David Applin

BIOLOGY

Northern Modular Science for GCSE

Stanley Thornes (Publishers) Ltd

First published 1997 by

Stanley Thornes (Publishers) Ltd,
Ellenborough House,
Wellington Street,
CHELTENHAM,
GL50 1YW

A catalogue record for this book is available from the British Library.

ISBN 0 7487 2608 X

97 / 98 / 99 / 00 / 01 / 10 / 9 / 8 / 7 / 6 / 5 / 4 / 3 / 2 / 1

Typeset by Ian Foulis & Associates and
Action Typesetting Limited, Gloucester
Printed and bound in Slovenia by Korotan

The author and publishers would like to thank the following for permission to reproduce photographs:
Allsport: Humans as Organisms title page; Allsport/Ralph Merlino: Environment A.13; Ander McIntyre: Maintenance of Life C.5; Animal Photography/Sally Anne Thompson: Inheritance & Selection A.5, A.6, G.6; Ann Ronan Picture Library: Maintenance of Life E.8; Audience Planners: Inheritance & Selection A.2a; Australian Information Service: Environment G.14; Barnaby's Picture Library: Environment K. 6b; Ben Pooley: Humans as Organisms B.2; BNFL: Inheritance & Selection H.8, H.9, H.10; Brian Armson: Environment O.6a, O.6b, O.6c; Biology Audio Learning: Inheritance & Selection H.2, I.1; Biophoto Associates: Humans as Organisms A4b, A4c, A8, C.6; Maintenance of Life B2, J.7; Environment D.1b, E.2, 1.3a, 1.3b, A1.f, A.1n; Blackpool Tourist Office: Environment M.6; Broom's Barn Experimental Station: Environment H.11; Bruce Coleman Ltd/Adrian Davies: Maintenance of Life G.9; Bruce Coleman Ltd/Jane Burton: Maintenance of Life G.10; Bruce Coleman Ltd/Gene Ahrens: Environment H.3; Bruce Coleman Ltd/Leonard Lee Rue: Environment I.7; Bruce Coleman Ltd/Michel Viard: Maintenance of Life F.1; Cambridge Laboratory/Plant Science Research Ltd, Norwich: Maintenance of Life G.4; CEGB: Environment O.5; Christopher Pillitz/Impact: Maintenance of Life J.1; Dalgety: Inheritance & Selection I.2; David Hoffman: Environment L.5b; D G Applin; Maintenance of Life title page, B.3; Environment A.2, B.3a, B.3b, B.3c, B.7a, D.1a, K.6a, K.6c, K.6d, K.7a, K.7b, K.8, K.9, F.1, A.1g, A.1h, A.1i, A.1j, A.1k, A.1l, A.1m, A.1p, A.1q, A.1s, A.1t, A.1u, A.1y, A.1z; FLPA/R Austing: Inheritance & Selection D.7; FLPA/Chris Mattison: Environment A.4a; FLPA/D Hall: Environment A.4b; FLPA/David Hosking: Environment F.6; FLPA/P Perry: Environment B.2; FLPA/Peter Reynolds: Maintenance of Life G.1; FLPA/Roger Wilmshurst: Maintenance of Life G.12a, G.12b; FLPA/W Broadhurst: Environment L.5a; Format/Jenny Matthews: Humans as Organisms D.9; Fran Harvey: Inheritance & Selection A.2b; Frank Spooner Pictures: Environment H.5; Graham Topping: Maintenance of Life G.13; Environment H.1, A.1w; Greenpeace/Pickaver: Environment M.3; Greenpeace/ Van der Veer: Environment M.4; Greenpeace/Greig: Environment M.5; Heather Angel: Inheritance & Selection I.6; Environment Q.4; Holstein Friesian Journal: Environment C.2; Holt Studios International/Nigel Cattlin: Inheritance & Selection D.1; Holt Studios Ltd: Environment F.2; Horticultural Research International, Littlehampton: Maintenance of Life D.3; Hulton Getty: Environment N.1, N.4; Hutchison Library/Nigel Smith: Inheritance & Selection D.3; I. Bowen: Inheritance & Selection A.12a, b; ICCE: Environment O.3; Images: Environment L.4c; Impact/Ben Gibson: Humans as Organisms D.6; Institute of Terrestrial Ecology/NERC: Environment I.8, I.9; John Arthur/Impact: Environment L.4a; John Gardiner Contractors Ltd: Environment N.8; John Innes Centre: Maintenance of Life G.11a, G.11b; John Radcliffe Hospital: Inheritance & Selection A.9, A.10; Katy Squire: Inheritance & Selection A.3, E.7; London Zoo: Environment A.1o; Mail on Sunday/Solo Syndication: Inheritance & Selection H.4; Martyn Chillmaid: Humans as Organisms C.1; Maintenance of Life E.12a, E.12b; Mary Evans Picture Library: Inheritance & Selection D.2, G.1, G.3, J.1a, J.1b; Massey-Ferguson: Inheritance & Selection C.3; Mike McQueen/Impact: Inheritance & Selection F.3; National Diabetic Society: Inheritance & Selection D.7; Oxfam (Ann Dalrymple-Smith): Humans as Organisms D.7, D.8; Oxford Scientific Films: Inheritance & Selection E.5, E.6, G.9; Oxford University Molecular Biophysics Dept: Environment J.3; P.J.Walters: Inheritance & Selection H.3; Panos/Paul Harrison: Inheritance & Selection D.6; Peter Gould; Humans as Organisms B.3, C.5, L.1a, L.1b; Popperfoto: Environment H.6; Planet Earth Pictures/Peter David: Environment B.7b; Planet Earth Pictures/Herwarth Voigtmann: Environment B.8; Remote Source: Environment A.1e; Rev. T.W. Gladwin: Environment B.7d; Ray Roberts/Impact: Maintenance of Life F.2; RNIB: Maintenance of Life H.1; Royal Geographical Society: Inheritance & Selection G.2; RSPB: Inheritance & Selection G.10, G.13; Science Photo Library: Humans as Organisms A.5, C.4, E.3, H.5; Maintenance of Life I.1, I.4, H.4, J.2; Inheritance & Selection title page, B.1, B.2, B.4, B.5, B.8, E.8, E.9, H.1, J.5a, J.5b, K.2, K.4; Environment A.1a, A.1b, A.1c, A.1d; Science Photo Library/Astrid & Hans-Frieder Michler: Humans as Organisms A4d; Science Photo Library/Claude Nuridsany & Marie Perennou: Humans as Organisms A.3; Science Photo Library/CNRI: Humans as Organisms J.1, J.2; Science Photo Library/David Scharf: Humans as Organisms J.4; Science Photo Library/Dr Tony Brain: Humans as Organisms I.1a; Environment A.1r; Science Photo Library/Dr Jeremy Burgess: Maintenance of Life E.10; Environment: B.7c, H.13, M.12; Science Photo Library/John Burbidge: Maintenance of Life J.8a; Science Photo Library/London School of Hygiene & Tropical Medicine: Humans as Organisms J.3, J7; Science Photo Library/Matt Meadows, Peter Arnold Inc.: Humans as Organisms: J.8; Science Photo Library/Prof P M Motta, G Macchiarelli, S A Nottola: Humans as Organisms I.1b; Science Photo Library/Patrick Lynch: Maintenance of Life J.8b; Science Photo Library/Philippe Plailly/Eurelios: Inheritance & Selection F.8; Science Photo Library/Sally Bensusen Maintenance of Life D.2; Inheritance & Selection F.5; Science Photo Library/Secchi-Lecaque/Roussel-UCLAF/CNRI Humans as Organisms A.4a; Science Photo Library/Sinclair Stammers: Maintenance of Life J.5; Science Photo Library/Space Telescope Science Institute/NASA: Inheritance & Selection F.1, Environment Q.6; Stanhay Webb/Laurie Martin: Environment H.7; Steven Newman: Environment A.1z; Tate & Lyle Sugars: Environment H.9; Tessa Carrick: Humans as Organisms C.2, C.3, C.7, D.1, D.4; Maintenance of Life C.1, E.4, E.7, G.3, G,8; Environment: G.3a, G.3b, G.3c, G.3d, G.5, H.2, H.10, H.12; Inheritance & Selection C.10, C.11; Thames Water: Environment J.2, J.3; Topham Picture Library: Environment M.9; UNFAO: Inheritance & Selection C.8; Unilever: Inheritance & Selection D.4; Warren Springs Laboratory: Environment N.3; Wyeth Laboratories: Inheritance & Selection M.1; ZEFA-Schimmelpfennig: Inheritance & Selection F.7; Zoological Society of London: Environment title page, A.2

Acknowledgements
The author and publishers would particularly like to thank Tessa Carrick, Morton Jenkins and Bob Lee for contributing material, originally published in the Blackwell Modular Science series, to this book.

Contents

Preface

Northern Modular Science: Biology is one of a set of three books written to prepare students for the Northern Examinations and Assessment Board's latest double award GCSE Modular Science syllabus. The other two books are *Northern Modular Science: Chemistry* by Ted Lister and Janet Renshaw and *Northern Modular Science: Physics* by Jim Breithaupt. These books are suitable for students aiming for either the higher tier grades (A* to D) or the upper grades of the foundation tier (grades C to E).

This book covers all the biology requirements of the course, as detailed in the syllabus, covering four biology modules: Humans as Organisms, Maintenance of Life, Environment, and Inheritance and Selection. The book is organised in four major sections, corresponding to the four biology modules. Sections are composed of Units based on topics in each module. Each Unit is clearly linked to the appropriate section of the module syllabus.

Within each Unit appropriate coursework is aimed at developing Sc1 skills and reinforcing essential concepts and knowledge. Some coursework activities are specifically designed and presented for assessment purposes.

A checklist is provided for each module to show students what they should know, understand and be able to do. Concepts that are exclusive for the higher tier (i.e. A*–B) are marked on the checklists and highlighted by extension questions in the book. Each Unit includes one or more sets of short questions to check progress and ends with a summary. Answers to questions and a detailed index are provided at the end of the book.

Modules end with specimen examination questions prepared for this book by the Chief Examiner and the NEAB examining team, to aid preparation for either module tests or terminal examinations, as appropriate. The modules on Humans as Organisms and Maintenance of Life are examined through module tests composed of multiple-choice questions, which may be taken at certain times during the two-year course. The other two modules are examined by terminal examination papers consisting of structured questions. The examination questions at the end of each section have been chosen to match the type of test for the module covered by that section.

Humans as Organisms

This module identifies the life processes that distinguish living things from non-living things. In it, we extend the idea from Key Stages 2 and 3: that cells group together to form tissues, which group together to form organs, which group together to form organ systems. Human organ systems and the functions they perform – in particular the digestive system, the breathing system and the circulatory system – are explored and the ways the body defends itself against infection are examined.

Humans as Organisms

Learning Objectives

What you should know:

- The characteristics of living organisms
- The structure of animal cells, the jobs of the nucleus, cytoplasm and cell membrance
- The differences between cells, tissues, organs and organ systems
- The sources and uses of carbohydrates, fats and proteins
- The digestive system breaks down insoluble food into soluble compounds and absorbs these into the blood
- The breakdown of food is catalysed by enzymes
- The sites of production and action of carbohydrates, proteases, lipases and hydrochloric acid
- The jobs of the small intestine and the large intestine
- The breathing system takes air into and out of the body so that oxygen from the air can pass into the blood and carbon dioxide can pass out of the blood into the air
- The breathing organs are situated in the thorax (chest).
- Respiration is the release of energy from sugars.
- The equation for aerobic respiration:
  glucose + oxygen → carbon dioxide + water + energy
- When there is a shortage of oxygen, cells may carry out anaerobic respiration, releasing lactic acid
- The uses of the energy released during respiration
- The circulation system transports substances around the body. The heart pumps blood around the body
- How, in the heart, the atria, ventricles and valves are involved in the circulation of blood to the lungs, then to the rest of the body
- The structure and jobs of arteries, vains and capillaries
- How oxygen, carbon dioxide, soluble foods and urea are transported around the body
- The structure and jobs of red cells, white cells and platelets
- The structure of bacteria
- The structure of viruses
- How we catch infectious diseases
- How bacteria infections may make us feel ill
- How the skin, the breathing organs and blood clots help to defend us against the entry of microbes
- How white cells defend us against microbes
- Diffusion as the spread of a gas or of a substance in solution from a higher to a lower concentration. Substances pass through cell membranes by diffusion

What you should be able to do:

- Examine cells from e.g. the mouth of a herring
- Relate the different types of human cells to their functions
- Label on a diagram: gullet, stomach, liver, pancreas, small intestine and large intestine
- Investigate what saliva does
- Use a model gut to investigate digestion
- Label on a diagram: rib muscles, diaphragm, lungs, trachea, bronchi, bronchioles and alveoli
- Investigate your own breathing
- Label on a diagram the parts of the heart
- Investigate how the rate of heartbeat changes wirh activity
- Explain how living conditions and lifestyle affect the spread of disease
- Investigate size and surface area

EXTENSION

- Where bile is made and stored. The role of bile in digestion, particularly in the emulsification of fats
- Energy from respiration may be used in active uptake
- Why anaerobic respiration leads to oxygen debt in muscles
- How the diaphragm muscles and the muscles between the ribs bring about inspiration
- How surface area for exchange is increased in humans

- Explaining how breathing takes place

What is special about living things?

Unit A
Features of living things

Figure A.1 What have all these living things in common?

What is an organism?

The scientific name used for a living thing is an **organism**. Humans are living, so they are also organisms. Two of the main groups or **kingdoms** of organisms are **plants** and **animals**. Plants are usually green and make their own food. Animals obtain their food from other organisms, alive or dead, or their products. As they cannot make their own food, all animals depend on green plants for food:

The plants are at the beginning of a **food chain** (see page 119). Humans belong to the **animal kingdom**. One hundred years ago many people found this idea very shocking, but nowadays most people are used to the thought and are not worried by it.

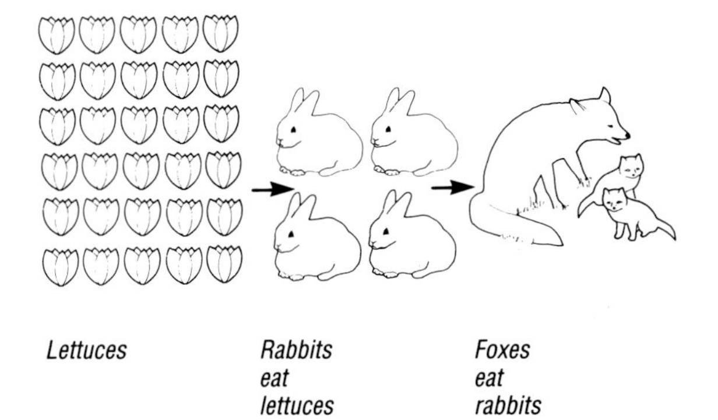

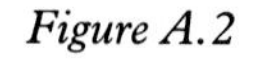

Figure A.2

The characteristics of living things

Everyone agrees that humans are living. But what exactly does this mean? Each of the living things shown in Figure A.1 has the same features which make it alive. These features are the characteristics of living things.

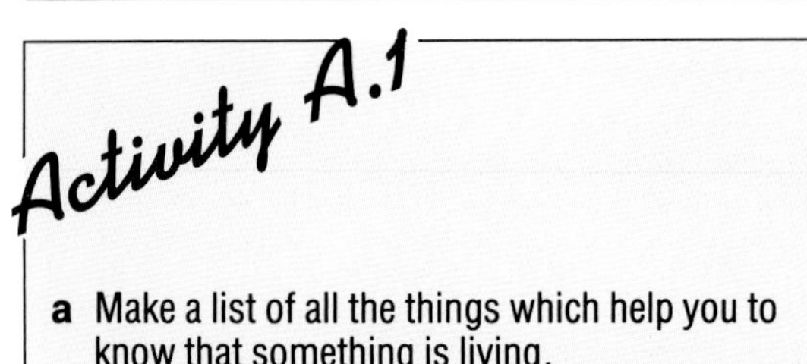

a Make a list of all the things which help you to know that something is living.
b Put a tick next to the things that you think apply to animals.
c Put a cross next to the things that you think apply to plants.
d Do plants and animals have exactly the same characteristics? If not, how are they different?

Turn to the summary on page 7. It lists eight features found in all living things, whether they are microscopic or elephant-sized, plant or animal. How many of these features did you include in *your* list?

As you can see, the first characteristic is rather different from all the others. In this module you will learn more about some of these characteristics in humans.

Cells

Living material is made up of tiny **cells**. The cells are invisible to the naked eye, but they can be seen with a microscope. The cells in different parts of the human body have different shapes and functions.

If you look at an *Amoeba*, a microscopic organism, you will have an idea of what cells are like. *Amoeba* is a single, large cell. It is about the size of a pinprick and lives on the mud in ponds and ditches.

Some of the simplest cells in the human body are inside the mouth, lining the cheek (Figure A.3). These cells fit together like paving stones, forming a covering. Because of the slight possibility of spreading disease, it is wiser not to examine living human cells in schools. We can, however, examine cells from other animals. Cells from the mouth of a herring can be used instead.

There are millions of cells in your body. Each has a **membrane**, or special layer which forms its surface. Substances pass into and out of the cell across the membrane. Inside is the jelly-like **cytoplasm**. Many of the chemical reactions occurring inside the cell take place in the cytoplasm. There are other small structures (organelles) in the cytoplasm. The **nucleus** is the most obvious. It controls the activities of the cell. Most of the other organelles are too tiny to see with a light microscope. They would show up under a much more powerful electron microscope.

There are different kinds of cells for different functions, or tasks, within the body. For example, some cover surfaces, some form muscles or nerves, and some are part of bone tissues. Figure A.4 shows some different human cells.

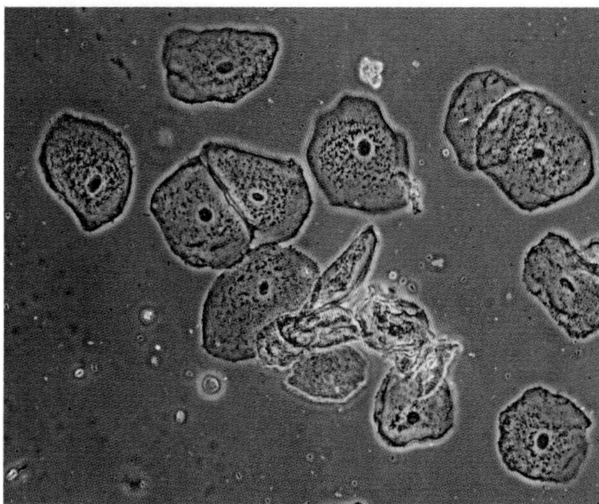

Figure A.3 This is a micrograph of cells from the lining of a person's mouth

a

b

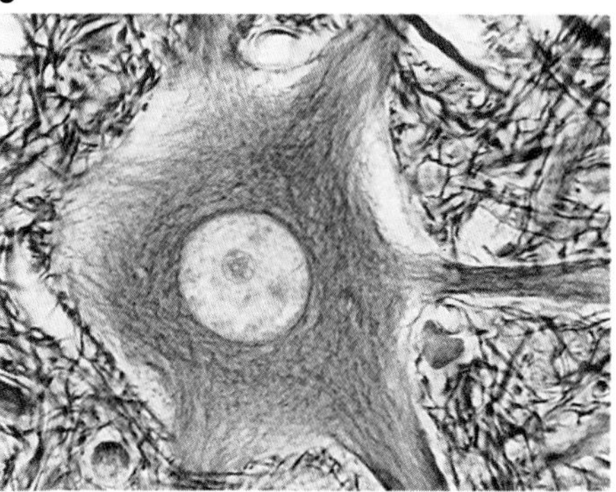

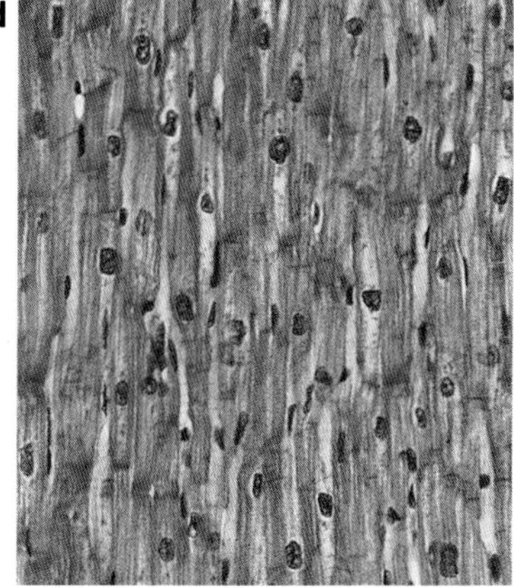

Figure A.4 Different kinds of cells: (a) white blood cell; (b) cells forming the lining of the small intestine; (c) a nerve cell; (d) muscle cells

Activity A.2

Looking at an amoeba

Look at a living amoeba under the microscope. (*If you cannot do this, look at Figure A.5 and begin at step number 3.*)

You need: fine pipette, slide, coverslip, mounted needle, microscope, amoeba culture

1. Use the pipette to draw up one of the tiny white specks from the amoeba culture. Place this on the slide in a drop of water. Cover the drop of water with the coverslip, lowering it as shown in Figure A.6.
2. Examine the slide under the low magnification of the microscope. Look for an amoeba (grey, with an irregular or bulging shape). When the water on the slide has settled, the amoeba will begin to change shape slowly. (You may need to alter the brightness of the light you are using with the microscope so you can see all the structures in the amoeba.)
3. Compare the amoeba or the photograph with Figure A.7 and identify as many parts as possible. Look for the **nucleus** (plural nuclei) and for the granules or specks in the **cytoplasm** around the nucleus. Make a drawing of your amoeba to show the important structures.

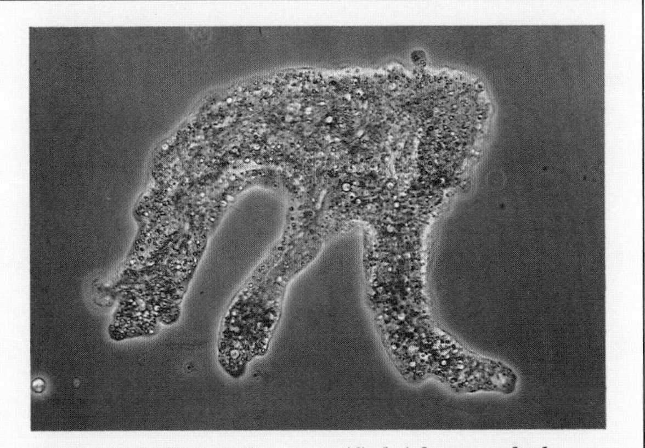

Figure A.5 An amoeba, magnified (photographed through a coloured filter – an amoeba is colourless)

Figure A.6 Lowering a coverslip on to a slide

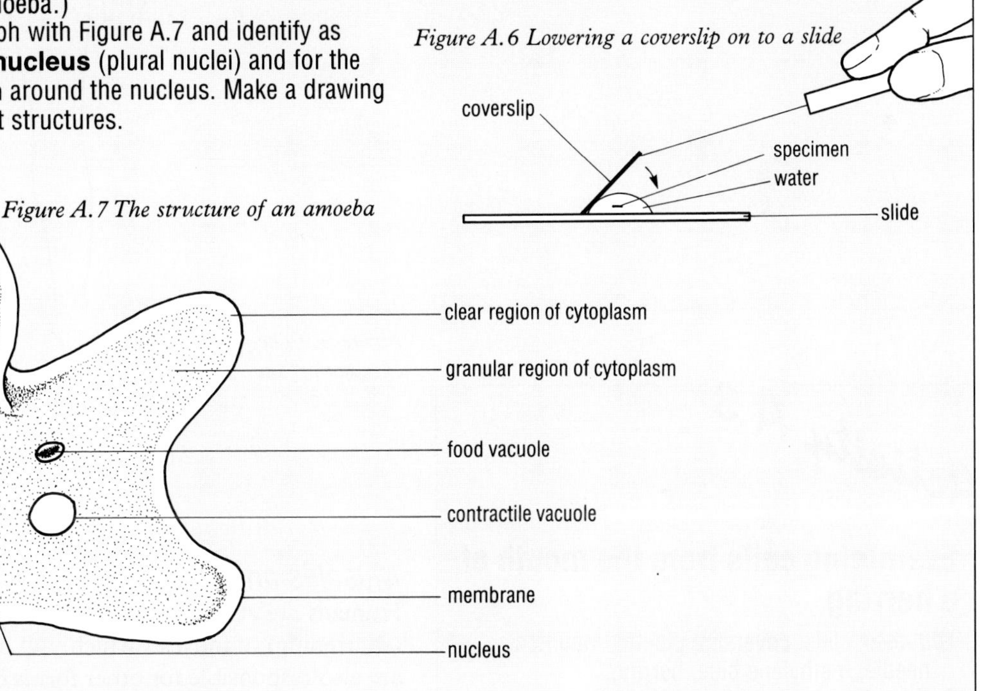

Figure A.7 The structure of an amoeba

Tissues, organs and organ systems of the body

We call groups of cells of a similar kind **tissues**. Muscle, different glands, cartilage and bone are all tissues. Tissues are grouped together to form **organs** like the heart or stomach. The **trachea** (windpipe) is an example of an organ. It is made up of tissues like cartilage and the **epithelium** or lining layer of cells. Figure A.8 shows a section across the trachea.

Figure A.9 shows the positions of some of the organs of the body. Organs are linked together into **systems** which carry out major activities of the body. For example, the **breathing** (respiratory) **system** consists of the trachea and other tubes branching into the lungs. The heart and the blood vessels form the **circulatory system**. The lungs and heart are both in the **thorax** (or chest). The **diaphragm**, a curved sheet of muscle, divides the thorax from the **abdomen**. You can see its position in Figure A.9.

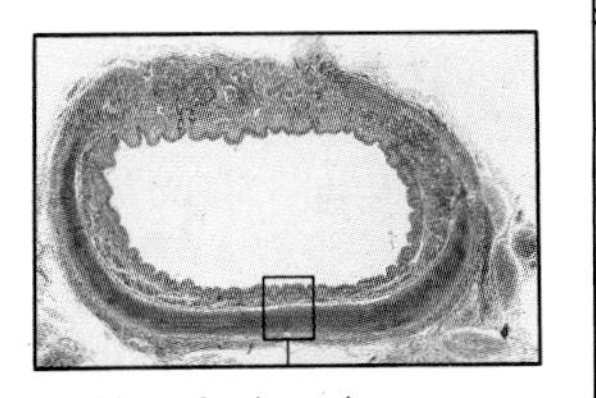

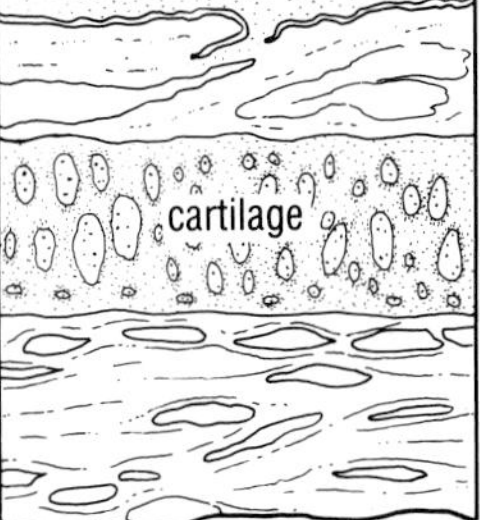

Figure A.8 Section through the trachea. Notice the cartilage and the lining cells

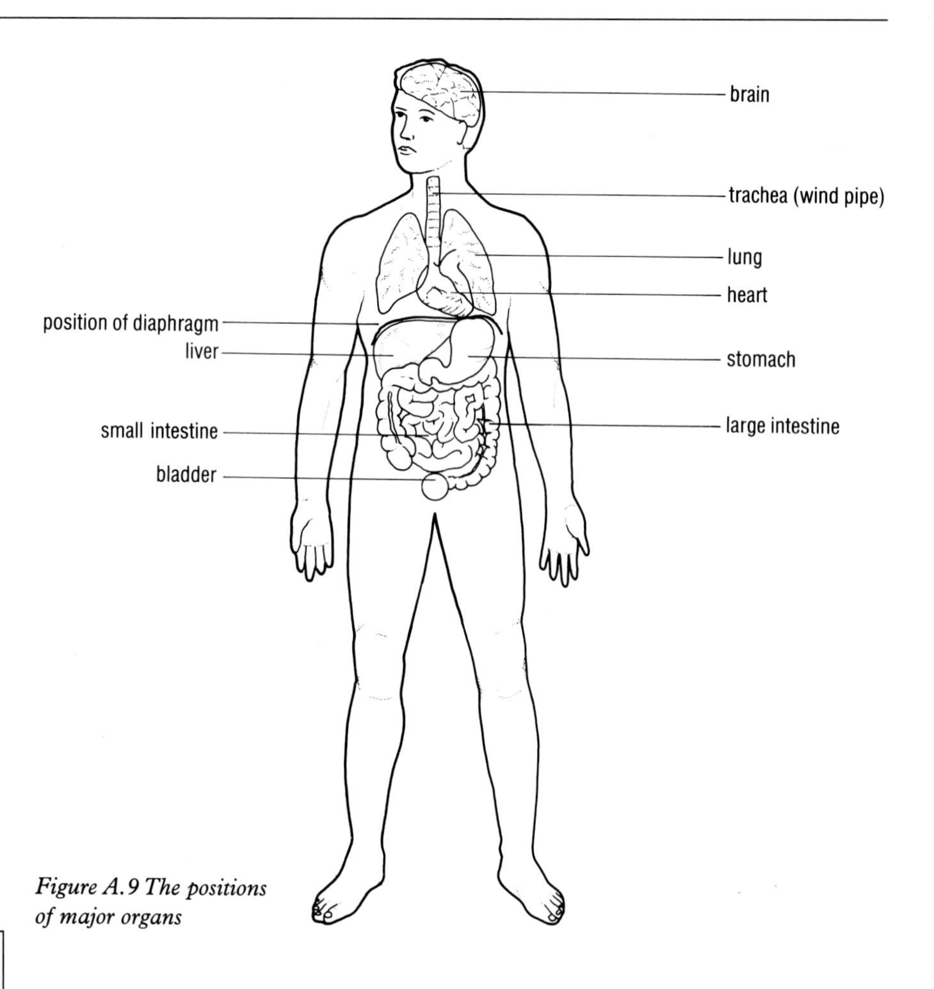

Figure A.9 The positions of major organs

Activity A.3

Examining cells from the mouth of a herring

You need: slide, coverslip, pipette, mounted needle, methylene blue, herring.

1. Open the herring's mouth and rub a wooden spatula firmly over the inside of the large scale at the side of the mouth. Smear the material on your spatula over the central area of a clean microscope slide. WASH YOUR HANDS.
2. Add one drop of methylene blue dye (0.5% concentration).
3. Lower the coverslip on to the slide gently, as shown in Figure A.6.
4. Place the slide on the stage of a microscope. Look for the small cells under low magnification. When you have found some, carefully turn to high power to see the detail of the cells. The dye will make the cells show up blue, and the nucleus will show up as a darker region in the centre. If you do not succeed in finding cells at first, try making a second slide.
5. Draw one of the cells.

Muscles and glands: examples of tissues

Humans are able to move from place to place because of the contraction (shortening) of muscles which pull on the skeleton. Muscular contractions are also responsible for other forms of movement. For example, as heart muscle contracts and relaxes rhythmically it propels blood through the blood vessels. Contraction and relaxation of the muscles in the wall of the intestine move food through the digestive system.

The endocrine glands are examples of glandular tissue which produce substances that the body needs – in their case, hormones. For example, groups of cells in the pancreas (see page 92) produce insulin and glucagon. These hormones help regulate glucose levels in the blood (see page 92).

Summary

Humans are animals. The body is made of cells. Humans have the same characteristics as all other living things:

- Made of complex substances – these are often arranged as cells.
- Nutrition – plants use energy from light to make food, animals obtain food by eating plants or by eating other animals.
- Respiration – the release of energy from food.
- Excretion – the release of waste products.
- Reproduction – the production of offspring.
- Growth – the growing of offspring to reach adult size.
- Sensitivity – the ability to react to the surroundings.
- Movement – the ability to move all or parts of the body.

Questions

1 Copy out the passage below and fill in the gaps with the correct words from the wordlist at the end of each paragraph. You may use the words in each list once, more than once, or not at all.

a All living things, or _______, are made up of _______. Each cell has an outer _______, enclosing _______, which surrounds a _______.

Wordlist
cells, cytoplasm, nucleus, water, organisms, membrane

b Living things are made of _______. Groups of similar _______ with similar function form _______, which can work together as _______. A group of _______ working together form _______.

Wordlist
an organ, organs, cells, an organ system, types, tissues

2 Try to work out the magnification of the cell in the drawing you made in Activity A.3. Write your answer under the drawing. There is more than one way to tackle this problem. Explain how you attempted to find the answer.

Unit B
What is digestion?

The food we take into our mouths cannot be used as it is. It has to be broken into small pieces and changed chemically. Large molecules are split chemically into smaller molecules by adding water.

Starch changes to simple sugars such as glucose.
Fats and oils change to fatty acids and glycerol.
Proteins change to amino acids.

Figure B.1 shows these changes.

The process of changing food so that it can be taken into the tissues of the body is known as **digestion**. It takes place in the **alimentary canal** or gut.

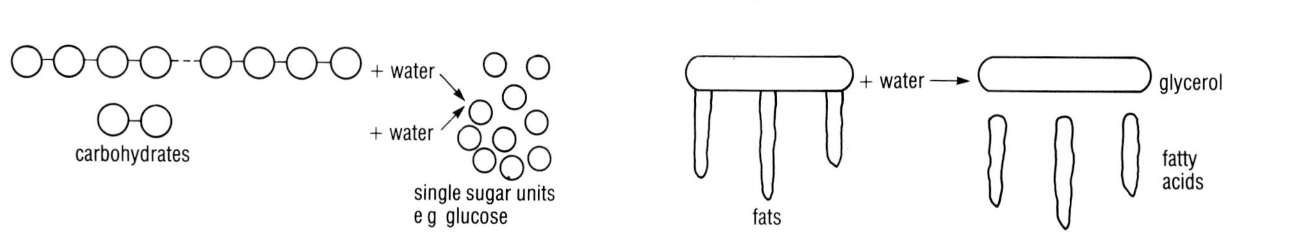

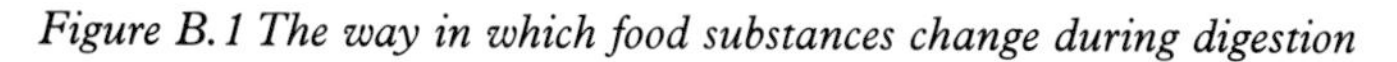

Figure B.1 The way in which food substances change during digestion

The diet and energy
The importance of energy
Whatever we do needs energy. Everyone knows that energy is necessary for moving about. But they forget that energy is required to send nerve impulses, to repair tissues, and to grow. It is also essential for transporting materials around the body. Without energy, living things die.

Energy for our activities
Our energy comes from food, which is used like a fuel. Oxygen in our bodies helps to release energy from food, just as oxygen enables a fuel like coal to burn and release heat energy.

Energy is measured in **joules**. One joule (J) will do a certain amount of work. The amount of work it will do is to move a force of 1 newton through a distance of 1 metre. One kilojoule (kJ) is 1000 joules.

A person doing nothing needs energy to keep the body working. The energy a person uses when resting varies according to the person's mass. Different amounts of energy are used when you move around (Table B.1).

Figure B.2 Where does his energy come from?

	kJ
Chapati with fat	1 420
Chips, fried	983
Cooking oil	3 696
Noodles, raw	1 646
Pitta bread	1 127
Potatoes	372
Rice, boiled	500
Spaghetti, raw	1 456
White bread	960

Table B.2 Energy in 100 g of some everyday foods

Activity	Energy used in kJ per hour
Sleeping	275
Sitting / watching television	about 360
Standing	420
Walking slowly	780
Climbing stairs	2 280
Household activities / driving	about 900
Carpentry / gardening	about 1 200
Moderately heavy work / cycling	about 1 500
Heavy work	about 2 000
Playing squash / football / swimming	over 1 800

Table B.1 Energy requirements for some activities for an average 65 kg man (aged 25)

Investigation B.1

Investigate the energy stored in various foods.

Figure B.3 Different foods provide different amounts of energy

Measuring energy in food

The foods we eat provide our energy needs. It is possible to measure and compare the energy stored in different foods by burning them in controlled conditions. Table B.2 shows the kind of results found for different foods using a **calorimeter**. More energy values are given on pages 32 and 33.

Release of energy in the body

When food reaches the cells, it is changed chemically and energy is released. This process of releasing energy is called **respiration**. Usually, oxygen is used during respiration. Carbon dioxide is a waste product. Some of the energy turns into heat, so you become warm when you exercise. The process can be written simply as:

Dissolved food + oxygen ⟶ carbon dioxide + water (+ energy)

Different kinds of energy-giving foods

A healthy diet includes foods of different types or classes. All the substances in one class of food are alike chemically. Most meals and snacks are mixtures of food substances from more than one of these classes. But one common food, household sugar, is a pure substance. It is made up of crystals of the sugar known as sucrose. Sucrose is obtained from sugar cane and sugar beet. The foods which give us energy usually belong to two groups, **carbohydrates** and **fats and oils**. A third class of foods, **proteins**, is also sometimes used as a source of energy. All these substances are made of complicated molecules.

Activity B.2

A more accurate method of measuring energy values

Read the passage below. Use Figure B.4 to help you fill in the gaps.

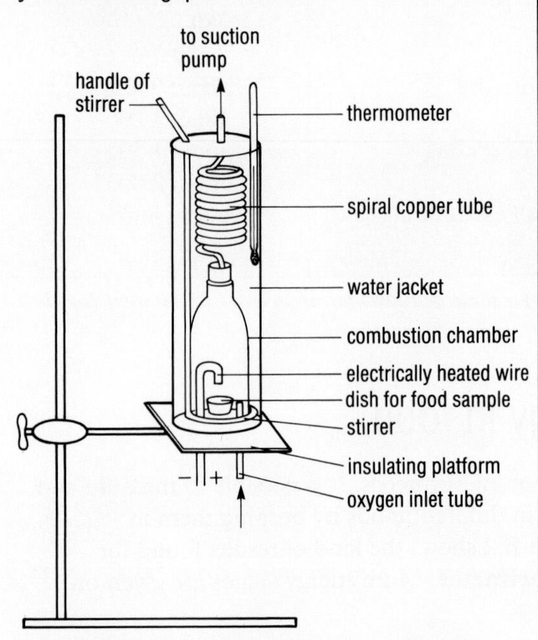

Figure B.4 A food calorimeter

The apparatus used for measuring energy values of food accurately is a **food calorimeter**. Food is burnt in the *combustion chamber* (burning chamber) of the calorimeter. This is surrounded by a water ______ with a *stirrer*. The ______ and ______ pass through the *lid* into the water.

The apparatus stands on an ______ platform. An *inlet* for ______ passes into the combustion chamber. At the bottom of the chamber is a *small* ______ to hold the food. An ______ *heated wire* can be lowered into the dish to set the food alight.

At the top of the combustion chamber a *spiral* ______ exchanges heat with the surrounding water. The top of the spiral tube passes through the ______. It can be attached to a ______ ______ which draws warm gases upwards from the ______ ______. The heat produced by the burning food is transferred to the water surrounding the ______ ______ and the ______ ______.

At the beginning of the experiment, the ______ of the water and mass of ______ are noted. The ______ ______ is lowered on to the food to ignite it. Then, ______ from a cylinder of ______ is switched on so that the food will ______ completely. The rise in water ______ is recorded.

Carbohydrates

Carbohydrates include sugars and starches. The word carbohydrate gives a clue to their chemistry. All of them are compounds of carbon, hydrogen and oxygen. In each carbohydrate molecule there are twice as many hydrogen as oxygen atoms. There are also twice as many hydrogen as oxygen atoms in water – H_2O – but there is no carbon in water.

Some carbohydrates are simple sugars: they are sweet and will dissolve. Glucose is an example.

Other carbohydrates, such as starch and cellulose, are insoluble substances made up of large molecules. Starch is a common substance in plants. When it is purified it is very like flour. Foods containing starch are an important part of most people's diets. They are called **staple foods.** Table B.3 shows how much starch some staple foods contain. Cellulose is a carbohydrate present in the walls around plant cells. It cannot be digested or absorbed by humans. It is important as roughage (see page 16) but cannot be used as a source of energy in the body.

Mass of starch / g	
Bread, white	45.8
Bread, wholemeal	39.8
Bread, pitta	57.9
Chapati	48.3
Gari, raw	82.7
Noodles, raw	76.1
Papadums	46.0
Potatoes, boiled	17.0
Rice, long grain, raw	86.8
Spaghetti, raw	74.1
Yams, boiled	29.6

Table B.3 The mass of starch in 100 g of some staple foods

More about carbohydrates

Glucose has the formula $C_6H_{12}O_6$. The atoms of the molecule form a single ring which is called a **monosaccharide**.

Disaccharides have two sugar units joined together in their molecules (see Figure B.5). Household sugar or sucrose has the formula $C_{12}H_{22}O_{11}$. Compare the number of carbon, hydrogen, and oxygen atoms in this molecule with the number in glucose or fructose molecules. You will see that it is almost twice as large, but has lost one molecule of water, H_2O. It is made up of two sugar units, which is why it is called a disaccharide. Sucrose (table sugar), maltose (malt sugar) and lactose (sugar of milk) are all disaccharides.

Starch and cellulose are **polysaccharides**, made up of many sugar units (see Figure B.5). Starch consists of chains of glucose molecules, which have lost a molecule of water wherever two sugar units are joined together.

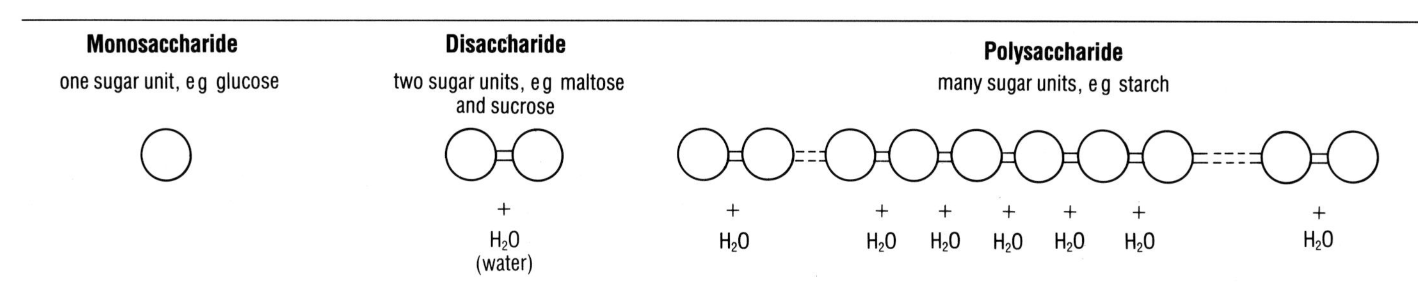

Figure B.5 Simple models of carbohydrates

Fats and oils

The molecules of fats and oils (or lipids) also contain carbon, with some hydrogen and a few oxygen atoms. Fats are solid and are usually found in foods coming from animals. Oils are liquid at room temperature. They come from plants. Both fats and oils make greasy marks on paper and will not mix with water.

Everyone needs some fat or oil in their diet. The body uses them to make the membranes of cells, to provide the covering material around nerve cells, and for various other substances in the body. Fats and oils can also be used to provide the body with energy.

Most people in Britain have a diet in which about 38% of the body's energy is supplied by fats and oils. Scientists now think that people in the developed countries ought to try to cut down their fat intake so that it gives only 30% of their energy. Also, some experts think that we ought to eat less of the kinds of fats called **saturated**. These are in animal and dairy foods. It is possible that these saturated fats cause deposits of fatty **cholesterol** in our blood vessels. This may cause heart and circulatory diseases in later life. Many people now use margarine and oils with **polyunsaturated** fats in them instead of butter and cream. Polyunsaturated fats come from plant oils, they do not lead to deposits of cholesterol. Skimmed milk has less fat than whole milk. If we change to drinking skimmed milk, less cholesterol will collect in our blood vessels. Our circulation will be healthier.

Storing carbohydrates and lipids

If the body has more carbohydrate than it needs, some of it can be changed to a polysaccharide called **glycogen**. Glycogen is stored in the liver and muscles. Any further surplus carbohydrate in the body is changed to fat.

Fat which is not used up immediately to give energy or to build important substances for the body is also stored. It collects around the organs and under the skin. Some fat provides useful insulation against losing heat from the body. It can also be used for energy when a person is really short of food. But too much fat makes a person overweight. A fat person finds it harder to move around and may become out of breath more easily. There may be more serious health risks, too. We all need *some* carbohydrates and fats and oils in our diet. But we must take care that we do not eat more than we need.

Proteins

Proteins are the third class of food which we need to eat. We use them for growth and repair of the body. Many of the activities which go on in the body

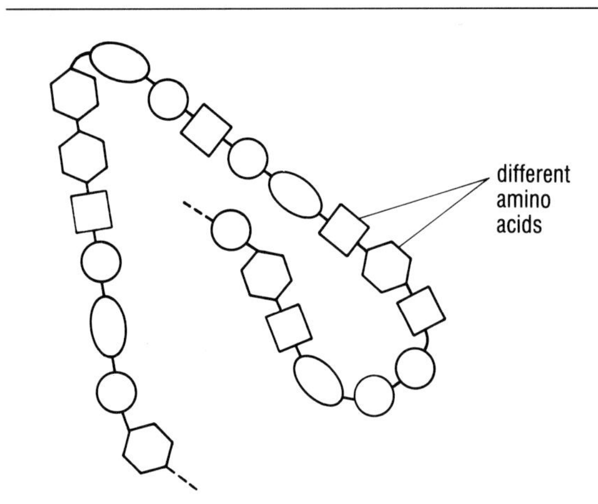

Figure B.6 A protein chain model

need special proteins. Common sources of protein are meat, fish, egg white, peas and beans.

If the body is short of carbohydrate and lipids for energy, it can use proteins. When humans are starving, some of the proteins in the body are used to supply energy, instead of for repairing damaged tissues, or growth. This makes the body less healthy. On the other hand, if the body has more proteins than it needs they are removed (excreted), and not stored.

Molecules of proteins are very large. They consist of long chains of different kinds of **amino acids** (Figure B.6). Each amino acid contains atoms of carbon, hydrogen, oxygen and nitrogen, arranged slightly differently.

Comparing foods as sources of energy

We need to balance our intake of carbohydrates and lipids against our need for energy. A girl of about 15 years needs about 9000 kilojoules a day, a boy nearly 12 000 kilojoules. The exact amount needed is affected by how active you are and by your size.

If we burn samples of carbohydrates, lipids and proteins in a calorimeter, we can compare their energy values (Table B.4).

Three examples of snacks or dishes of food are given in Table B.5.

	Kilojoules in 1 g of pure substance
Carbohydrate	16
Lipid	37
Protein	17

Table B.4 Energy values of carbohydrates, lipids and proteins

	Potato crisps	**Steak and kidney pie**	**Vegetable curry**
Carbohydrates	49.3	22.2	10
Lipids	35.9	17.1	15.2
Proteins	6.3	9.3	1.8

Table B.5 Quantities (grams) of food classes in 100 g of the food

Summary

- Energy comes from food and is released for the body's activities during respiration.
- Carbohydrates, fats and oils and proteins can all be used by the body as sources of energy.

Questions

1. Draw a bar graph showing the energy a 65 kg man uses for his activities (see Table B.1). Put the activities along the horizontal axis. Leave gaps between the bars. Label the vertical axis 'Energy used / kJ per hour'. Give your graph a title.
2. Find out how to cut down on the amount of fatty foods you eat. Make a poster or write a short magazine article recommending a less fatty diet.
3. Estimate the energy you have used since you got up this morning. Choose the activities in Table B.1 that are closest to what you have done to give you an idea of the energy you needed for each activity. Explain how you obtained your estimate. What sources of error are there in your estimate?
4. Collect ten food labels and use them to complete this table:

	In 100 g			
Food	Energy / kJ	Carbohydrate / g	Fats and oils / g	Proteins / g

Unit C
What do we need food for?

Bulky foods

We have seen that there are three main classes of nutrients which can be used to give the body energy. The three kinds of food are:

- carbohydrates
- fats and oils (or lipids)
- proteins

We need fairly large amounts of these foods, so they are called the bulky foods. In Britain almost everybody can get enough to eat. In fact, many people eat too much of the sugary carbohydrates (Figure C.1) and fatty foods (Figure C.2). Everyone is slightly different and each of us has to learn how much food and drink we need to remain healthy, to grow properly, and then to keep a steady weight. The advice of nutrition experts will help us to remain fit and to avoid illness.

Figure C.1 Foods rich in carbohydrates. Which kinds of carbohydrates should you eat less of?

Getting the balance right

Choosing the right foods to eat is complicated, so it is useful to build up some guidelines. We all need to eat foods that will give us energy. But, as we have just seen, energy can come from carbohydrates, proteins or fats and oils. Which sort of food is best?

Think about the following information:

- nutritionists say that too much fatty or oily food in the diet is unhealthy
- protein foods are expensive
- too many sugary carbohydrates are bad for us
- starchy carbohydrates are healthier than sugars and cheaper than proteins

Figure C.2 Foods containing fats and oils. Should you eat more or less of these?

Starchy carbohydrates are clearly the best source of most of the energy we need. The fairly cheap, staple foods (see Table B.3 on page 10) are all starchy carbohydrates, so it is sensible to keep on eating them.

Nutrition experts tell us that most people in the developed world eat too much sugar. The carbohydrate sucrose (table sugar) is very common in our diet. At present an average person in Britain eats over 100 g of sucrose each day, and experts recommend that we cut down sugar intake to 55 g. Many people stir sugar into their tea or coffee, but it is also in many everyday foods. If you look at the labels on packets and tins you will see that sugar is in baked beans, soups, tinned fruit and breakfast cereals, as well as in 'sweet' products like cakes and biscuits. We have to be careful not to eat more sugar than is good for us.

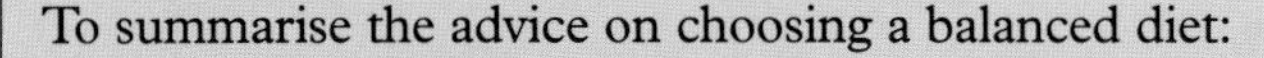

> To summarise the advice on choosing a balanced diet:
> - Eat a mixed diet containing some fat, some protein and some carbohydrate.
> - Keep the proportion of fats and sugars fairly low, and meet most of your energy needs by eating starchy foods such as bread, potatoes or rice.

In the pages that follow you can find out about other things that must be considered in choosing food wisely.

More about proteins

Figure C.3 shows foods containing proteins. Meat, fish, eggs, cheese and milk are good sources. Many other foods contain small amounts. Peas and beans (sometimes called pulses or legumes) also contain plenty of protein. Proteins cannot be stored in the body so we must have some protein every day if we are to remain healthy.

Table C.1 shows how much protein nutritionists recommend we should eat.

Figure C.3 Foods containing plenty of protein. Why do you need proteins?

	Males	**Females**
1 year	30	27
5-6 years	43	42
12-14 years	66	53
15-17 years	72	53
Young adult, moderately active	72	54
Young adult, very active	84	62
When pregnant		60
When breast-feeding baby		69
Middle age	60	47
Over 75 years	54	42

Table C.1 Recommended daily intake of protein (in grams)

The table on pages 32–33 tells you how much protein is in some everyday foods. Proteins are very important for the body. They are used for growth and repair and form important parts of all cells. They also keep many of the body's activities working properly. For example, they are part of the bone and cartilage.

Properties of proteins

Egg white or **albumen** is a protein mixed with water. When dried albumen or any substance which is rich in protein is burnt it gives an irritating smell, like burning hair.

A protein is a very large molecule made up of a chain of many smaller molecules called **amino acids**. Each amino acid is built up from the elements carbon, hydrogen, oxygen, nitrogen (and sometimes, sulphur). There are about 20 different kinds of amino acid. You can imagine a protein molecule as a chain of different beads which represent the amino acids (see Figure B.6 on page 12). This chain may be twisted in a complicated way (Figure C.4). It may even have other molecules attached to it.

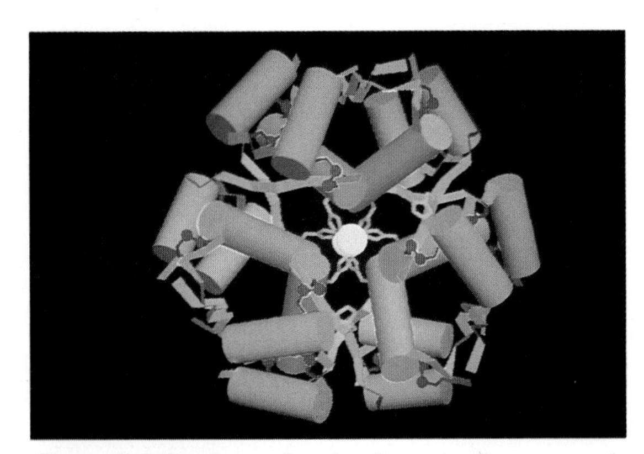

Figure C.4 The shape of an insulin molecule, very much enlarged

Essential and non-essential amino acids

One kind of protein food may not contain all the different kinds of amino acids. The amino acids are described as **essential** and **non-essential**. Essential amino acids cannot be made in the human body. They must all be present in the proteins humans eat. Non-essential amino acids are also important, but they can be manufactured by the body from other amino acids. Nine amino acids are essential for infants; everyone else needs eight. If we eat a mixture of protein foods we should take in all the essential amino acids.

Foods which contain animal protein supply the essential amino acids required by humans in about the right proportions. No single plant food has the right amounts of all essential amino acids. For this reason vegetarians must eat a variety of protein foods. A mixture of nuts and seeds, cereals, potatoes, peas and beans helps to provide a better balance of essential amino acids. Many vegetarians also eat eggs, cheese and milk, which makes it easier to obtain all the essentials.

Soya beans are a source of protein, but they do not have the correct balance of amino acids. Manufacturers convert soya beans into textured vegetable protein (TVP). TVP contains essential amino acids in about the right proportions for the human diet. It is often made to look and taste like meat, but it can be eaten by vegetarians (Figure C.5).

Figure C.5 Foods containing textured vegetable protein

Minerals and vitamins

A healthy diet also includes **minerals**. Small amounts of many different minerals or chemical elements are essential for life. Compounds of calcium, phosphorus, iron and iodine are important. Some people's diets do not contain enough calcium, iron or iodine.

Calcium is necessary for bones, teeth and the clotting of blood. Iron is part of the red substance **haemoglobin** which carries oxygen in the blood (see page 39). Iodine is part of the hormone **thyroxine** produced (secreted) by the **thyroid gland** in the throat. This substance helps to maintain the activity of cells. If there is a shortage of iodine the thyroid gland may swell into a **goitre** (Figure C.6).

Some magnesium and very small traces of zinc must also be present in our food.

Our bodies also need sodium, but many people eat too much sodium (as sodium chloride, or salt) rather than too little. On average, we use two teaspoonfuls (10 grams) of table salt a day. A small amount of salt is added in cooking foods, some is added at the table, but most of the salt is added when food is processed. In fact we need only one gram of salt each day, not the 10 grams most of us eat. Worse, for some people too much salt can lead to high blood pressure. People with high blood pressure are more prone to heart disease and strokes. It is impossible to know who will develop high blood pressure, so it is a good idea for everyone to cut down on salt.

If one of the essential minerals is missing from the diet, a person develops a **deficiency disease**.

Vitamins are also needed for healthy living. They are fairly small **organic molecules** (containing carbon). Humans need only tiny amounts of each vitamin, but if one of them is missing a deficiency disease occurs. For example, deficiency of vitamin C causes **scurvy** – bleeding, especially from the gums, is a symptom of this disease.

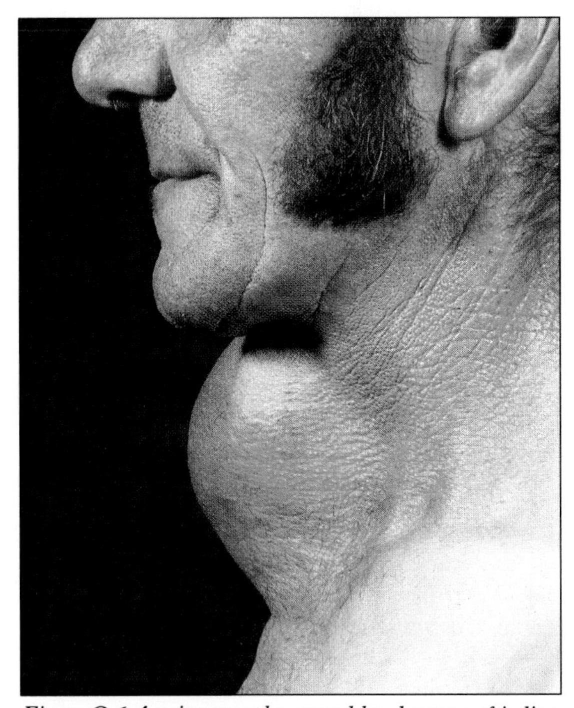

Figure C.6 A goitre may be caused by shortage of iodine in the diet. This photograph shows a severe case

Figure C.7 Some foods containing fibre

The need for fibre in the diet

Dietary fibre (fibre in the diet) is sometimes called **roughage**. Although we need fibre in our food, we do not absorb it into our bodies. It passes through the digestive system unchanged. Fibre provides bulk during digestion and helps to prevent constipation. It protects against diseases of the lower part of the digestive system. Doctors think that people who eat very little fibre are more likely to suffer from cancer of the bowel or **colon**.

Fibre is made up of cell walls and other fibrous material in plants. Wheat bran, wholemeal foods, prunes, apricots, peas and beans are all good sources of fibre (Figure C.7).

Activity C.1

Looking at fibre under the microscope

You need: slides, cover slips, 2 mounted needles, small dish, iodine stain, microscope, boiled celery, bran, rabbit pellet in 90% ethanol

1. Use two mounted needles to tease out short pieces of the tough strands from the celery. Place them in a drop of iodine stain on a slide. Lower a coverslip gently on to the drop. Examine the strands under a microscope. They have walls which provide fibre.
2. Place a few pieces of bran on a slide. Add a drop of iodine and cover with a coverslip. Examine the thinnest part of the bran under a microscope. Bran is a good source of fibre.
3. Tease apart the rabbit pellet in the dish. Place some small pieces in a drop of water on a slide. Add a coverslip. Examine them with the microscope. These fragments passed through the rabbit's digestive system. What are they? NOW WASH YOUR HANDS

Water

Water is not a nutrient but it is certainly needed by the body (see page 93). You would die without water much faster than if you had no food. You must take in enough water to balance the water you lose.

Questions

Table C.2 compares the amount of energy and some nutrients that an adult human needs each day with the energy and nutrients in milk. Study the table and answer the following questions:

1. On a diet of milk only, how much would an adult have to drink to satisfy daily energy needs?
2. Milk is a 'balanced' diet for a baby. What is meant by a 'balanced' diet?
3. Give one reason why milk is particularly important for development of bones and teeth in babies.
4. How much milk would an adult need to satisfy the daily need for calcium?
5. Collect food labels. Draw up a chart to show which foods contain the following: starch, sugar, fats and oils.

Energy or nutrient	Daily needs of adult	Content of 100 g cows' milk
Energy	12 000 kJ	272 kJ
Protein	72 g	3.2 g
Calcium	0.5 g	0.1 g
Vitamin C	30 mg	1.5 mg

Table C.2

Summary

- Humans need to eat proteins, minerals and vitamins, as well as a small amount of fat.
- They must have enough energy-giving foods, usually in the form of polysaccharide carbohydrates.
- Water is essential.
- There must also be fibre in the diet.

Unit D
What is a good diet?

Testing foods

The next stage in the work on our diet explains how to test everyday foods to find out what is in them. Once you have tried out the tests for starch, sugars, fats, proteins and vitamin C on pure substances, you can then discover for yourselves what is in some of the foods you eat.

Figure D.1 What foods will give you a balanced meal?

Activity D.1 Testing foods WEAR SAFETY GOGGLES

You need: test tubes in rack; spatula; 2 pipettes; Bunsen burner; test tube holder; mat; tripod; gauze; 250 cm^3 beaker for use as water bath; iodine solution; Benedict's solution; ethanol; 10% sodium hydroxide; 0.5% copper sulphate solution; paper; and starch – a suspension of starch; glucose powder; oil; egg white.

1 Try out each of the tests shown in Table D.1 on the pure substances provided, so that you know what result to look for.

Test for	Method	Heat?	Positive result
Starch	Add few drops of iodine in test tube	No	Black or blue-black colour
Glucose	Add equal volume of Benedict's solution. Boil in water bath carefully (Figure D.2)	Yes	Orange colour
Fat or oil test 1	Smear small quantity on paper	Warm gently	Translucent (clear) greasy mark
Fat or oil test 2 (Emulsion test)	Add substance to 2 cm depth of warm* ethanol in test tube. Shake. Drop by drop pour liquid into water in another test tube.	NO - away from flames	Milky emulsion forms
Proteins (Biuret test)	Add 2 cm depth of sodium hydroxide mixed with water. Add few drops of copper sulphate. Examine carefully.	No	Violet or purple

*Warm by standing in water bath at 40 °C.

Table D.1 Food tests

water
Benedict's solution
food
heat

Figure D.2 Apparatus for Benedict's test for simple sugars

2 Choose one food substance and test a small sample of it for starch. You will need about half a spatula of food, broken up into small pieces in about 1 cm^3 of water, or about 0.5 cm^3 of the food if it is liquid.
3 Make a table to record your results. Use the headings: *Food*; *Test used*; *What I saw happen*; *What class of food is present*.
4 Using new samples of the same kind of food each time, test the food for sugar, for fats and oils, and for proteins.
5 Now test other foods in the same way. It is easier to see the results of tests if you choose foods which are not very strongly coloured. Make a table to show all your results.

Testing for vitamin C

You need: Two 1 cm^3 syringes; test tube; 0.1% solution of DCPIP (freshly prepared); 0.1% solution of vitamin C; fresh fruit juices; carton of fruit juice; water in which cabbage has been cooked.

1. Use a syringe to measure out exactly 1 cm^3 of DCPIP. Place it in the test tube.
2. Draw up 1 cm^3 of vitamin C solution into the other syringe.
3. Release the vitamin C solution 1 mm^3 at a time into the DCPIP.
4. Stop when all the colour disappears and note how much vitamin C solution you have used.
5. Repeat steps 1 to 4 using one of the following liquids instead of vitamin C solution: fresh fruit juice, fruit juice from a carton, water in which cabbage has been cooked. Add 1 mm^3 of the liquid at a time and record the number of mm^3 needed for all the colour to disappear.
6. Compare the results you obtain for the liquid with those for the 0.1% solution of vitamin C. Make a table to show the results for the 0.1% solution of vitamin C and the other liquids.

Testing for vitamin C

The test for vitamin C is carried out using a chemical known as DCPIP. DCPIP stands for dichlorophenol-indophenol but you do not have to remember that name!

Summary

- Iodine is used as a test for starch.
- Benedict's solution reacts with sugars when boiled.
- Fats and oils leave a greasy mark on paper.
- The test for proteins is the Biuret test.
- DCPIP reacts with vitamin C.

Questions

1. Design an experiment to discover whether boiling lemon juice for 5 minutes has any effect on the amount of vitamin C in it. Explain what apparatus you would use, how you would carry out the experiment and how you would record your results.
2. Make a food snakes and ladders board (like the one in Figure D.3). Put pictures of unhealthy foods at the top of snakes. Draw healthy foods at the bottom of ladders.
   Rules: You need a die and counters for up to 6 players.
   Throw a 6 to start. Then take it in turns to throw the die and move the number of squares it shows. If you land on a 'bad' food square you go down a snake. On a 'good' food you go up a ladder. The first person to reach 100 wins.

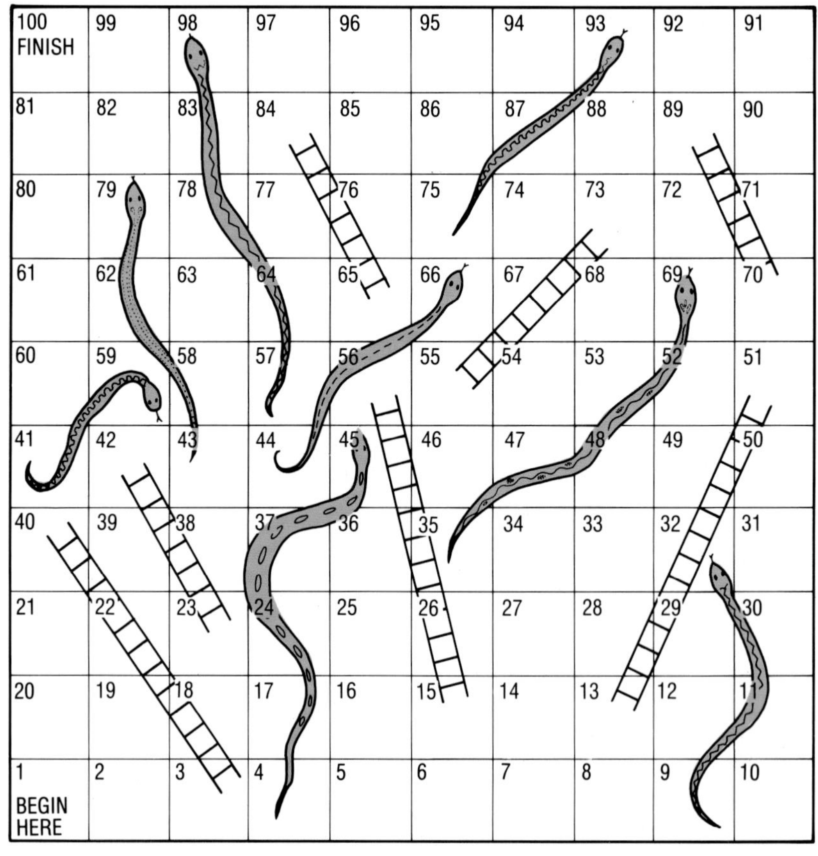

Figure D.3 Snakes and ladders board

Balanced diet

What is a balanced diet? If you look back at pages 13–16 you should be able to make a list of all the types of nutrients you need each day. Can you give examples of everyday foods containing each of these nutrients? Having a balanced diet means eating the nutrients in the right proportions for your needs. Which should you eat most of? Which should you eat only in small amounts?

Table D.2 shows the recommendations for approximate daily intake of food for a teenage girl and boy.

	*Energy /kJ	Protein /g	Calcium /mg	Iron /mg	Vitamins							Fibre /g
					A /mg	B_1 /mg	B_2 /mg	Ni /mg	B_{12} /mg	C /mg	D /mg	
Girl	9600	58	700	14	0.7	0.9	1.4	19	trace	25	.002	30
Boy	11700	70	700	14	0.7	1.1	1.4	19	trace	25	.002	30

*** Energy needs will come from carbohydrates and fats.**

Table D.2 Recommended daily intake

From Table D.2 you can see that males and females of similar age require different amounts of food. Food needs are also influenced by age, body weight, build and activity (see Tables B.1, page 9 and C.1, page 14).

Older, less active people need less energy than lively, growing teenagers. Older people generally need to eat smaller meals. Some find it difficult to chew or digest certain foods, and this may affect what they eat. It's very important for an older person to eat a balanced diet which includes all the essential nutrients. Many old people cannot go out very much. This means that they do not obtain vitamin D from sunlight, so they must eat foods which are rich in vitamin D, such as herring, mackerel or cod-liver oil. Some old people cannot afford to buy fruit or meat frequently, so their diet may lack vitamin C, fibre and iron. Schemes like 'Meals on Wheels' help ensure that old people have a balanced diet.

Activity D.3

Design a poster to tell people what foods to eat to obtain a balanced diet. Your poster is to be displayed in one of these places:

- the school dining hall
- a mother and toddlers group
- a factory canteen

Eating during pregnancy

A pregnant mother has to take in enough food to feed the developing baby (or fetus) as well as her own body. If she does not take in enough food, some nutrients, such as calcium, may move from her body to the baby. Severe or long shortage of foods puts mother and baby at risk.

If the mother does not eat enough energy-giving food the baby may be small and the mother may not be able to provide enough breast milk. It is healthy for a mother to put on about 12 kg in weight during pregnancy.

During pregnancy, the mother should have extra protein daily. She should eat more foods containing calcium while pregnant or breast feeding (**lactating**). Milk will provide both extra protein and calcium. If calcium intake is really low during pregnancy, or if one pregnancy follows another too quickly, calcium can be lost from the mother's bones. The new baby may be born with rickets, too. Pregnant women whose diet is lacking in vitamin D and who do not receive much sunlight can take cod-liver oil to prevent the baby getting rickets.

If a mother is breast-feeding her baby she needs to take even more care with her diet. A good, mixed diet with more milk than usual, some extra iron and possibly vitamin D keeps both mother and baby fit.

Vegetarianism

Vegetarians do not eat meat. Vegans consume nothing from animal sources, that is no meat, no milk or its products like butter or cheese, and no eggs. People choose to be vegetarians for various reasons. They may not like the taste of meat, they may believe a vegetarian diet is healthier, they may not approve of killing animals, or their religion may not allow them to eat meat.

Vegetarians have to take care to eat all the essential nutrients. Vitamin C and fibre are common in plant foods, so vegetarians are unlikely to be deficient in them. Meat is high in saturated fats, which nutritionists say we should cut down on. However, if a vegetarian relies on eggs and cheese as an alternative, the level of saturated fat in the diet may be just as high as for a person who eats meat. Vegetarians, and especially vegans, need plenty of outdoor activity so that their skin can make vitamin D in sunlight.

It is particularly difficult for strict vegetarians and vegans to obtain vitamin B_{12}. Yeast products and vitamin B_{12} supplement can be taken to avoid deficiency. Iron and calcium may also be lacking in some strict vegetarian diets.

As long as vegetarians are careful to eat a varied diet, they should have no problems. Evidence suggests that a varied vegetarian diet is a healthy one.

Figure D.4 Spices and herbs contain important vitamins and minerals

Examination of food contents of meals

You can use tables giving nutrients in different foods to calculate whether your daily diet provides enough of each class of food. Table D.3 shows three common foods. The table on pages 32 and 33 gives much more information.

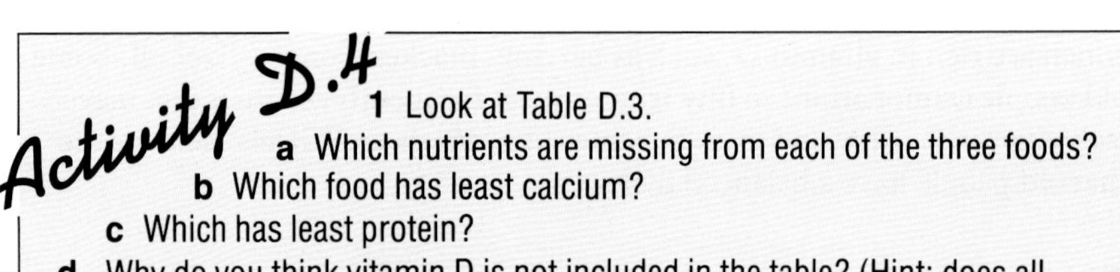

1 Look at Table D.3.
  a Which nutrients are missing from each of the three foods?
  b Which food has least calcium?
  c Which has least protein?
  d Why do you think vitamin D is not included in the table? (Hint: does all our vitamin D come from food?)

2a Write down the foods in your favourite meal. Use the information on pages 32–33 to make a table showing the nutrients in 100 g of each of the foods in your meal.
  b Write down which of the nutrients are low or missing from your meal.
  c Are any nutrients too plentiful in the meal?

Second World War rations: a controlled diet

During and just after World War II food was very scarce. The government of the time knew it was important to have a healthy population. Government scientists worked out the basic food requirements for health and people were given ration books containing coupons for different kinds of food. Vegetarians had special ration books giving extra cheese and no meat.

It was a dull diet. But it was healthy: it provided everything that was needed and no-one had too much food. Many people grew vegetables in their gardens or even kept chickens to supply eggs. People collected rose hips and used them to make a drink containing vitamin C. Recipes tried to make the

<table>
<tr><th rowspan="2">Nutrients</th><th colspan="3">Amount of each nutrient in 100 g of the food</th></tr>
<tr><th>Potato crisps</th><th>Steak and kidney pie</th><th>Vegetable curry</th></tr>
<tr><td>Carbohydrates</td><td>49g</td><td>22g</td><td>10g</td></tr>
<tr><td>Fats and oils</td><td>35g</td><td>17g</td><td>15g</td></tr>
<tr><td>Proteins</td><td>6g</td><td>9g</td><td>1g</td></tr>
<tr><td>Calcium</td><td>37mg</td><td>47mg</td><td>39mg</td></tr>
<tr><td>Iron</td><td>2.1mg</td><td>1.8mg</td><td>0.8mg</td></tr>
<tr><td>Vitamin A</td><td>0</td><td>0</td><td>0.6mg</td></tr>
<tr><td>Vitamin B_1</td><td>0.19mg</td><td>0.12mg</td><td>0.07mg</td></tr>
<tr><td>Vitamin B_2</td><td>0.07mg</td><td>0.25mg</td><td>0.05mg</td></tr>
<tr><td>Niacin</td><td>6.1mg</td><td>4.9mg</td><td>1.1mg</td></tr>
<tr><td>Vitamin C</td><td>17mg</td><td>0</td><td>12mg</td></tr>
</table>

(Niacin is used in the body to give nicotinic acid (B_{12} group)).

Table D.3 Nutrients in different foods

rations more interesting. Can you imagine cakes made without fat, carrot jam, or sweets made with dried milk and banana flavouring?

Diets in different cultures and countries

Traditional eating habits often provide a sensible diet and avoid some health risks. For example, pork is not eaten by people of some religions. This helped people of those faiths avoid the tapeworm and trichinosis parasites which may have been in pork. Today, meat is inspected before it is sold and pork is cooked thoroughly to ensure that it does not carry live parasites.

Over many generations, people discover the kind of diet which will keep them healthy. They learn about the foods that are most suitable for the climate they live in and their way of life. However, poverty, famine, changes in lifestyle or pattern of agriculture and trade can disturb traditional eating customs. If people migrate to another part of the world they may find that their customary foods are not available or suitable. It may be necessary to learn to eat new foods.

The effect of environment – Eskimos

Eskimos survive almost entirely on protein and fat in meat and fish. In summer they eat a few berries. Their diet is rich in calcium, iron, and vitamins A and D, but it tends to be low in vitamin C and fibre. Over thousands of years, the bodies of Eskimos have evolved so that they can live healthily on this very specialised diet. What is more, their metabolism enables them to use this diet to survive in the severe Arctic climate and lead a strenuous hunting life.

Malnutrition in Britain

A poor diet causes malnutrition (bad nutrition). Malnutrition may be due to:

1 Not having enough food
2 Missing out on protein, or on some minerals and vitamins
3 Eating too much, especially of fatty or sugary foods

Figure D.5 Wartime rations provided the basic requirements for people to stay healthy

Figure D.6 What differences are there between the nutrients in an Eskimo diet and your food?

Cause 1 is not often a problem in Britain, where most people have enough food. Some people may suffer the second type of malnutrition if they have an unbalanced diet – perhaps because they do not know enough about nutrition, or because they can't afford good food. But the main form of malnutrition in Britain is number 3, eating too much of the wrong foods. The only way to overcome this problem is to increase people's knowledge about nutrition and the importance of a well-balanced diet.

Getting the balance wrong

A person who eats more than they need stores fat and becomes obese. People who are seriously obese have a shorter life expectancy (the average number of years a person can expect to live) than people of a normal weight. Some diseases are more common in people who are overweight. These include heart disease, high blood pressure, strokes, diabetes, gallstones and varicose veins. Being overweight is also a problem for anyone with arthritis (disease of the joints).

Although it is better not to be very overweight, it is unwise to try to slim too fast or too much. People vary a great deal in build; the right weight for you may be much lighter, or heavier, than that for someone else of your age. Instead of accepting what is right for them, many people try to slim unnecessarily. Sometimes this can be dangerous; a person may develop **anorexia nervosa**. Sufferers from this illness (usually girls and young women) refuse to eat. They become thinner and thinner and their bodies cease to function normally. Anorexia nervosa *can* be cured, but treatment may take a long time and the patient needs a great deal of support.

The best way to slim is gradually. Eat less sugar and fat, but make sure you get a good balance of essential minerals and fibre. This is very important if you are still growing. Taking more exercise can also help you lose weight and improve your general fitness. If you need to lose a lot of weight it is sensible to ask your doctor for advice.

Figure D.7 A child suffering from marasmus

Shortage of food

In some parts of the world people suffer from malnutrition because they are very poor, or because food is not available, as a result of famine. Millions of people in the developing world do not have enough to eat, through no fault of their own. As well as being dangerous in its own right, malnutrition increases the risk of serious illness or death from infectious diseases.

If there is not enough carbohydrate in the diet, the body can obtain energy from protein. But a diet without carbohydrate is also likely to be short of protein. If this happens, the body begins to use up the protein needed for growth and repairs. Growth stops and the muscles begin to waste away. People suffering from such a shortage of food, or starvation, become very thin; their skin is loose and dry, their muscles shrivel and they are sluggish and lacking in energy. If the starvation lasts for some time, there will be more serious damage to the body, until eventually the person dies from lack of food.

Two kinds of disease are linked with undernourishment in children. **Kwashiorkor** occurs in older infants and children. Face, hands, feet, and liver become swollen and the children become sluggish. Kwashiorkor may be caused by undernourishment, particularly shortage of protein. However, new research suggests that it may be linked with **toxins** (poisons) from mould on food. **Marasmus** is suffered by younger infants who lack both energy-giving foods and protein (Figure D.7). The infant becomes very thin and irritable. Often marasmus occurs in a child who has been weaned from the mother's milk on to food which does not provide enough energy or protein.

Figure D.8 The same child, after treatment

Food for the world

There is enough food in the world to feed everyone. But political problems and transport costs make it difficult to carry this food from country to country. World population growth, changes in lifestyle and wars all add to the problem of the unequal distribution of food throughout the world. In some areas, agricultural land has become desert. People in poorer nations cannot afford fertilisers or tractors; they have to rely on simple, cheaper methods. They also need help in controlling major pests such as locusts. Where plants will grow, the governments of poor countries often decide to plant crops which will sell overseas rather than those which might provide food for the local people.

Although emergency help can be given, the long-term solution is for people to become self-sufficient (that is, able to provide enough food for their own needs). New methods of farming that are *appropriate* for local people will provide surplus for drought years. Organisations like Oxfam and Cafod help with emergencies and also assist people with long-term schemes to grow food for the future. Computers are used to store information about crops, and to help with planning (Figure D.9). Hard work, suitable agricultural methods and careful planning are needed if the world is to solve the very complicated problems of providing food for all.

Figure D.9 Using a computer to keep records of crops and to plan for the future

Summary

- A varied diet will usually fulfil most kinds of food requirements. It should contain some fibre, but not too much fat, sugar or salt.
- Bad diets cause health problems.
- Malnutrition arises from eating too much, too little or the wrong proportions of classes of food.

Questions

1 If you were shipwrecked on a remote island, explain what foods you would need to keep alive until you were rescued.
or
2 You are in charge of the food for an expedition to a wild, isolated place. What kinds of foods must you take with you? Explain your answer.

Unit E
The teeth

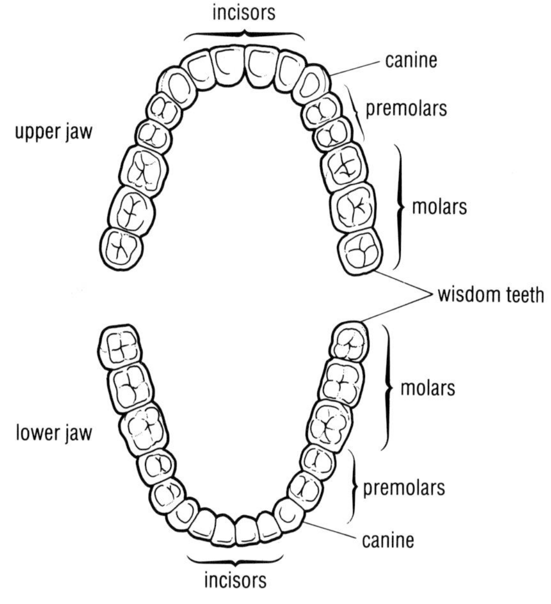

Figure E.1 A full set of teeth. Which of these teeth have you got?

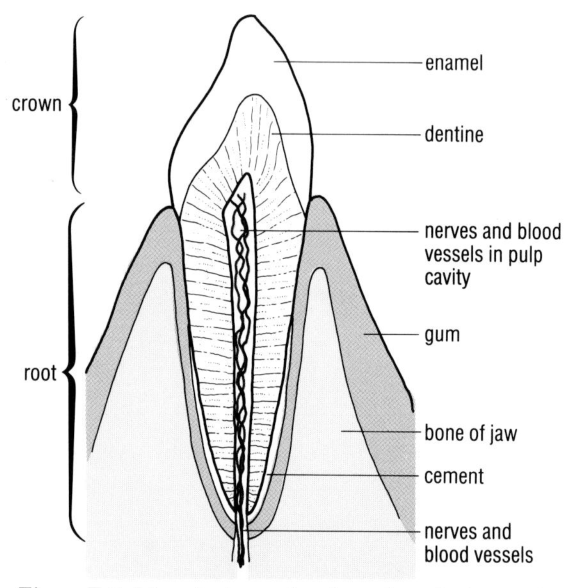

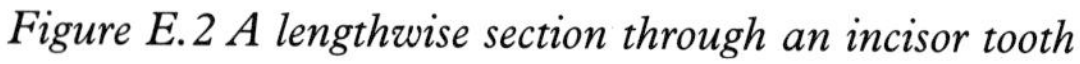

Figure E.2 A lengthwise section through an incisor tooth

Food has to be made *soluble* before it can be used in the body. First it is broken up, by chewing, into many small pieces. This increases the total surface area which the digestive juices can act upon.

The teeth – the food-fragmenters

Humans have four kinds of teeth: **incisors, canines, premolars** and **molars** (Figure E.1). Activity E.1 explains how you can find out what each kind of tooth does.

Young children have 20 **milk teeth**. These fit easily into the child's small mouth. Gradually the milk teeth are lost and replaced by adult teeth. The molars only grow in the adult set of teeth. The back molars, known as **wisdom teeth**, often appear a long time after the others – sometimes when a person is over 20 years old.

Figure E.2 is a section through a front tooth or incisor. It shows the regions of the tooth and its structure. The outside enamel is very hard indeed. It protects the dentine enclosing the pulp cavity. This contains nerves and blood vessels. The nerves and blood vessels run through the jaw to the brain and circulatory system.

Activity E.1 Your own teeth

a Examine your own teeth with a mirror (or ask a friend to do so for you). Record your observations on a table like the one below:

	Upper jaw				Lower jaw			
	Incisors	**Canines**	**Pre-molars**	**Molars**	**Incisors**	**Canines**	**Pre-molars**	**Molars**
Total number								
Number of fillings								

b Describe the shape of each type of tooth.

c At home, experiment to find out what each kind of tooth does. As you follow these instructions record what each kind of tooth does and what the tongue does. Take half a raw carrot. Nibble or gnaw off a small piece. Then, bite off a piece. Tear a piece off in the same way that dogs tear at meat. Finally, chew a piece and grind it up for swallowing. Record your results in a table under the headings 'Type of tooth', 'Shape of tooth' and 'Function'.

The dental formula

The **dental formula** gives the number of teeth of each kind found in one half of the mouth. Dentists use the dental formula to record what they do to your teeth. The number in the upper jaw is placed over the number in the lower jaw. The formula for an adult with a full set of teeth is:

$$i\,\frac{2}{2}\;c\,\frac{1}{1}\;pm\,\frac{2}{2}\;m\,\frac{3}{3}$$

Tooth care

Teeth have to last a long time. If neglected they can develop decay or **dental caries**. Bacteria form a film over the surface of the teeth and between them. These bacteria act on sugary food and change it to acid. The acid first attacks the enamel, then the dentine, and finally breaks through to the pulp. Toothache follows. Gum or **periodontal disease** is caused by **plaque**. Plaque is a whitish, soft film which contains bacteria. It forms between the teeth, and where they meet the gums. Sometimes plaque causes the gums to become inflamed and infected. Eventually the jaw bone itself may be destroyed. Fortunately, there are several ways of reducing tooth troubles.

- Make sure your diet includes sufficient calcium (found in dairy products) and avoid sweet foods.
- Clean teeth and gums daily with a fluoride toothpaste. Brush all surfaces of the teeth with a circular movement. (It takes about three minutes to do this thoroughly).
- Have 1 part per million (ppm) of fluoride in drinking water (this is decided by the local water authority).
- Make regular visits to a dentist, who will clean the parts of your teeth you cannot reach and check for signs of decay.

Activity E.2

Are your teeth clean?

You need: your toothbrush and toothpaste, Vaseline, a disclosing tablet, mirror.

1. Smear Vaseline on your lips to prevent them being dyed.
2. Use the disclosing tablet, following the instructions on the packet.
3. Examine your teeth in a mirror. The dye in the tablet will have stained the plaque on your teeth. How clean are they? Where is there most plaque?
4. Clean your teeth thoroughly. Which parts of your teeth are the most difficult to keep clean? How could you clean your teeth more thoroughly?

The case for fluoride

Fluoride occurs naturally in the water of some parts of the world. Evidence from scientific research suggests that low concentrations (up to 1 part per million) of fluoride help reduce tooth decay. Now, it is added in low concentrations to some water supplies as a preventive measure, to protect against decay. Table E.1 shows the results from several studies of pairs of towns; half with fluoride added to the water, and half without.

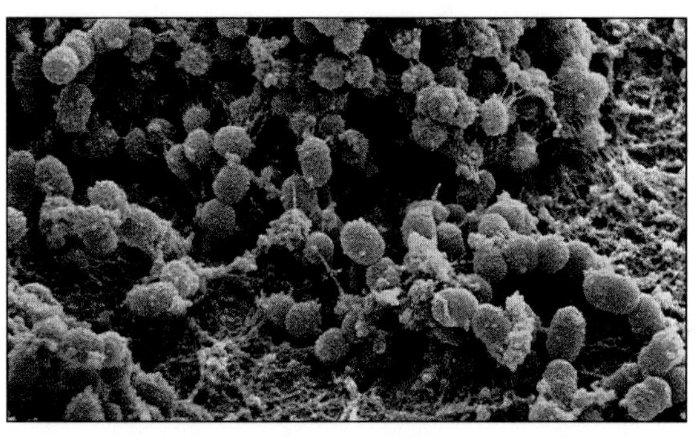

Figure E.3 Plaque under a microscope ▲

▼*Table E.1*

Town	Is fluoride added?	No. of extractions per 100 children at 5	No. of extractions per 100 children at 12	No. of treated teeth per 100 at 5	% free of decay at 5	Average no. teeth decayed at 5
A1	Yes	6				
A2	No	40				
B1	Yes	1	13			
B2	No	45	74			
C1	Yes			130		
C2	No			330		
D1	Yes				62	
D2	No				30	
E1	Yes				About 50	1.5
E2	No				Less than 20	4

A survey showed that over 70% of the population believed fluoridation of water is beneficial, but some people objected. Why do you think people object? They may be concerned about possible side-effects such as mottling of the teeth. It is also known that higher concentrations of fluoride can affect bones and joints. But much research has been done to discover possible risks from adding fluoride to the water supply, and most scientists, doctors and dentists believe that the amounts used are safe. In 1985 the government passed the Water (Fluoridation) Act. This Act makes it legal for Health Authorities and water suppliers to add fluoride to water.

Fluoride can be added to toothpaste, too. In places where fluoride occurs naturally or is added to the water supply, no more needs to be added and fluoride toothpaste is unnecessary.

In some areas fluoridation of the water has greatly reduced the amount of decay in children's and young people's teeth. This saves dental costs and means that teeth will last longer. Where no fluoride is added many children still develop dental caries and have to have their teeth filled or pulled out. Artificial teeth on a plate are often uncomfortable and never as good as a person's own teeth. If the teeth have been removed, the jaw bones gradually change shape and so hold the artificial teeth less well.

Summary

- The four kinds of teeth – incisors, canines, premolars and molars – have different functions.
- Regular cleaning is important to keep teeth and gums healthy.
- Very small quantities of fluoride added to the water supply protect teeth against decay.

Questions

1 Write the dental formula for a) the first set of teeth, and b) the second set of teeth before the wisdom teeth come through.

2 List the arguments for and the arguments against adding fluoride to the water.

3 *Either* a) make a poster on 'Caring for your teeth', to appear in a dentist's waiting room.

or b) write a pamphlet for parents on 'Looking after your children's teeth.'

Unit F
Digesting our food

Figure F.1 shows the human digestive system (alimentary canal, or gut). It is a long, twisting tube running from the mouth to the **anus**. On Figure F.1, follow the path of the food from the mouth, past the opening of the trachea, down the **oesophagus** (gullet), through the **stomach**, and out – via the **pyloric sphincter** – into the **duodenum**. The pyloric sphincter is a circular muscle which is normally contracted. It relaxes to allow food into the intestine.

The duodenum is the first part of the **small intestine**. The whole of the small intestine is about six metres long (nearly as long as a cricket pitch). The next part of the digestive system, the **large intestine**, is wider (but shorter). The **appendix** is like a short worm attached to the **colon** or first part of the large intestine. In humans the appendix has no known function. The colon runs up the right side of the abdomen, across and down again. It then joins the short **rectum** which opens at the **anus**.

The digestive system changes food so that it can be absorbed by the body. In the mouth, the teeth break up the food mechanically or **masticate** it (see Unit E). This gives the food a larger surface area for juices to act upon. The food is moved around by the tongue and mixed with saliva from three pairs of **salivary glands**. The saliva moistens the food.

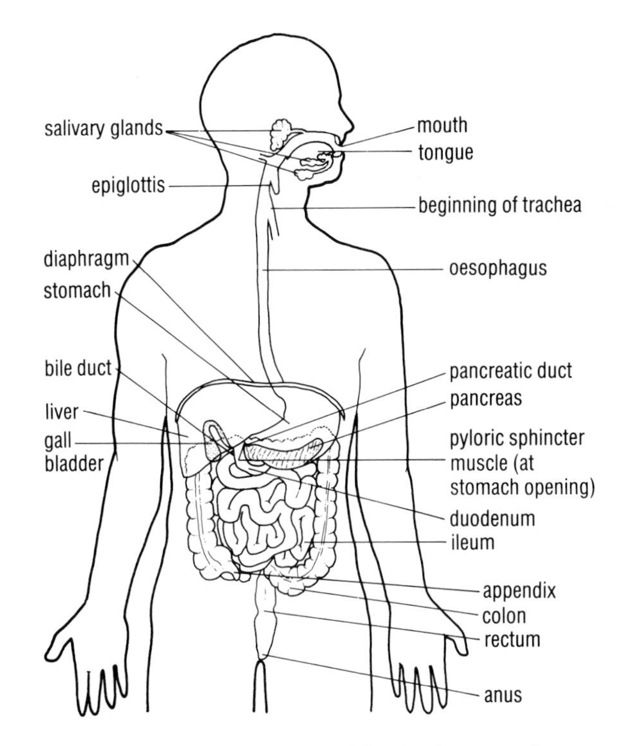

Figure F.1 The digestive system. The duodenum and ileum together make up the small intestine. The appendix, colon and rectum comprise the large intestine

Enzymes – biological catalysts

Saliva contains an active substance or **enzyme**. An enzyme is a special protein which acts like a catalyst in chemistry and *speeds up chemical reactions without itself being changed*. Enzymes are always produced by living cells. Each enzyme catalyses one special reaction – it is *specific* in its action. Some enzymes build up new chemicals in the body, some break them down. During digestion a number of enzymes change the food. Saliva contains the enzyme **salivary amylase**. It catalyses the reaction:

starch + water ➡ malt sugar

Malt sugar has two sugar units. It is soluble and, like glucose, produces an orange colour when heated with Benedict's solution (see Figure D.2 on page 17).

Swallowing

After food has been mixed with saliva and chewed in the mouth, it is rolled into a small ball or bolus. This is swallowed by a quick, muscular movement. Place your finger on your throat and swallow. You will feel the muscle action. While you swallow the back of the **palate** (roof of the mouth) closes the nose cavity. A flap called the **epiglottis** seals the windpipe (**trachea**) so that you do not choke. The bolus is pushed along the oesophagus by circular muscles in its walls. These contract behind the food (Figure F.2) and push it

Activity F.1

What does saliva do? (A homework activity)

1. Chew a piece of bread (about 2 cm cube) slowly, keeping it in your mouth for as long as possible.
2. Note down any changes in texture and taste. Is it still solid? How does it taste?
3. Read the section about the digestive system on this page. How does it help explain what happened to your bread? Write down your explanation. Suggest an experiment you could do to test it.

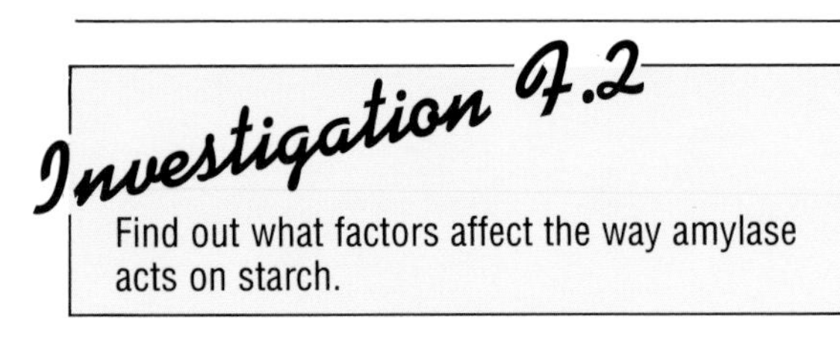

Find out what factors affect the way amylase acts on starch.

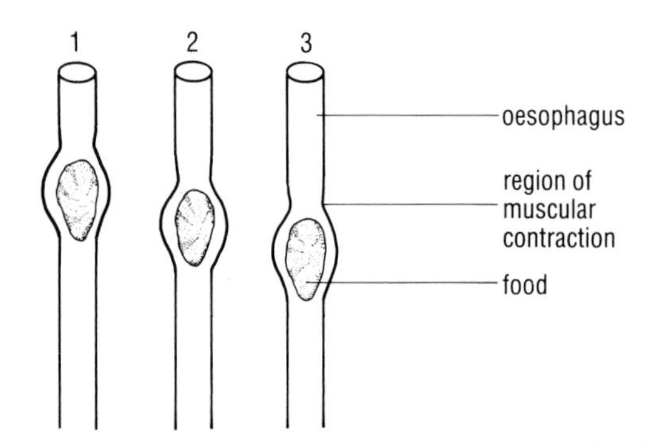

Figure F.2 Peristalsis in part of the oesophagus. What is happening?

downwards. This muscular action is called **peristalsis**. The oesophagus carries food through the thorax and the diaphragm to the stomach. The food is pushed along rapidly and reaches the stomach in under ten seconds.

In the stomach

The stomach is a large, muscular bag. Food remains in it for about three to four hours. **Gastric juice** is secreted (produced) by glands in the stomach wall. The juice is very acid. Slimy mucus is also produced. This protects the stomach wall from the action of the acid. The acid kills any bacteria in the food. The muscular stomach wall moves and churns the food up with the gastric juice until it becomes fairly liquid. This mixture, known as **chyme**, then passes through the pyloric sphincter to the duodenum.

The small intestine

In the small intestine three juices are mixed with the chyme:

1 **Bile** – a very alkaline, dark green liquid which comes from the **liver**. It is stored in the **gall bladder** and passes along the **bile duct** to the small intestine.
2 **Pancreatic juice** – produced by the **pancreas**.
3 **Intestinal juice** – which comes from glands in the wall of the small intestine.

In the small intestine, bile emulsifies fat in the food, that is, it makes the fat form many small droplets. The surface area of these fat droplets is very large, so it is easier for digestive juices to work on the fat. Bile is alkaline: it neutralises the acid from the stomach and makes the correct conditions for enzymes of the pancreatic and intestinal juices to work. Food takes about eight hours to pass through the small intestine. During that time digestion is completed and the dissolved, useful materials are absorbed.

More about digestive enzymes

In the stomach and small intestine food is bathed in juices. These contain several enzymes which change food to soluble substances. Enzymes can be grouped according to the reactions they catalyse. In digestion three groups are involved, **carbohydrases**, **lipases** and **proteases**. Table F.1 shows their reactions. Salivary amylase is an example of a carbohydrase.

Food	Enzyme group	Final product
Carbohydrates	Carbohydrases	Sugars
Fats and oils (lipids)	Lipases	Fatty acids and glycerol
Proteins	Proteases	Amino acids

Table F.1 The groups of digestive enzymes

Enzymes have special properties. For instance, they work well at 37 °C, the temperature of the body. Above 40 °C the proteins of enzymes begin to break down chemically. Nowadays, enzymes are used in many ways in industry.

Table F.2 shows the whole process of digestion. For each region of the digestive system the secretions (juices) and mechanical activity (movements) are given.

Region of digestive system	Acid or alkaline	Mechanical activity	Secretions present	What happens
Mouth	Neutral	Chewing	Saliva	Food broken up Beginning of starch digestion
Oesophagus	Neutral	Peristalsis	None	Movement downwards
Stomach	Acid	Churning	Gastric	Coagulates milk proteins Begins protein digestion
Small intestine	Alkaline	Peristalsis	Bile from liver Pancreatic juice from pancreas Intestinal juice	Makes conditions alkaline Emulsifies fat Continues digesting starch Digests fats and proteins Completes digestion of proteins and carbohydrates Absorption of sugars, amino acids, fatty acids and glycerol
Large intestine				Absorbs water Produces faeces

Table F.2 Stages in digestion

Questions

1 Copy out the beginning of this sentence and complete it with the best answer from **a–e**.
Enzymes are said to be catalysts because they
a are proteins.
b are affected by temperature.
c are affected by pH.
d change the speed of chemical reactions.
e alter food substances during digestion.

2 **a** Use the table on pages 32–33 to work out what kinds of major food classes are present in a cheese sandwich.
b What happens to the sandwich as it passes through the digestive system?

3 Some washing powders contain enzymes. Imagine that you have been given two washing powders: *Washwhite* and *Sudso*. One of them contains enzymes. Explain what you would do to find out which of them contained enzymes. Suggest one reason why enzymes are used in washing powders.

Summary

- As food passes along the digestive system it is broken up mechanically and chemically.
- Enzymes are proteins which act as catalysts.
- During digestion enzymes break down large food molecules to smaller ones.

Unit G
What happens to digested food?

Activity G.1

A model of part of the wall of the small intestine can be made using a piece of transparent Visking tubing. Inside is a mixture of starch and sugar molecules representing partly digested food. Outside the tubing is water, which represents blood.

You need: 15 cm of Visking tubing soaked in water; thread; beaker; thermometer; small funnel; pipette; test tubes; Bunsen burner; iodine solution; Benedict's solution; starch and sugar mixture (1 g starch made into a paste and 5 g glucose made up to 100 cm^3).

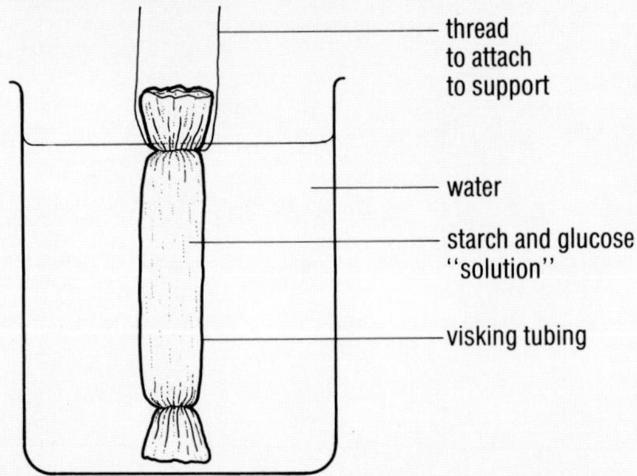

Figure G.1 A model of the small intestine made of Visking tubing

1. Tie one end of the tube in a knot (see Figure G.1).
2. Using the funnel, fill the tubing with starch and glucose mixture.
3. Tie the opening of the tubing tightly with thread. Rinse the outside of the tubing gently under the tap.
4. Stand the tubing in a beaker of water (Figure G.1).
5. Leave the tubing for 30 minutes.
6. Place 1 cm depth of water from the beaker in each of two test tubes. Test one for starch, the other for sugar (see Table D.1, page 17).
7. Has the starch or sugar passed from the tubing into the water in the beaker?

Why is digestion necessary?

We all know that food has to be digested. But can you explain why this is? As you have seen, starch is common in our food. Starch does not dissolve, although it will make a suspension with water. Glucose, which is produced during digestion, *is* soluble. The next activity uses a model of the small intestine to show the difference in how starch and glucose behave. It will suggest one reason why digestion is necessary.

The tubing acts like the wall of the small intestine. It allows small molecules to pass through but is a barrier to large molecules.

Inside the small intestine

Food remains in the small intestine until most of it has been digested. Digestible carbohydrates are changed to simple sugars. Protein chains become amino acids. Fats and oils produce fatty acids and glycerol. The small molecules of vitamins and minerals are also present.

The small intestine is narrow and very long. Inside it are minute finger-like projections or **villi** (singular villus). There are about 30 tiny villi on each millimetre square of surface (making it rather like velvet) (Figure G.2). Altogether there are about 5 million villi. The total inside area of the small intestine is about 10 m^2! This provides a huge surface for absorbing food.

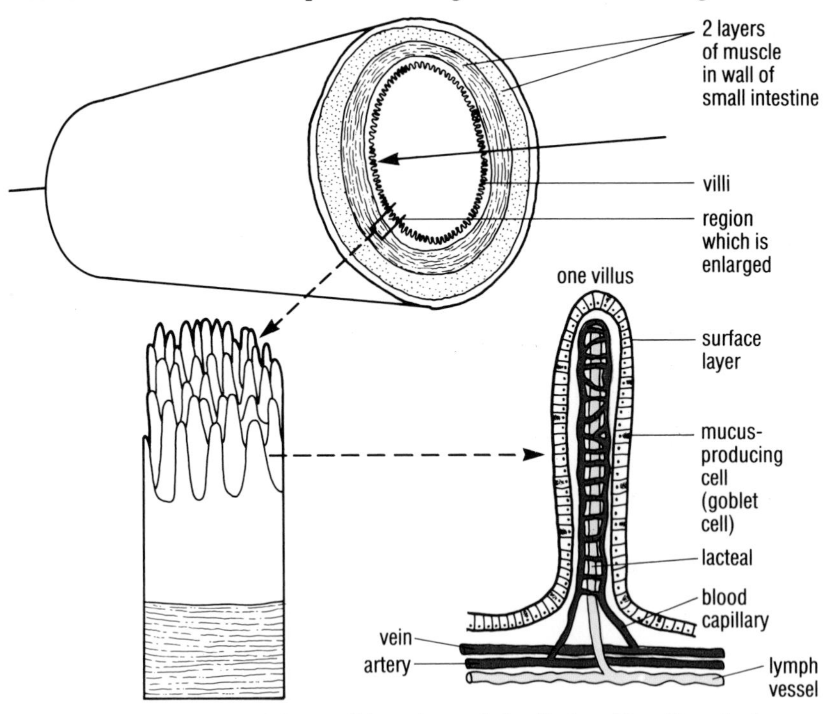

Figure G.2 A cross-section of the small intestine and detail of a villus. How do the villi improve absorption of digested food?

Absorption of digested food

Each villus has a thin layer of cells which form its surface or walls (see Figure G.2). Inside is a network of blood capillaries surrounding a single tube (**lacteal**).

Digested food is absorbed through the walls of the villi. Sugars, amino acids, vitamins and minerals enter blood capillaries and are carried in the blood back towards the heart via the liver. Fatty materials pass into the lacteal in the middle of each villus. This connects with lymph vessels which collect materials from around the body. They join the blood system in the neck.

Assimilation

After dissolved food has been absorbed, it is **assimilated** or used by the body.

Useful amino acids are made into new proteins. Any unwanted amino acids are changed or **deaminated** in the liver. Deamination produces carbon-containing molecules (which can be used as a source of energy) and urea, containing nitrogen. The urea is excreted as waste through the kidneys.

Some glucose is stored in the liver or muscles as **glycogen**, a substance rather like starch. Some is used in respiration to give energy. The remainder is changed to fat and stored with the surplus fatty material.

What happens in the large intestine?

Food which has not been digested is moved along by peristalsis into the large intestine. Water from the food is absorbed into the blood through the colon walls. The remains pass through the rectum and are **eliminated** through the anus as **faeces**.

Summary

- The surface area inside the small intestine is increased by projections called villi.
- Once substances have been absorbed they are assimilated or used by the body.

Activity G.2

Looking at tripe

Tripe comes from a cow's stomach. When you buy tripe, it is white because it is partly cooked. The projections on it are similar to villi but larger.

You need: 1 cm^2 tripe; 2 mounted needles; pair of forceps.

1. Look carefully at the projections on the surface.
2. Use a mounted needle to lift off this inner surface layer. You will uncover a layer of muscle with all its fibres lying in one direction. Note the direction of the fibres. What would happen if these fibres contracted?
3. Lift off this layer of muscle. Underneath it you will find another muscle layer. Which way do the muscle fibres run in this layer? Compare the muscle layers with Figure G.3.
4. Copy Figure G.3. Indicate the inside surface and complete the three labels. Add notes on how the two layers of muscles enable the wall to change shape.

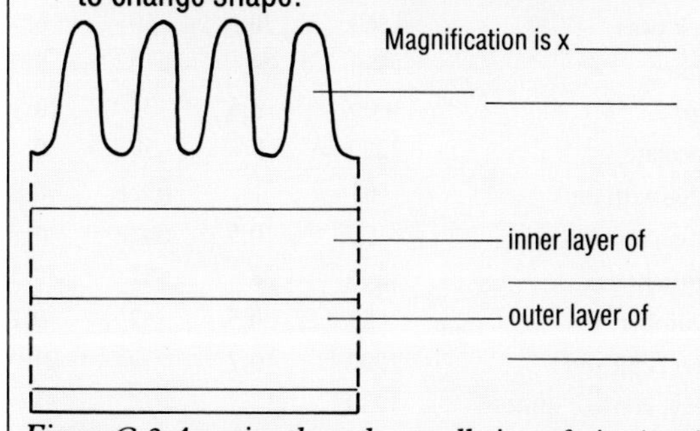

Figure G.3 A section through a small piece of tripe (cow's stomach wall). Remember that tripe has larger projections than the small intestine

Nutrients in 100 grams of food (approximate values)

Food	Energy /kJ	Protein /g	Carbo-hydrate /g	Fat /g	Water /g	Sodium /mg	Calcium /mg	Iron /mg	Vitamin A /mg	Vitamin B_1 /mg	Vitamin B_2 /mg	Niacin /mg	Vitamin C /mg	Vitamin D /mg	Fibre /mg
Apple	200	0.5	12	0	85	2	5	0.3	0.005	0.04	0.02	0.1	5.0	0	2
Bacon, grilled	1650	28	0	31	35	2400	15	1.3	0	0.57	0.27	12.5	0	0	0
Banana	350	1	19	0.3	70	1	5	0.4	0.02	0.04	0.07	0.8	10	0	3
Beans, baked	350	5	15	0.5	75	550	50	1.4	0.01	0.08	0.06	1.3	0	0	7
Bean sprouts	150	4	6	0.2	90	5	20	1.3	trace	0.13	0.13	0.8	19	0	3*
Beans, runner	80	2	3	0.2	90	1	20	0.7	0.06	0.03	0.07	0.8	5	0	25
Beef, stewed	930	31	0	11	55	360	15	3.0	0	0.03	0.33	10.2	0	trace	0
Bhajia (Pakora)	450	2.5	5	9	75	530	60	2.9	0.07	0.04	0.08	0.8	15	trace	3
Biscuits, chocolate	2200	5.5	67	28	2	160	110	1.7	0	0.03	0.13	2.7	0	trace	3
Biscuits, plain, digestive	2000	6	69	21	3	600	90	3.2	0	0.14	0.11	2.4	0	trace	7
Bounty bar	1950	5	58	26	10	180	100	1.3	0	0.04	0.1	0.3	0	0	2*
Bread, white	950	8	50	2	40	540	100	1.7	0	0.2	0.03	1.4	0	0	4
Bread, wholemeal	900	9	42	3	40	540	25	2.5	0	0.3	0.08	4.0	0	0	9
Butter	3050	0.5	0	82	15	870	15	0.2	0.98	0	0	0.1	0	trace	0
Cabbage, boiled	70	1.5	2	0	95	4	40	0.4	0.05	0.03	0.03	0.5	20	0	2
Cake, sponge	1900	6.5	53	27	15	350	140	1.4	0.3	0	0.1	0.5	0	trace	1
Carrots	100	0.5	6	0	90	100	48	0.6	2	0.06	0.05	0.7	6	0	3
Cauliflower	40	1.6	1	0	95	4	20	0.4	trace	0.06	0.06	0.8	20	0	2
Chapatis	1400	8	50	13	30	130	70	2.3	0	0.25	0.05	1.7	0	0	5
Cheese	1670	26	0	34	35	600	800	0.4	0.4	0.04	0.5	6	0	trace	0
Chick peas	1360	20	50	6	10	40	140	6.4	0.03	0.5	0.15	4.2	3	0	15
Chicken	750	29	0	7	65	80	0	1.2	0	0.06	0.2	7	0	trace	0
Chips	1000	3.5	34	10	45	40	15	0.8	0	0.2	0.02	1.5	10	0	2
Chocolate	2200	8	59	30	2	120	220	1.6	0	0.1	0.2	0	0	0	0
Coffee with milk	65	1	trace	0.8	96	20	30	0.1	0.02	0.02	0.04	0.02	0.2	0	0
Coffee, no milk	12	0.5	trace	0	100	1	5	0.1	0	0	0.01	0.6	0	0	0
Cornflakes	1500	8	82	0.5	5	1200	0	7	0	1	1.5	16	0	trace	11
Cucumber	50	0.5	2	0.1	95	13	25	0.3	trace	0.04	0.04	0.3	8	0	0.4
Curry, chicken	1000	10.7	2	21	64	475	25	1.8	0.22	0.03	0.09	4	1	0	0.4
Curry, lamb or mutton	1550	17	2	33	45	770	25	1.9	0.18	0.07	0.18	6.3	6	0.2	0
Custard	500	4	17	4	75	80	140	0.1	0.04	0.05	0.20	1.0	0	trace	0
Dahl (lentils)	450	5	13	5	75	280	20	1.4	0.04	0.07	0.03	0.5	1	trace	4*
Egg, boiled	650	12	0	11	75	140	50	2.0	0.14	0.1	0.5	3.5	0	trace	0
Egg, fried	950	14	0	20	65	220	65	2.5	0.14	0.17	0.42	4.2	0	trace	0
Fish fingers	950	14	17	13	55	350	45	0.7	0	0.1	0.1	1.4	0	trace	0
Fish, fried in batter	850	20	8	10	60	80	80	0.5	0	0.1	0.1	1.2	0	0	0
Guava	250	1	9	1	85	4	25	0.9	trace	0.05	0.05	1.2	240	0	1.5
Ham	500	20	0	5	75	1200	0	1.2	0	0.5	0.3	4.0	0	N	0
Hamburger, fried	1100	20	7	17	55	900	35	3.1	0	0	0.2	4.0	0	N	0
Ice cream	700	3.5	21	8	65	70	120	0.3	0	0.04	0.15	trace	0	trace	0
Jam	1100	0.5	69	0	30	14	18	1.2	trace	0	0	0	10	0	0
Kebab, Indian	1500	30.1	1	26	40	550	40	3.3	0.06	0.19	0.38	12.9	2	trace	0
Kheema, lamb or mutton	1300	20	2	25	50	820	22	1.8	0.15	0.07	0.2	7.1	5	trace	0
Lamb, roast	1100	26	0	18	55	65	8	2.5	0	0.12	0.31	11.0	0	trace	0
Lemonade	80	0	6	0	95	0	0	0	0	0	0	0	0	0	0
Lettuce	40	1	1	0	95	0	25	0.9	0.2	0.07	0.08	0.3	15	0	15
Liver, lamb's, fried	1000	30	0	13	55	83	8	10.9	30.5	0.38	5.65	24.7	18.6	trace	0
Mandarin oranges, tinned	250	0.5	14	trace	85	9	18	0.4	trace	0.07	0.02	0.3	14	0	0.3
Mango	250	0.5	15	trace	80	7	10	0.5	0.2	0.03	0.04	0.4	30	0	1.5
Margarine	3050	0	0	80	15	800	0	0	1.0	0	0	0	0	trace	0
Marmalade	1100	0.1	70	0	30	20	35	0.6	trace	0	0	0	10	0	0
Marrow (Dudi)	70	0.06	4	0	95	1	17	0.2	trace	0	0	0.3	5	0	2

Nutrients in 100 grams of food (approximate values)

Food	Energy /kJ	Protein /g	Carbohydrate /g	Fat /g	Water /g	Sodium /mg	Calcium /mg	Iron /mg	Vitamin A /mg	Vitamin B_1 /mg	Vitamin B_2 /mg	Niacin /mg	Vitamin C /mg	Vitamin D /mg	Fibre /mg
Milk	250	3.3	5	4	90	50	120	0	0.05	0.05	0.2	0.1	1	trace	0
Milk shake	300	3.0	60	3	90	200	100*	0	0.03*	0.03*	0.01*	trace	trace	0	0
Mooli (white radish)	100	1.0	4	trace	95	25	25	0.4	0	0.02	0.03	0.7	40	0	2*
Muesli	1550	13	66	8	6	180	200	4.6	0	0.33	0.27	5.7	0	0	7
Mushroom	50	1.8	0	1	92	9	3	1.0	0	0.1	0.4	4.6	3	0	3
Okra (Lady's fingers)	70	2.0	2	trace	90	7	70	1.0	0.02	0.10	0.10	1.3	25	0	3
Onion	100	0.9	5	trace	93	10	30	0.3	0	0.03	0.05	0.4	10	0	1.3
Orange	150	1	9	0	85	0	40	0.3	0.05	0.10	0.03	0.2	50	0	2
Orange juice	170	0.5	9	0	87	0	0	0.3	0.05	0.08	0	0.2	50	0	0
Papadum	1150	21	46	2	10	2850	80	13.0	0	0.21	0.12	1.2	0	0	12
Peanuts, fresh	2400	24	9	50	5	0	60	2	0	1.0	0.1	16.0	0	0	8
Pears	175	0.5	10	0	85	2	8	0.2	trace	0.03	0.03	0.3	3	0	2
Peas, fresh or frozen	300	6	11	0.4	80	0	35	2	0.05	0.3	0.09	1.6	12	0	5
Pineapple, fresh	200	0.5	12	0	85	2	12	0.4	trace	0.08	0.02	0.2	25	0	1
Pitta bread	1150	9	58	2	35	520	90	1.7	0	0.24	0.05	1.4	0	0	4
Pork chop, grilled	1400	29	0	24	45	80	10	1.2	0	0.66	0.20	11.0	0	trace	0
Potato, baked	350	2	20	0	60	0	0	0.6	0	0.1	0.03	1.0	10	0	2
Potato, boiled	350	1.5	20	0	80	0	0	0.3	0	0.1	0.03	1.0	10	0	2
Raisins	1050	1.0	65	0	20	50	60	1.5	0	0.1	0.08	0.5	0	0	7
Rice, boiled	500	2	30	0.3	70	2	1	0.2	0	0.01	0.01	0.8	0	0	1
Rice pudding	550	4	20	4	70	60	30	0.1	0.03	0.04	0.14	1.1	0	trace	0.5
Samosa	2400	6	19	54	25	40	35	1	0.02	0.09	0.06	2.7	3	0	0.5
Sausage roll	1950	8	38	32	20	580	80	1.5	0.1	0.1	0.04	2.0	0	0	1
Soy sauce	250	5	8	0.5	70	5720	65	4.8	0	0.04	0.17	1.8	0	0	0
Spaghetti	1450	12	74	2	10	3	25	2.1	0	0.22	0.03	3.1	0	0	6
Spinach	100	3	4	0.5	90	70	95	3.1	1	0.1	0.2	1.1	50	0	6
Sugar	1650	0	100	0	trace	trace	trace	0	0	0	0	0	0	0	0
Sweetcorn kernels	550	4	24	2.5	65	1	4	1.1	0.04	0.15	0.08	2.2	12	0	4
Tea with milk	50	1	1	1	97	20	26	0	0.02	0.02	0.04	0.02	0.2	0	0
Tea without milk	trace	trace	trace	0	98	trace	2	0	0	0	trace	trace	0	0	0
Toast, white	1250	10	65	2	24	640	110	2.2	0	0.2	0.04	1.8	0	0	4
Toffees	1800	2	70	17	5	300	100	1.5	0	0	0	0	0	0	0
Tomatoes	60	1	3	trace	93	3	13	0.4	0.1	0.06	0.04	0.8	20	0	1.5
Tomato ketchup	400	2	24	0	65	1120	25	1.2	0	0.06	0.05	0.3	0	0	0
Yam, boiled	500	2	30	0.1	66	17	9	0.3	trace	0.05	0.01	0.8	2	0	4
Yoghurt	200	5	6	1	86	80	180	0.1	trace	0.05	0.26	1.2	0.4	trace	0

* Estimated value

Unit H
Breathing and respiration

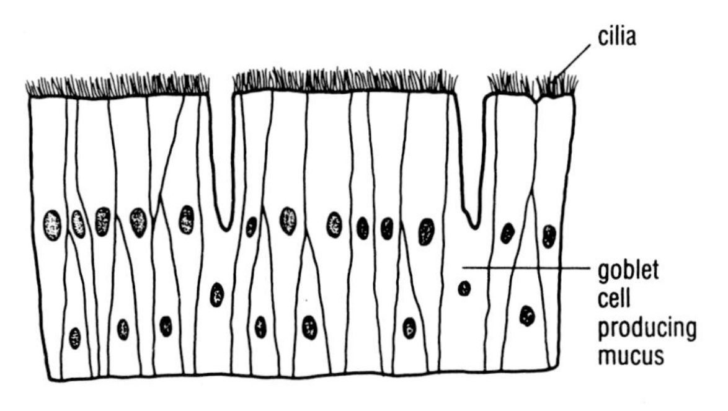

Figure H.1 Cells lining the trachea

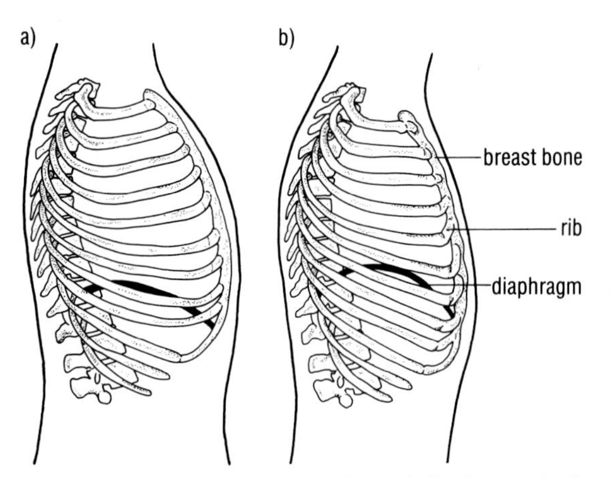

Figure H.2 The position of the ribs and diaphragm during breathing: a) inhalation; b) exhalation

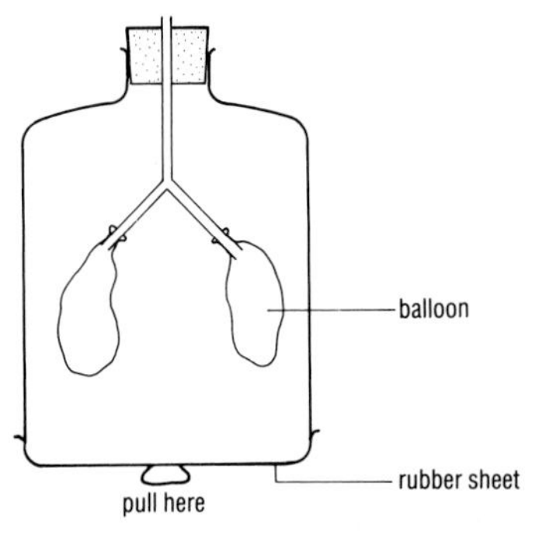

Figure H.3 A bell-jar model of the chest. What does each part of the model represent? What happens when the rubber diaphragm is pulled downwards? What criticisms of the model can you make?

Every day we breathe about 2160 times. Each time we breathe in and out at least 500 cm^3 of air, which amounts to at least 1 080 000 cm^3 of air daily. Any dust in the air is filtered out in the nose. In the trachea and bronchi there is a second line of defence against dirt and infection. They are lined by a single layer of special cells (Figure H.1). A few of these, **goblet cells**, produce slimy mucus which traps particles in the air. The other cells have a 'carpet' of tiny hairs on their surface (see Figure H.1). These move together, driving the mucus up towards the throat. The mucus is then swallowed and digested.

Breathing brings air containing oxygen to the lungs, where the oxygen passes into the blood. Breathing in humans is a sign of life. If we wanted to know if someone was alive we would probably ask whether they were breathing.

Breathing: the sign of life

The chest wall moves gently in and out as you breathe. When you become more active the movement becomes faster and more vigorous. More air is breathed in and out. Investigation H.1 asks you to examine your own breathing movements.

The breathing movements are caused by muscles which contract to pull the ribs up and out, and then relax again, letting the rib cage drop (Figure H.2). The ribs are hinged to the vertebrae at the back and attached to the flexible cartilage of the breastbone at the front. As the ribs move out, the muscular, arched **diaphragm** contracts and moves downwards. This movement of the ribs and diaphragm leads to **inhalation** or **inspiration** of air into the lungs.

During **exhalation** or **expiration** of air, other rib muscles (the **intercostal muscles**) bring the ribs back to the resting position. The diaphragm moves back to its arched position as it relaxes (see Figure H.2).

The lungs and tracheal system

Normally, we breathe in and out through our noses, with our mouths closed. The nasal cavity is lined with a thin film of mucus and contains many fine hairs. These catch particles of dust in the air as it passes through the nasal cavity. At the same time the air is warmed and made moist. From the nose the air passes through the throat into the windpipe or **trachea**. This is a short tube, kept open by rings of cartilage in its wall (see Figure A.8 on page 5). The first part of the trachea is the **larynx** or 'voice box'. This can be seen as the Adam's apple on the outside of the throat. The larynx (Figure H.4) contains the vocal cords, used for speech.

The trachea branches into two **bronchi** (singular bronchus), which lead to the lungs. Each bronchus splits into many branches or **bronchioles** which take air to all parts of the lung (see Figure H.5). At the ends of the branches there are clusters of **alveoli** (singular alveolus). The walls of an alveolus are moist and consist of a single layer of flattened cells. Capillary branches from the pulmonary artery form a network around the alveoli.

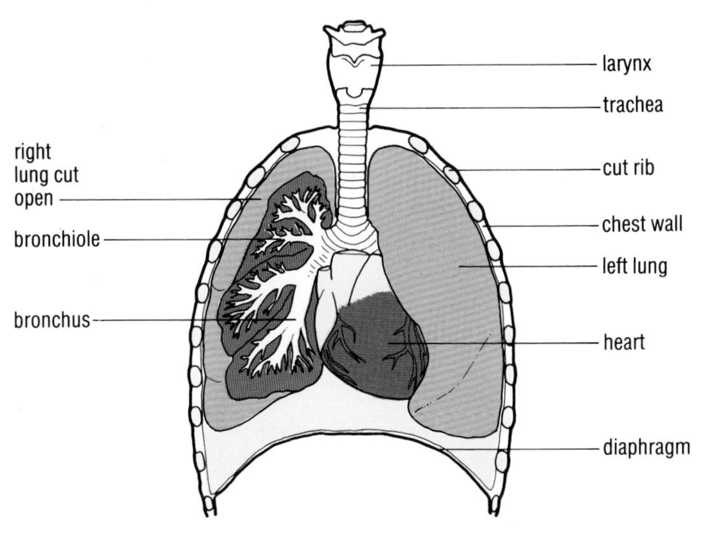

Figure H.4 The lungs and tracheal system. Can you list all the structures the air passes through?

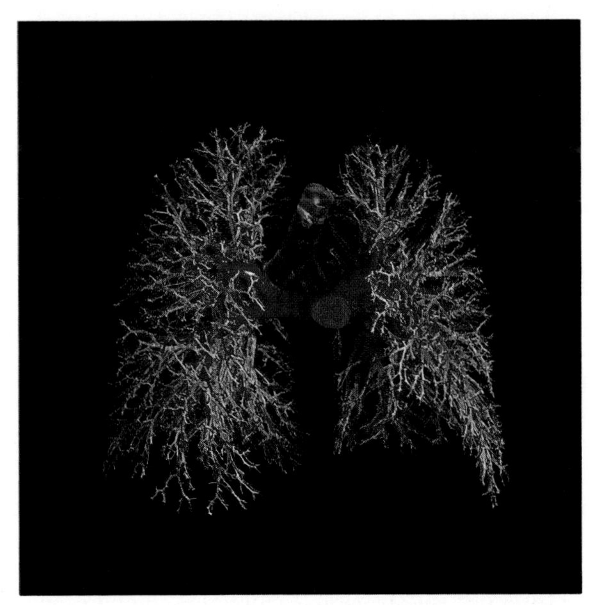

Figure H.5 A latex injection of the lungs which shows the bronchioles and blood vessels

Two thin walls, each one cell thick, lie between the alveolus and the blood inside the capillary (Figure H.6).

When you breathe in, normally about 500 cm^3 of air enters the lung. If you breathe in very hard you will take in more air; if you breathe out hard, you can expel more. The total air you can move in and out is normally about 3500 cm^3. There is always about 1500 cm^3 of air in the lungs which hardly moves at all.

Investigation H.1

Looking at your own breathing

You need: a mirror, a piece of blue cobalt chloride paper, a tape measure (or string and a ruler), a clock.

Work in pairs.

1. Place the mirror in the refrigerator. When it is cool, breathe on it. Note down what happens.
2. Place a strip of cobalt chloride paper on the mirror. Note down what happens (cobalt chloride turns pink if water is present).
3. Place the measuring tape around your partner's chest and ask her/him to breathe out. Record the result.
4. Repeat the measurement while your partner is breathing in deeply. Record the result.
5. What things affect how fast you breathe?
6. What is the change in the circumference of your/your partner's chest when you breathe in and out? What effect does exercise have on your breathing rate?

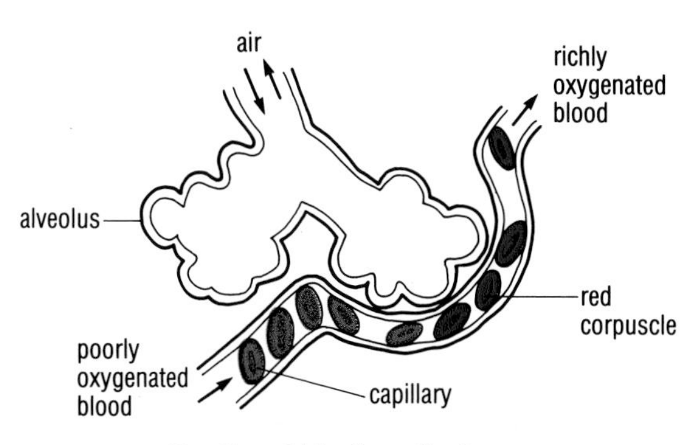

Figure H.6 Alveoli and blood capillaries

Activity H.2

Explaining how breathing takes place

Work in pairs. Read the next two paragraphs and look at Figures H.2 and H.3. When you have both read the section, one person should try to explain what happens when you breathe in and out. The other partner should ask questions when the explanation is not clear.

Breathing in

The chest works rather like a pair of bellows. As the ribs move outwards and the diaphragm downwards, the volume of the chest cavity increases. The air pressure in the air outside is greater than the pressure in the chest. Air flows into the lungs, equalising the pressure again. The lungs are inflated as this happens.

Breathing out

When the ribs move inwards and the diaphragm upwards the chest cavity is reduced. This increases the pressure in the chest and forces air outwards again. The mechanism can be shown by a simple model (Figure H.3). Try to explain how it works.

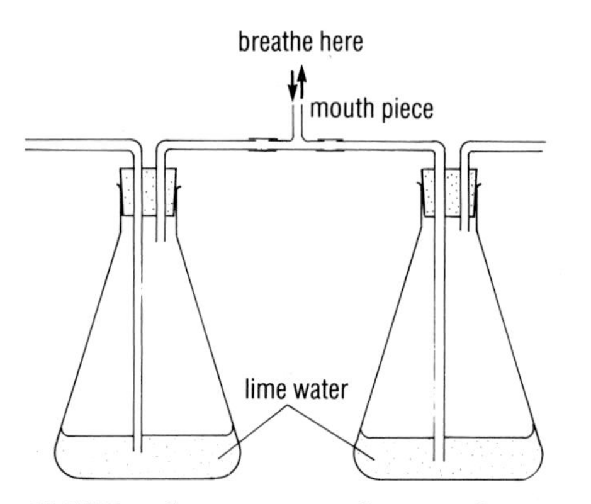

Figure H.7 What do you expect to happen when you breathe in and out of the mouthpiece? (The mouthpiece must be disinfected between different people.)

Gaseous exchange

Which do you think contains the most carbon dioxide, the air you breathe in or the air you breathe out? You can show that there is a difference between the air breathed in and breathed out using the simple apparatus in Figure H.7. Limewater turns milky when carbon dioxide passes through it. Can you explain how the apparatus works?

If you did Investigation H.1 you will have shown that the air we breathe out contains water droplets. It is essential that the walls of the alveoli are moist for gases to pass through them.

In the alveoli the oxygen is in higher concentration than in the blood in the capillaries. So, oxygen diffuses through the thin walls of the alveoli and capillaries and combines with haemoglobin in the red corpuscles. The haemoglobin changes from dark red to scarlet oxyhaemoglobin. At the same time, carbon dioxide which is mostly in the blood plasma, passes out of the blood into the air in the alveoli.

Table H.1 shows some of the other differences between the air breathed in and out. The percentages will vary slightly, depending on how active the person is.

	Air breathed in	Air breathed out
Nitrogen	79%	80%
Oxygen	21%	16%
Carbon dioxide	0.03%	4%
Temperature	Cool	Warm
Moistness	Dry	Moist
Cleanliness	May be dirty	Clean

Table H.1 Differences between inhaled and exhaled air

The processes of breathing and gaseous exchange in the lungs are sometimes called **external respiration.**

Respiration in the cells

The blood leaving the lungs is rich in oxygen. It travels along the blood vessels to all parts of the body. When this blood reaches the capillaries in the tissues it contains more oxygen than the surrounding cells. There is a concentration gradient with a greater concentration of oxygen in the red corpuscles and a lower concentration of oxygen in the cells. Oxygen is released from the oxyhaemoglobin and diffuses down the concentration gradient into the cells. There the oxygen is used in respiration and carbon dioxide is produced. Respiration in the cells is sometimes called **tissue respiration** or **internal respiration**. Figure H.8 compares what happens in the lungs with what happens in the tissues.

The chemistry of respiration

Respiration takes place in every cell of the body. Energy is released from glucose. The overall chemical reaction is:

Glucose + Oxygen ⟶ Carbon dioxide + Water (+ Energy)

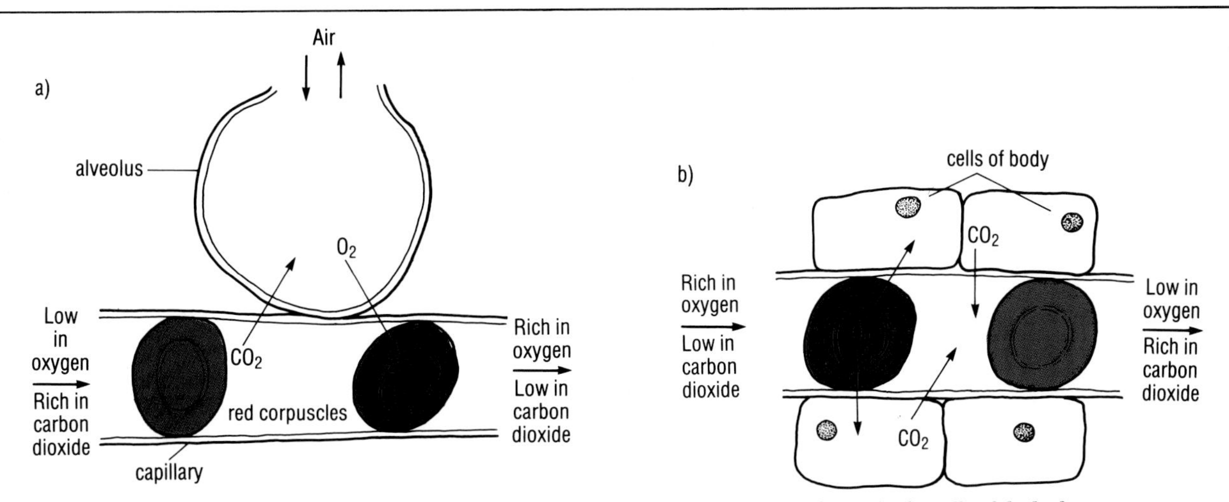

Figure H.8 Exchange of oxygen and carbon dioxide: a) gaseous exchange in the lungs; b) gaseous exchange in the cells of the body

Because oxygen is used, this is called **aerobic respiration.** The energy released from one gram of glucose is about 16.1 kJ. Activities such as the contraction of muscles require the energy released by the chemical reactions of aerobic respiration. Figure H.9 shows the active transport of glucose molecules. The process is another example of a biological activity that needs the energy released from glucose by aerobic respiration. The energy is needed to move glucose molecules across the cell membrane from a region where they are present in low concentration to a region where they are present in high concentration. By active transport cells can build up stores of substances (in this case, glucose) which would otherwise be lost by diffusion (see page 47). The energy released by aerobic respiration also warms us and helps us to maintain a steady body temperature, even in cold weather.

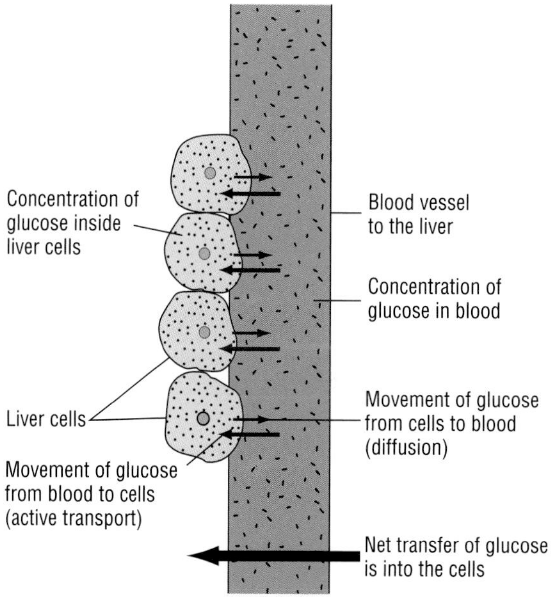

Concentration of glucose inside liver cells
Blood vessel to the liver
Concentration of glucose in blood
Liver cells
Movement of glucose from cells to blood (diffusion)
Movement of glucose from blood to cells (active transport)
Net transfer of glucose is into the cells

Figure H.9 Glucose is carried into cells by active transport, which uses energy released during respiration

Anaerobic respiration

If you are exercising, oxygen may not reach the muscles fast enough to provide the energy you need. Then **anaerobic respiration** occurs, which does not use oxygen. Lactic acid is produced and collects in the muscles. The overall chemical reaction is:

$$\text{Glucose} \longrightarrow \text{Lactic acid}$$

The energy released by this reaction is 0.83 kJ per gram of glucose – less than in aerobic respiration. This causes an **oxygen debt** to build up – after the muscles have respired anaerobically for a few minutes, the lactic acid they have built up stops them from working. During the recovery period, lactic acid is removed from the muscles. Vigorous panting brings a rush of oxygen to the muscles. Aerobic respiration re-starts and some lactic acid is oxidised to carbon dioxide and water. The rest is converted into glucose. In this way the oxygen debt is re-paid and the muscles start to work again.

Summary

- Breathing brings air in and out of the lungs.
- Gaseous exchange occurs in the lungs, respiration takes place in the cells.
- Oxygen is used in aerobic respiration. When no oxygen is available in muscles, anaerobic respiration gives less energy and produces lactic acid.

Questions

1 Draw up a table or write a paragraph comparing the following:
a gaseous exchange in the lungs and gaseous exchange in the cells of the body (see Figure H.8).
b aerobic and anaerobic respiration.

2 Find out how you would carry out mouth-to-mouth resuscitation. (Important: this can be dangerous. You must only practise on a model.)

Extension Questions

3 Explain the difference between diffusion and active transport. Give one example of the importance of active transport.

4 Complete the following paragraphs with words from the wordlists provided. Each word might be used once, more than once or not at all.

a During a sprint, your leg ______ will need oxygen faster than it can be supplied by your blood. To compensate, the muscle cells ______ anaerobically. This process forms ______, which stops ______ from working. The oxygen needed to oxidise the ______ is called the ______.

Wordlist
respire, repaid, oxygen, muscles, oxygen debt, liver, lactic acid, aerobic

b The ______ of oxygen and removal of carbon dioxide occur in the ______ of the lungs. These provide a large surface area (about 90 m^2) for efficient gas ______. They are thin walled, have an excellent blood supply, are ______ and are kept well supplied with air by breathing. Air is taken into the lungs by ______ and removed by ______.

Wordlist
uptake, respire, inhalation, oxygen, alveoli, energy, exchange, moist, exhalation

How are substances transported around the body?

Unit I
Transporting materials

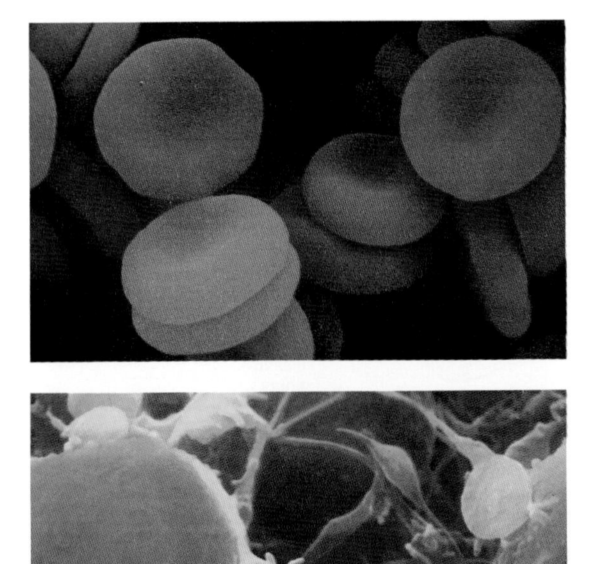

Figure I.1 Blood cells. The false colour scanning electron micrographs show red cells coloured red, white cells coloured violet and platelets coloured yellow

Blood

Blood is an almost colourless fluid (**plasma**) containing cells (Figure I.1). The red cells or **corpuscles** contain a red pigment, **haemoglobin**. Haemoglobin combines easily with oxygen in the lungs to produce scarlet **oxyhaemoglobin**. When the blood reaches parts of the body that lack oxygen, oxyhaemoglobin releases oxygen equally easily. This is a reversible reaction:

haemoglobin + oxygen (deep red)  oxyhaemoglobin (scarlet)

Haemoglobin is a protein with iron attached to it. If the diet is short of iron the red cells contain less haemoglobin. This means that the blood cannot transport enough oxygen for the body's normal activity. People with too little haemoglobin suffer from **anaemia**. Anaemia can usually be treated by taking iron tablets. **Sickle cell anaemia** is a special form of anaemia which is inherited. In people who have sickle cell anaemia the haemoglobin is different chemically and some of the red cells become shaped like sickles or crescents.

Blood transports food, oxygen and other chemicals around the body and helps to protect against infection. Table I.1 gives more information about the parts of the blood and their functions.

Component	Description	Size in 1/1000 mm	Numbers in 1 mm^3	Function
Plasma	Almost colourless liquid containing proteins and dissolved substances			Transports carbon dioxide, nutrients, urea
Red cells	Biconcave; no nuclei; contain red haemoglobin	8	5 million	Transport oxygen combined with haemoglobin
White cells	Various shapes, with nuclei	8–20	8000	Attack disease-causing organisms in various ways
Platelets	Disc-shaped; no nuclei	3	3–500000	Help clot blood at the site of a wound. A series of chemical reactions which require different substances, including factor VIII, eventually stops the bleeding

Table I.1 The composition and functions of blood

Blood vessels

Blood travels around the body in three kinds of blood vessels, **arteries, veins** and **capillaries**.

Arteries carry blood away from the heart to all regions of the body. They are quite large in diameter and have thick muscular walls which help them stand up to the pressure of blood coming from the heart (Figure I.2). The thick walls prevent the blood showing through as red.

Veins take blood back to the heart and have **pocket valves** in them to stop blood from flowing backwards. They are quite large in diameter, but have thinner walls than arteries, and appear purple. Veins are often visible through the skin on the inside of the wrist.

Capillaries form a network of fine tubes running through the tissues. They are fed by arteries and drain into veins (see Figure I.2). Capillaries have very thin walls, which means that oxygen and food needed by cells can diffuse out of them and the urea and carbon dioxide (wastes) produced by cells can diffuse into them. The flow of blood to all parts of the body follows the pathway shown:

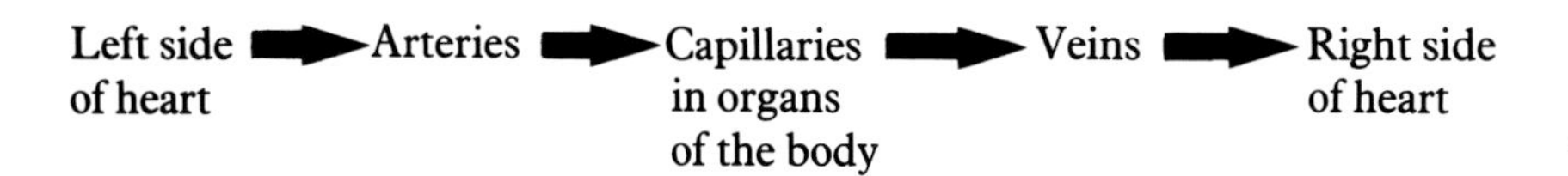

This pathway around the body is called the **systemic system**.

The heart pumps blood into the arteries at high pressure. The main artery carrying blood away from the left side of the heart to the body is the **aorta**. Branches from this go to capillaries in the head, arms, alimentary canal, liver, kidneys and legs. Veins carry the blood back from the capillaries in each of these organs to the two **venae cavae** which enter the right side of the heart. The blood pressure in veins is quite low.

A special **hepatic portal vein** runs from the capillaries in the wall of the small intestine to capillaries in the liver. This vein carries blood containing nutrients coming from the intestine to the liver, where they are processed.

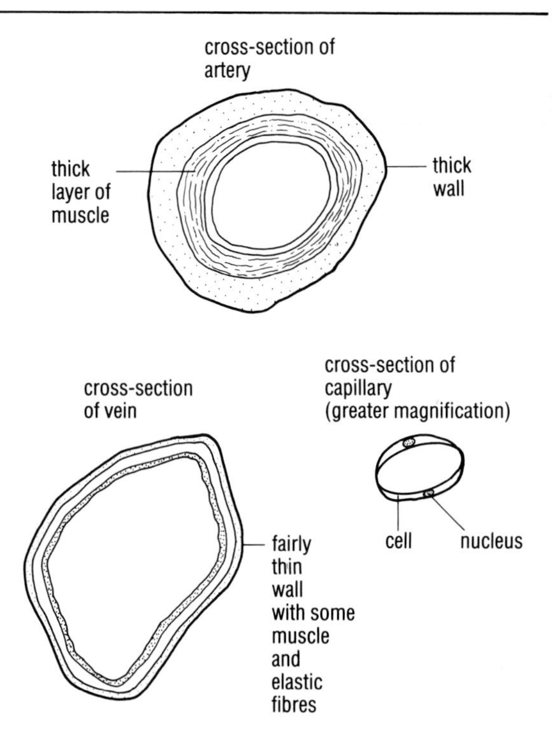

Figure I.2 Artery, capillary and vein compared

The pulmonary system to the lungs

The **pulmonary artery** carries blood away from the right side of the heart and branches to each lung. In the lungs carbon dioxide dissolved in the plasma diffuses out and oxygen diffuses in through the thin walls of the capillaries. The oxygen combines with haemoglobin, forming oxyhaemoglobin. This scarlet, richly oxygenated blood is carried back to the left side of the heart along the **pulmonary veins**. The pathway through the lungs is the **pulmonary system**:

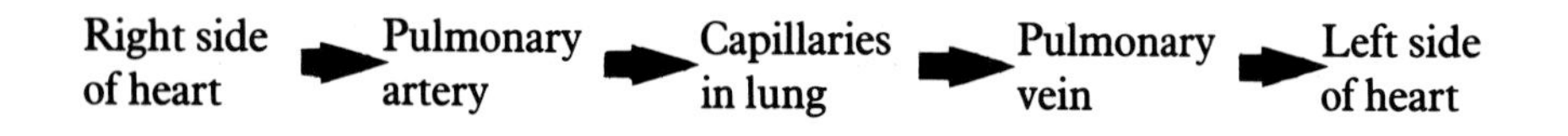

The richly oxygenated blood is pumped out of the left side of the heart again (see above), through the aorta to all parts of the body. Blood flows alternately through the pulmonary and the systemic systems. The pathway of the circulating blood can now be completed as shown on the next page.

Investigation I.1

Investigate the factors that affect the rate of activity in humans.

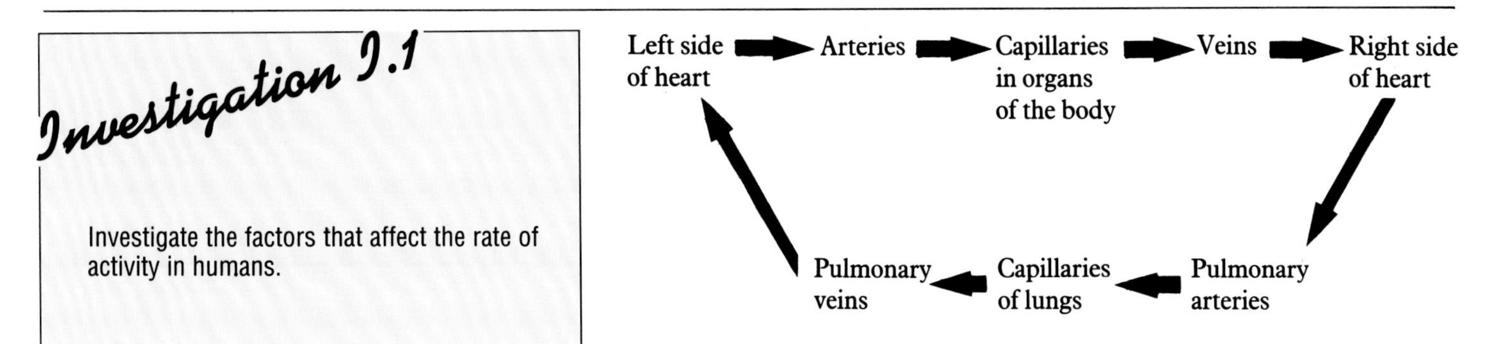

The systemic and pulmonary systems together make up the **double circulation** (Figure I.3).

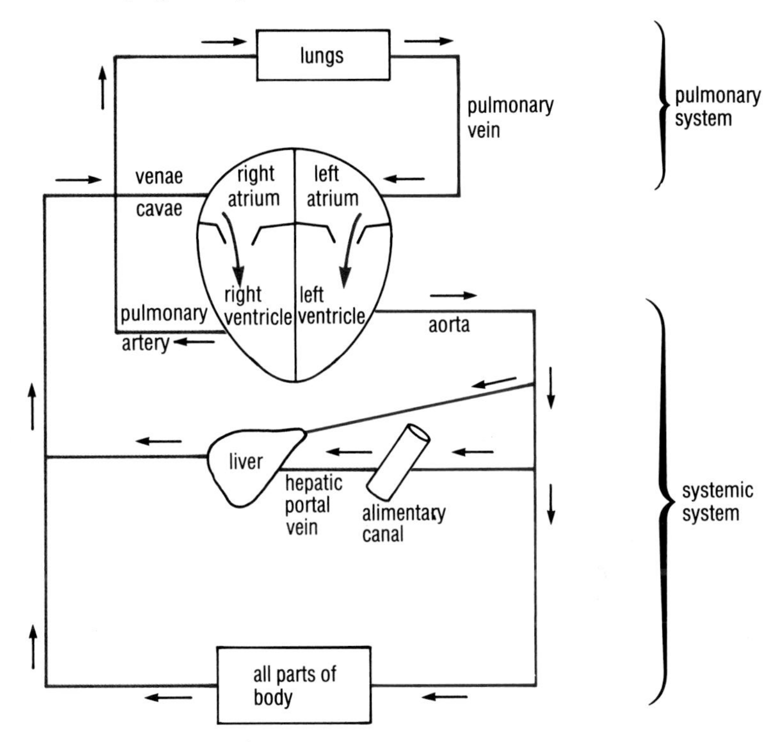

Figure I.3 The double circulation. Can you follow the systemic system and the pulmonary system?

Summary

- Blood transports materials around the body and protects against disease.
- The heart pumps the blood around the circulatory system.
- Arteries carry blood away from the heart. Veins carry blood to the heart. Capillaries link arteries and veins.

Questions

1. Describe the path of a red corpuscle from the lung to the liver.
2. Explain how the blood carries a) oxygen and b) dissolved nutrients.

The heart – a muscular pump

The heart is a pump that propels blood through the arteries and veins. It is made of a type of muscle called cardiac muscle, which contracts and relaxes rhythmically for a lifetime.

The heart is divided into four compartments or chambers. The **left atrium** and the **left ventricle** are separated by a muscle wall from the **right atrium** and the **right ventricle**. The two atria are smaller and have thinner muscular walls than the ventricles. The thickest walls are in the left ventricle which forces blood all round the body when it contracts.

Unit J
Diseases

For most of us the word 'disease' conjures up the idea of feeling unwell. The normal functions of the body are disrupted, either because cells are damaged, or because the disease-causing agent releases a toxin, making us feel ill. The human body is an ideal environment for a range of organisms causing different diseases. We blame **bacteria** (Figure J.1) for many of our ailments, but **viruses** (Figure J.2) are the most important disease-causing agents. **Protists** (Figure J.3) and **fungi** (Figure J.4) are culprits as well, and different animal **parasites** (Figure J.5) also cause disease as a result of their activities inside our bodies.

Not all viruses and bacteria cause disease: those that do are called **pathogens**. Some protists, fungi and parasites are also pathogens. Diseases are said to be **infectious** if the pathogens can be passed from one person to another. Not all diseases are infectious. Many **non-infectious** diseases develop because the body is not working properly. The way we treat our bodies and increasing age affects the onset of non-infectious disease. Many diseases can be avoided or at least delayed according to lifestyle.

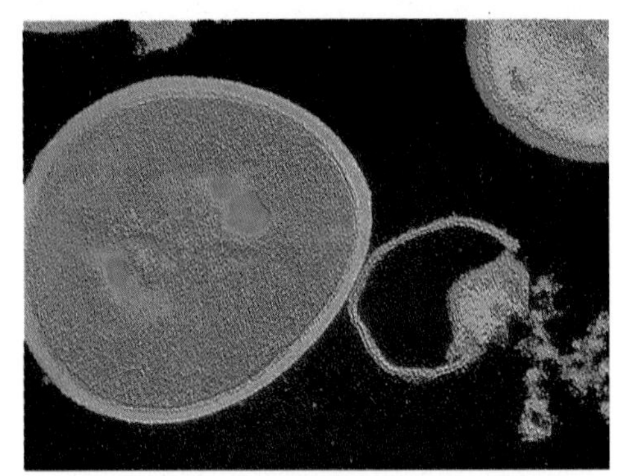

Figure J.1 Bacteria are simple single-celled organisms, which lack a distinct nucleus and many of the other organelles in animal and plant cells. This electron micrograph shows Staphylococcus aureus, *which can cause stomach upsets. The bacterium on the left is complete: the cell on the right has burst because of the action of an antibiotic. Some bacterial cells propel themselves along using a tail-like* ***flagellum***

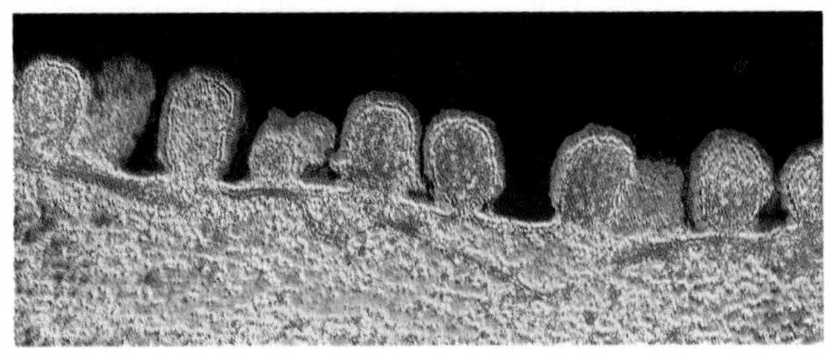

Figure J.2 Viruses are not cells. Consisting of a strand of nucleic acid enclosed in a protein coat, viruses cannot reproduce independently. They take over the biological machinery of living cells to reproduce new viruses. This electron micrograph shows many influenza virus particles budding from the surface of an infected cell

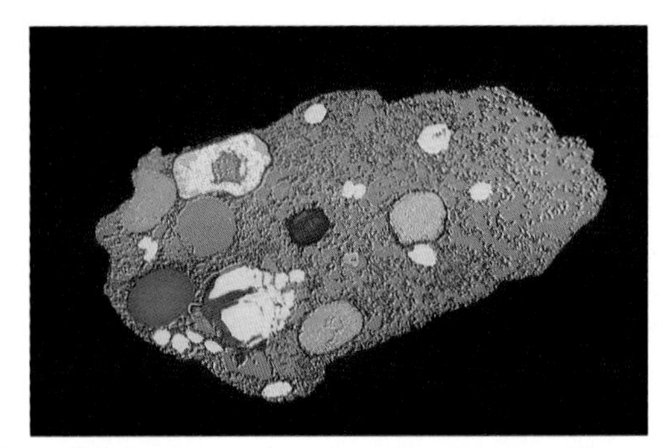

Figure J.3 Protists, such as Amoeba *(see page 5), are single-celled organisms which have a distinct nucleus.* Entamoeba histolytica *causes amoebic dysentery*

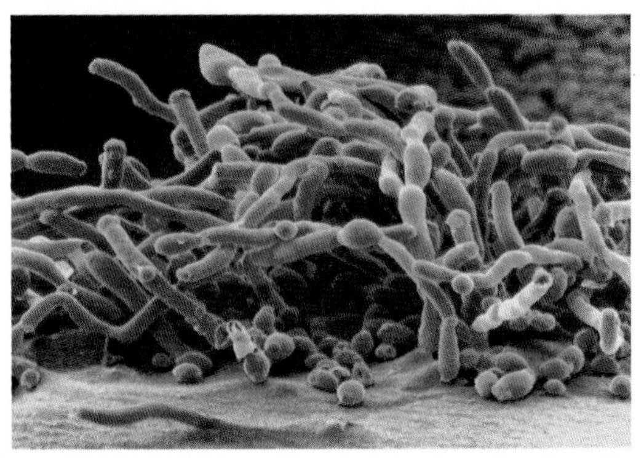

Figure J.4 Fungi are made up of slender tube-like cells called ***hyphae****. This electron micrograph shows hyphae of the fungus* Candida albicans, *which causes thrush. Moulds and yeasts are also examples of fungi*

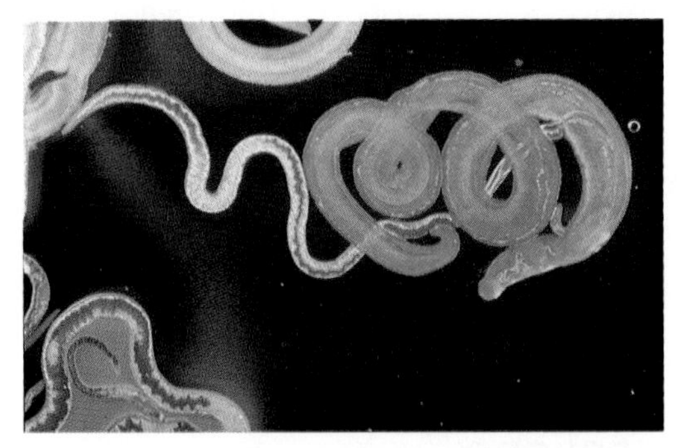

Figure J.5 This micrograph shows the parasitic blood flukes that cause bilharzia. Humans can become infected with their eggs while bathing in contaminated water. The eggs develop in intestinal blood vessels, causing severe dysentery

Spreading disease

Figure J.6 Clean drinking water is vital to prevent the spread of disease, but for many communities this basic resource is very hard to obtain

In the developed countries of Japan, Australasia, Europe and North America, where there are good medical services and housing with modern sanitation and water services, few people die of infectious diseases. Instead, unhealthy lifestyles are partly responsible for the early onset of diseases such as cancer and heart disease. These and other non-infectious diseases are major killers. In sharp contrast, in the developing countries of Africa and Asia, infectious diseases are responsible for 40% of all deaths. Poor housing, lack of proper sanitation and contaminated drinking water create the unhygienic environment which is ideal for the spread of infectious diseases.

Water-borne diseases

In the 19th century, Britain's cities were rapidly expanding following the Industrial Revolution. Insanitary and overcrowded living conditions meant that infectious diseases spread quickly. **Cholera**, which is spread by contaminated drinking water, was particularly common. Although at the time people were not at all clear about the true cause of disease, many realised that drinking water contaminated with raw sewage was a danger to health.

Today, three out of five people in developing countries have difficulty in obtaining clean water (Figure J.6). For many communities the local river is a source of drinking water as well as a way of disposing of sewage. Bacteria present in faeces infect the water, causing not only cholera but also other diseases such as **typhoid**, **diarrhoea** and **dysentery**. Around five million people each year are killed by water-borne diseases.

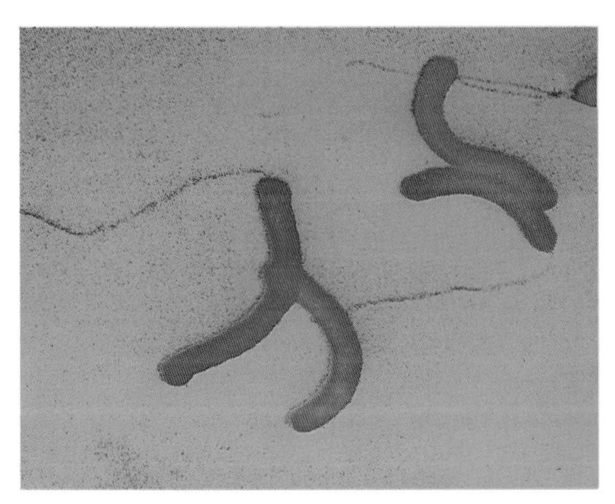

Figure J.7 This electron micrograph shows the bacteria that cause cholera, which move through water using a single flagellum. Cholera causes severe vomiting and diarrhoea, and can lead to death due to dehydration as so much water is lost from the body

Fly-borne diseases

Flies are a health hazard! The bacteria that cause typhoid fever, cholera and dysentery can be transferred to human food by houseflies. The diseases appear wherever people are crowded together in unhygienic conditions. For example, during the Second World War (1939–1945) the British Army in North Africa kept outbreaks of dysentery to a minimum by strict attention to hygiene and the control of flies. The picture was different in the Crimean War (1854), when filth provided ideal breeding conditions for flies and 8000 British soldiers suffered from dysentery. More than 2000 died of the disease.

Improvements in treatment, health, personal hygiene and sanitation have helped to reduce the extent and effects of fly-borne diseases. However, remember that flies carry pathogens. Infections are reduced by covering food to prevent flies from landing and feeding.

Figure J.8 When you sneeze or cough, you send an invisible mist of water droplets into the air that could carry infection to others

Air-borne diseases

Figure J.8 shows how a sneeze produces a jet of moisture droplets that shoots out of the nose and mouth. Pathogens infecting the **mucous membranes** lining the tracheal system (see page 34) are carried in the droplets and spread to other people nearby. This is how the viruses that cause diseases such as colds and influenza ('flu) pass from person to person, especially in crowded places like schools and hospitals.

Pneumonia is caused by a particular type of bacterium which passes easily from person to person in air-borne droplets. Air-borne bacteria can also infect the **pleural membranes** which line the ribcage and cover the surfaces of the lungs. The infection, called **pleurisy**, roughens the membranes causing pain when they rub together.

Summary

- The human body is an ideal environment for pathogens, which either produce toxins or damage cells, making us feel ill.
- Many bacteria and viruses are pathogens, which cause disease as a result of their activities within our bodies.

Questions

1 'Oral rehydration treatment' is a low-cost way of preventing death from diarrhoea. How do you think the treatment works?
2 Copy out the list of diseases below. Identify the cause of each disease by writing either B (for bacterium) or V (for virus) next to each one.

Cholera
Typhoid
Cold
'Flu
Pneumonia

Unit K
Fighting disease

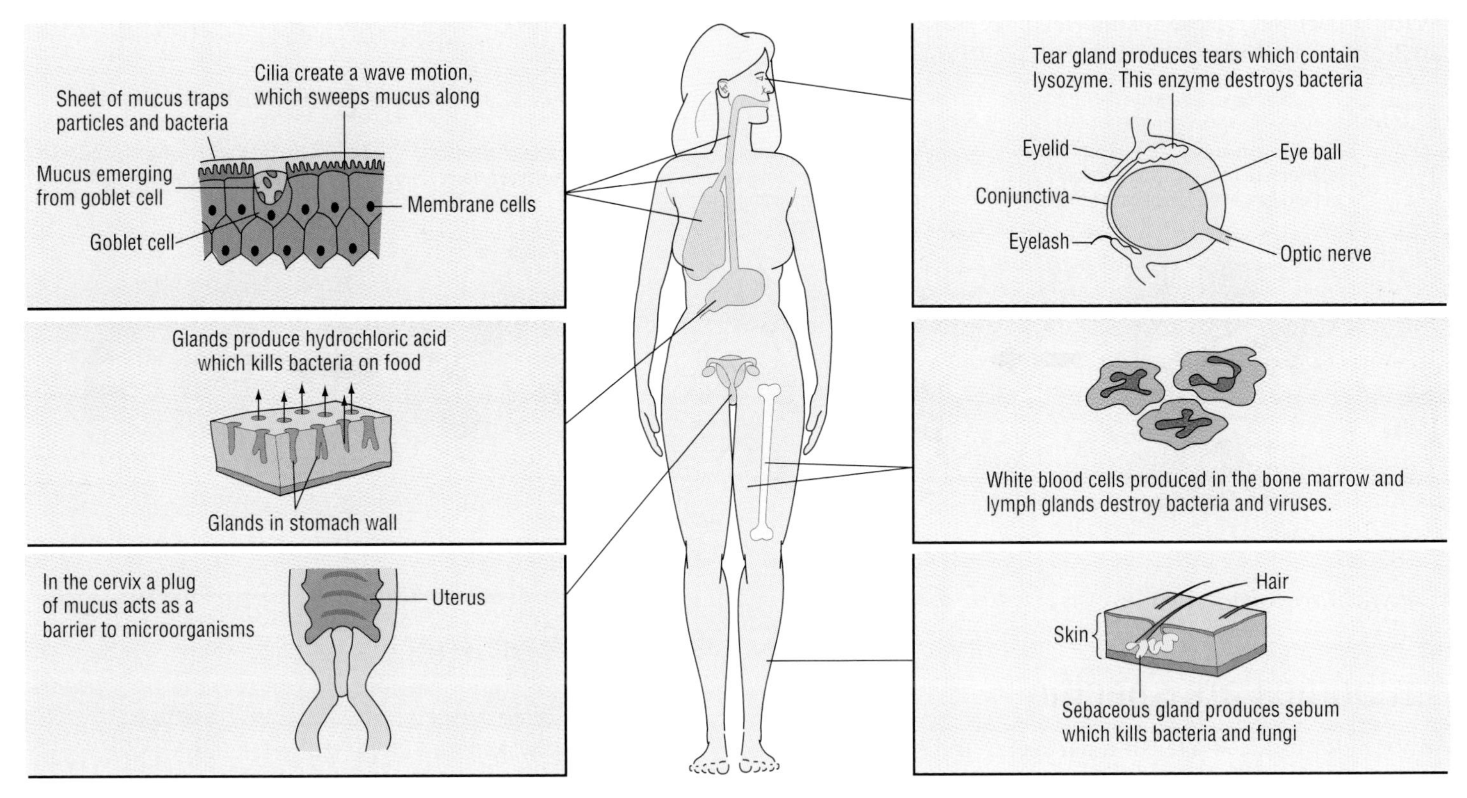

Figure K.1 The body's natural defences

Figure K.1 shows the body's natural defences against disease. Notice the physical and chemical barriers to infection. For example, inside the body, mucus covers internal surfaces and plugs openings, trapping particles and bacteria before they can infect tissues. Blood platelets help to seal cuts and wounds (see page 39), preventing the entry of pathogens. All these different defence mechanisms help to keep most of us healthy most of the time. Looking after ourselves and maintaining good hygiene help the defences do their job.

Role of the skin

The skin is one of the main barriers against infection. Washing regularly reduces the chances of infection because the pathogens resting on the skin are washed away. Some soaps and skin cleansers have antiseptic properties, which stop bacteria from multiplying. They are particularly effective against the bacteria that cause body odour. They also help prevent fungal infections such as ringworm and athlete's foot. Sweaty feet are an ideal environment for fungi – so make sure that you wash and dry carefully between your toes, and try not to borrow other people's footwear. The fungi are spread from person to person by physical contact.

White blood cells protect the body

There are two main types of white blood cell: **lymphocytes** and **phagocytes**. They protect the body from disease by quickly destroying viruses and bacteria. They also destroy any other cells or substances that the body does not recognise as its own.

Materials foreign to the body are called **antigens**. When antigens come into contact with lymphocytes, the lymphocytes produce proteins called **antibodies**, which begin the process of destruction. The phagocytes finish the job. Figure K.2 shows what happens. Some invading bacteria release toxins (see page 42), with which **antitoxin antibodies** combine to prevent damage to tissues (Figure K.3).

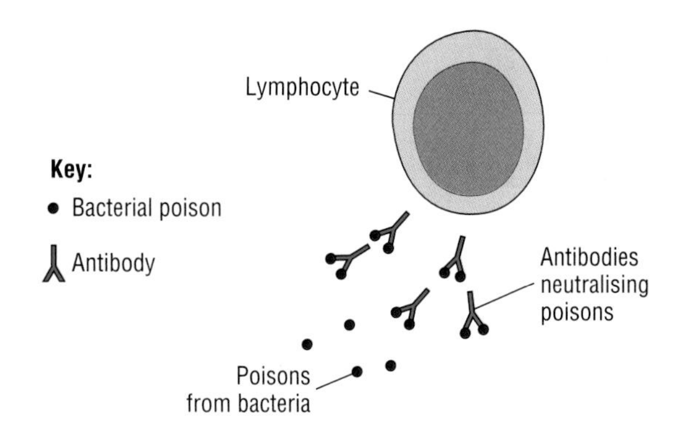

Figure K.3 Antibodies protect the body from poisons

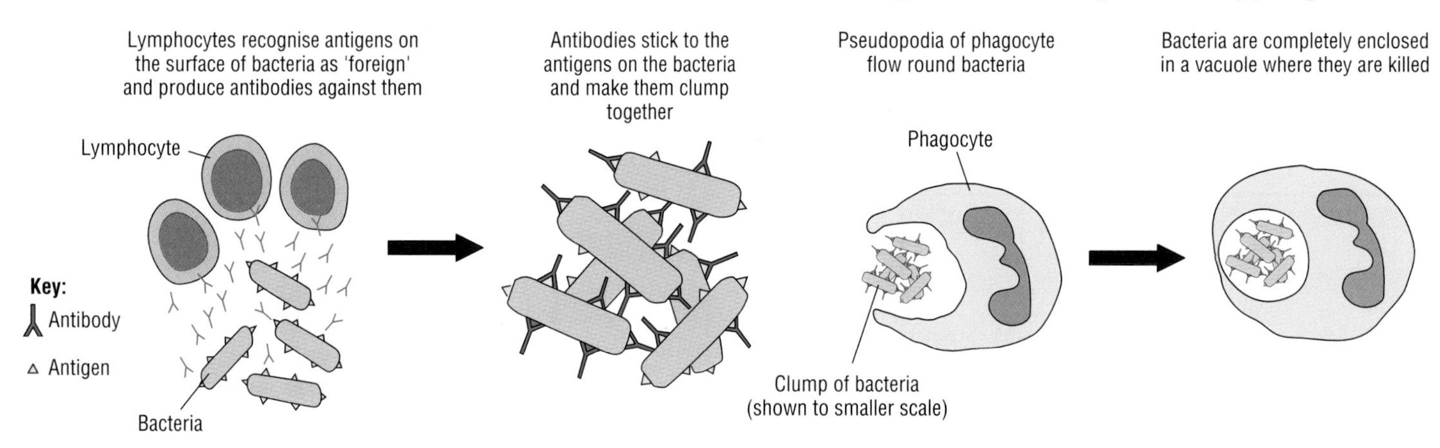

Figure K.2 Lymphocytes and phagocytes protect the body from disease

Antibodies are specific

Antibodies produced against a particular antigen will attack only that antigen. Why? The antigen in question stimulates the production of antibodies whose shape will fit that antigen and no other. We say that the antibody is **specific** to that antigen. This is why, for example, antibodies produced against typhoid bacteria will not attack pneumonia bacteria. Antibody and antigen will fit together only if their shape and structure match. Each of us can produce tens of millions of different antibodies to deal with all of the antigens we are ever likely to meet in a lifetime.

Active immunity

The action of lymphocytes and phagocytes against invading bacteria, viruses and other antigens is called an **immune response.** The antibodies produced can stay in the body for some time, ready to attack the same antigens when they next invade the body. Even if the antibodies do not stay in the body for long, they are soon made again because the first-time battle between antigen and lymphocytes primes the lymphocytes to recognise the same antigen next time. This means that re-infection is dealt with by immune responses that are even faster and more effective than the first reaction. In other words, you become **resistant**. This is why you rarely catch diseases like chicken pox or measles more than once.

Summary

The body's natural defences provide physical and chemical barriers to infection:

- The skin prevents entry of micro-organisms.
- Mucus covers internal surfaces of the body, trapping micro-organisms.
- Blood clots seal cuts.
- White blood cells destroy pathogens.

Questions

1. Discuss how the body's natural defences keep most of us healthy for most of our lives.
2. List the different types of white blood cell. Explain the role of each type of cell in the defence of the body against disease.
3. Why are diseases like chicken pox and measles rarely caught more than once?

Unit L
Diffusion

(a)

(b)

Figure L.1 The ink slowly spreads out, by diffusion, until the concentration is the same throughout the solution

Figure L.1 (a) captures the moment just after drops of ink have been dripped into a beaker containing water. Figure L.1 (b) shows the appearance of the ink/water solution several hours later. The ink molecules have slowly become evenly distributed throughout the volume of water.

The molecules of ink and water are in constant motion. If it were possible to track a single ink molecule, we would see that its movements are random and independent of the motion of all of the other molecules in the solution. How do ink molecules get from one side of the beaker to the other? Think of it like this. To begin with, there is more ink in one place in the beaker than in any other (Figure L.1 (a)). Even though there is an equal probability that any one ink molecule will move in a particular direction, there is a better than even chance that some molecules will spread from where they are highly concentrated to where they are fewer in number.

This kind of movement of a substance through a solution or through a gas is called **diffusion**. Diffusion happens when there is a higher concentration of a substance in one part of a solution or gas than in another. In other words, substances move from a region of higher concentration of their own molecules to a region of lower concentration, down a **concentration gradient**. The greater the difference between the regions of high and low concentration, the steeper the concentration gradient and the faster the rate of diffusion of that substance. A substance moving in the opposite direction, towards a higher concentration of its own molecules, is said to move *against* a concentration gradient. This process is called **active transport** (see page 37).

Cells and diffusion

Water, oxygen, carbon dioxide and some other small, simple molecules freely diffuse across cell membranes. Diffusion is therefore very important for movement of substances between cells. It is also important for the movement of substances within cells. However, dependence on diffusion is one of the factors that limits the size of cells. The slowness of the process, except over short distances, means that delivery of molecules to the parts of the cell that need them becomes increasingly inefficient as the distance covered by the diffusing molecules increases.

Efficient diffusion requires not only a relatively short distance but also a steep concentration gradient. Cells maintain steep gradients through their metabolism. For example, aerobic respiration constantly produces carbon dioxide (see page 37). As a result, there is a higher concentration of carbon dioxide inside the cell than outside it, and a gradient is produced. Carbon dioxide diffuses out of the cell down the

Activity L.1

Figure L.2 shows three cubes of different sizes. A cube has six faces.
Copy and complete Table L.1:

	Cube A	Cube B	Cube C
Surface area of one face	$1\ \text{cm} \times 1\ \text{cm} = 1\ \text{cm}^2$		
Surface area of a cube	$6 \times 1\ \text{cm}^2 = 6\ \text{cm}^2$		
Volume of a cube	$1\ \text{cm} \times 1\ \text{cm} \times 1\ \text{cm} = 1\ \text{cm}^3$		
Ratio: surface area/volume	$6\ \text{cm}^2/1\ \text{cm}^3 = 6/\text{cm}$		

Table L.1

State whether the surface area/volume ratio of a cube increases or decreases as the cube increases in size.

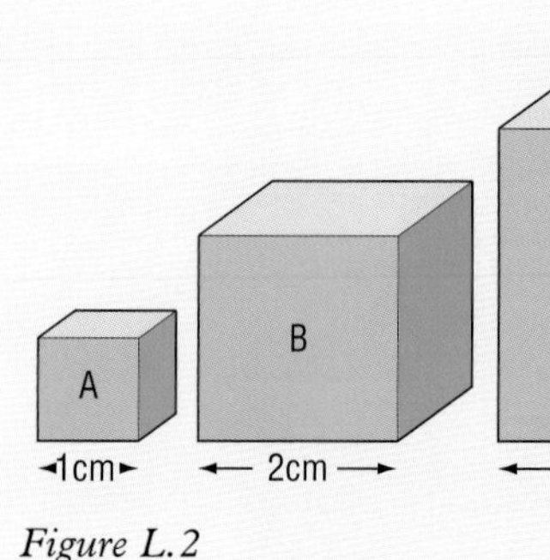

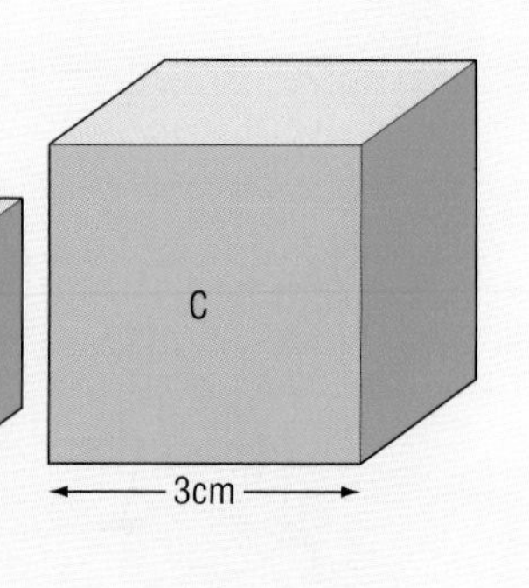

Figure L.2

gradient. Aerobic respiration also consumes oxygen, which is therefore at a higher concentration outside the cell than inside. Oxygen diffuses into the cell, again down a concentration gradient.

Specialised exchange surfaces

Each of us exchanges gases, food and other materials between our body and the environment. The exchanges occur slowly by diffusion across body surfaces, and supply tissues and organs with the materials they need to do their different jobs.

Activity L.1 should help you understand some of the problems associated with exchanging materials across surfaces. Your calculations should tell you that as a cube increases in size its surface area increases more slowly than its volume.

Of course, tissues and organs are not cube-shaped, but the arithmetic applies to structures of any shape. Imagine that the surfaces and insides of the cubes represent the surfaces and insides of different sized bodies. Increasing the size means that the surfaces across which materials pass into and out of the body decrease in area relative to the volume of the tissue and organs which the surfaces supply. Since diffusion is a slow process, tissues at a distance from the surfaces where exchanges take place might not receive sufficient materials to do their jobs efficiently.

Humans have large bodies. The problems discussed above are overcome by specialised structures that massively increase the area across which materials are exchanged. The two examples below show how these increased surface areas allow materials to be exchanged at a fast enough rate to serve all the cells in the human body.

Example 1

Figure F.1 (page 27) shows a part of the small intestine called the ileum. In the ileum, digested food is absorbed into the bloodstream. The small intestine is highly folded and coiled, packing as much gut as possible into the restricted space available. The extra length increases the surface area for absorption.

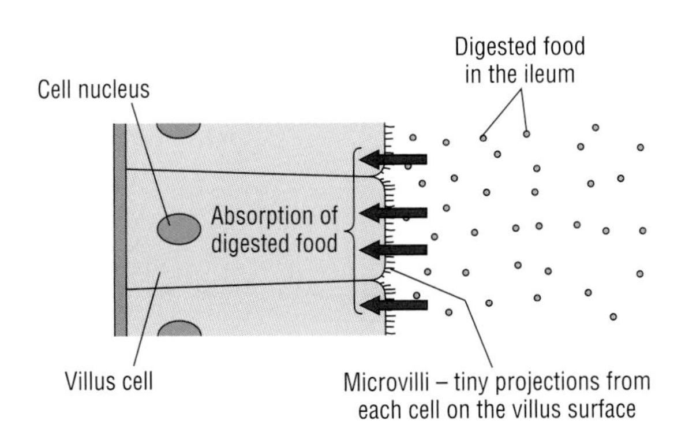

Figure L.3 The cells lining the small intestine are specialised for absorption of digested food materials

Figure G.2 (page 30) shows inside the small intestine. The closely packed finger-like villi increase the surface area further. Figure L.3 shows one of the cells lining a villus. Notice the tiny projections along the cell's surface. The projections are called **microvilli**, and they increase the surface area for absorption about 20 times more. If the villi provide an area about 10 m^2, how large a surface area do the villi and microvilli make altogether for the absorption of digested food?

Example 2

Look at Figure H.8(a) on page 37. It shows that oxygen and carbon dioxide are exchanged between the lungs and blood across the thin walls of the alveoli. The alveoli are like clusters of grapes at the end of the small tubes which branch and extend from the bronchioles (see Figure H.4, page 35). They honeycomb the lungs with millions of air sacs. Together they give a surface area of approximately 90 m^2 for gaseous exchange.

Summary

- Diffusion is the movement of a substance through a solution or gas from a region where the substance is in high concentration to a region where its concentration is lower.
- Diffusion occurs down a concentration gradient of the substance. The steeper the concentration gradient, the faster the rate of diffusion.
- Materials move across body surfaces by diffusion. Organs are specialised to increase the surface area across which the movement of materials occurs.

Questions

1 Explain the difference between diffusion and active transport.
2 How does a cell maintain diffusion processes between itself and its environment?
3 As you blow up a balloon, how does its surface area change, *relative to its volume*?

Extension questions

4 List the specialisations of the small intestine which increase its surface area for the absorption of food.
5 Explain how the alveoli increase the surface area of the lungs for the exchange of gases.
6 Emphysema is a lung disease that results in the fine walls between individual alveoli breaking down (see page 98). Explain why a person suffering from emphysema becomes short of breath easily.

End-of-module questions

QUESTION ONE

This is a diagram of the digestive system. Choose words from the list for each of the labels on the diagram.

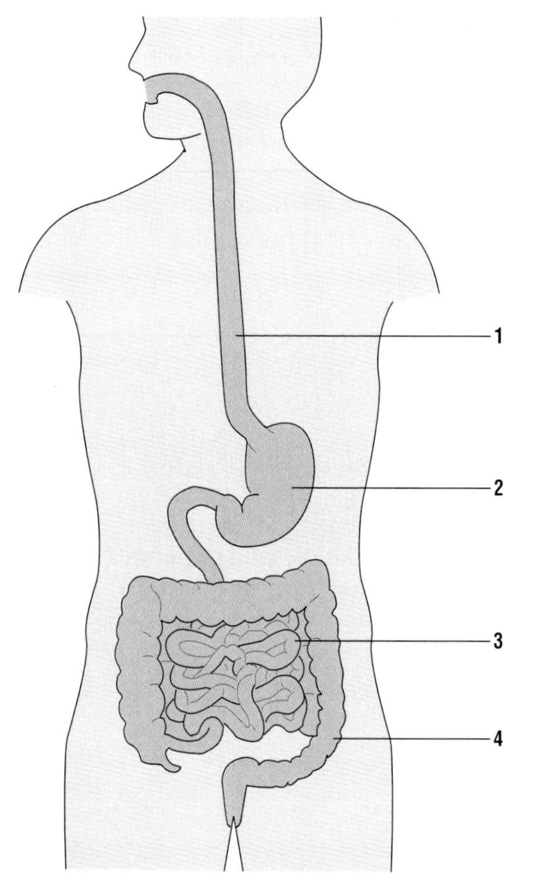

large intestine **small intestine**
gullet **stomach**

(E/F)

QUESTION TWO

The table is about life processes. Choose words from the list for each of the items 1–4 in the table.

reproduction **excretion**
nutrition **respiration**

1	obtaining food to eat
2	production of offspring
3	release of energy from food
4	release of waste products

(E/F)

QUESTION THREE

The table gives the jobs of parts of the digestive system. Choose words from the list for each of the items 1–4 in the table.

salivary glands **stomach**
small intestine **large intestine**

1	only carbohydrase enzymes are produced here
2	the blood absorbs most of the water here
3	the blood absorbs the products of protein digestion here
4	hydrochloric acid is produced here

(E/F)

QUESTION FOUR

The table gives information about cells and tissues. Choose words from the list for each of the items 1–4 in the table.

cell membrane **tissue** **nucleus** **cytoplasm**

1	controls the activities of a cell
2	solution in which reactions take place
3	a group of cells of similar structure carrying out a particular function
4	allows passage of substances in and out of cells

(E/F)

QUESTION FIVE

The table is about the jobs of parts of the blood. Choose words from the list for each of the items 1–4 in the table.

red cells **platelets** **plasma** **white cells**

1	have a nucleus, and defend against microbes
2	transports substances
3	fragments of cells, which form clots
4	transport oxygen

(C/D)

QUESTION SIX

Arteries are blood vessels. Choose from the list **two** characteristics of arteries.

they have thin walls
they have muscular walls
they have valves
the walls allow substances in and out
the walls contain elastic fibres

(C/D)

QUESTION SEVEN

White cells are found in the blood. Choose from the list **two** characteristics of white cells.

they ingest microbes
they carry oxygen
they produce sugars
they produce antibodies
they cause disease

(C/D)

QUESTION EIGHT

Breathing involves movement of the ribs and diaphragm. Choose from the list **two** of the following that enable breathing in to take place.

the pressure in the chest increases
muscles between the ribs contract
the ribs move downwards
the diaphragm muscles contract
the diaphragm moves upwards

(A/B)

QUESTION NINE

Choose from the list **two** correct facts about anaerobic respiration.

carbon dioxide and water are produced
energy is released
sugar is completely broken down
oxygen is required
lactic acid is produced

(A/B)

QUESTION TEN

Choose from the list **two** statements about diffusion that are true.

the surface area of the lungs is increased by villi
ions cannot diffuse through membranes
sugar can diffuse through membranes
sugar passes through alveoli by diffusion
the greater the difference in concentration the faster the rate of diffusion

(A/B)

QUESTION ELEVEN

The diagram shows some parts of the heart.

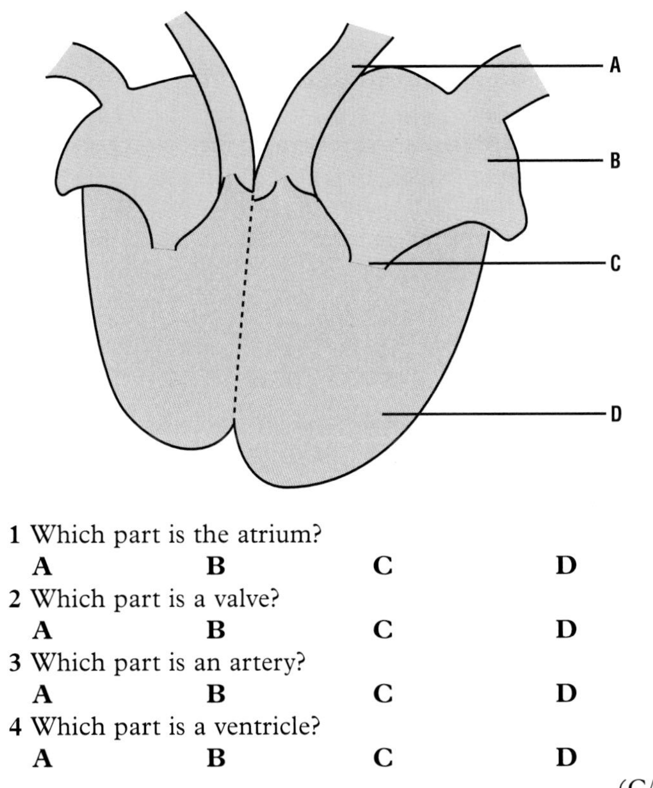

1 Which part is the atrium?
A **B** **C** **D**

2 Which part is a valve?
A **B** **C** **D**

3 Which part is an artery?
A **B** **C** **D**

4 Which part is a ventricle?
A **B** **C** **D**

(C/D)

QUESTION TWELVE

The diagram represents the human circulation system.

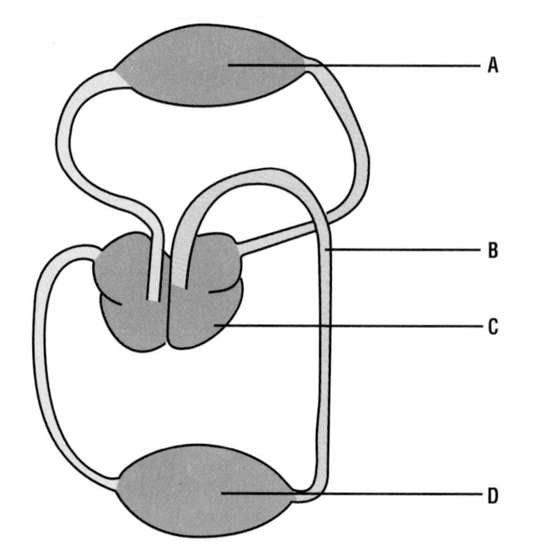

1 Which part represents the heart?
A **B** **C** **D**
2 Which part represents an area where the blood picks up a lot of carbon dioxide?
A **B** **C** **D**
3 Which part represents an area where the blood picks up a lot of oxygen?
A **B** **C** **D**
4 Which part represents an artery?
A **B** **C** **D**

(C/D)

QUESTION THIRTEEN

The diagram shows part of the breathing system.

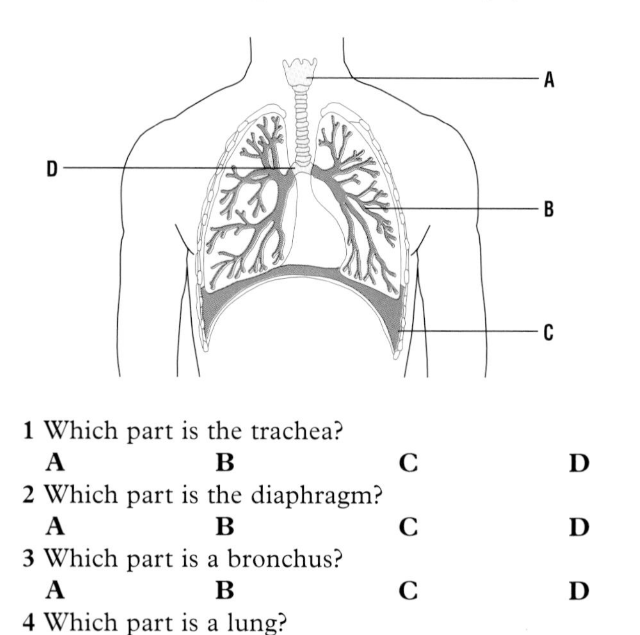

1 Which part is the trachea?
A **B** **C** **D**
2 Which part is the diaphragm?
A **B** **C** **D**
3 Which part is a bronchus?
A **B** **C** **D**
4 Which part is a lung?
A **B** **C** **D**

(E/F)

QUESTION FOURTEEN

An athlete has a resting heart rate of 55 beats per minute. The athlete does some training and checks her heart rate afterwards. The results are shown below.

Heart rate	Time
55	before exercise
145	immediately after exercise
116	15 seconds after exercise
97	30 seconds after exercise
83	45 seconds after exercise
76	60 seconds after exercise

1 Plot a graph of these results.
2 Estimate how long it will take for the athlete's heart rate to return to normal.

(C/D)

3 Why is her heart rate higher after exercise?

(A/B)

4 How do red cells carry oxygen around the body?

(C/D)

QUESTION FIFTEEN

At rest, a girl's breathing rate is 14 breaths per minute. She goes for a 20 minute run and measures her breathing rate immediately upon stopping, and every 30 seconds thereafter until it returns to normal. Some of her results are shown in the table.

Time	Breaths per minute minutes
at rest	14
immediately after exercise	48
30 seconds after exercise	46
60 seconds after exercise	39
90 seconds after exercise	35
270 seconds after exercise	14

1 Draw a graph of these results.
2 Estimate the breathing rate after 180 seconds.
3 What is the equation for respiration?
4 Why does the breathing rate increase during exercise?
5 Why does the breathing rate remain high after exercise has been completed?

(A/B)

Maintenance of Life

This module considers processes in the lives of plants and animals, and adds to our knowledge and understanding of human biology. In it we investigate how plants obtain water, nutrients and carbon dioxide, make food by photosynthesis, transport materials to where they are needed and respond to their surroundings. The ways in which humans co-ordinate their behaviour in response to stimuli from their environment and maintain a stable internal environment are studied. We also examine the effects drugs have on the human body.

Maintenance of Life

Learning Objectives

What you should know:

- The structure of plant cells, the jobs of the nucleus, cytoplasm and cell membrane, cell wall and chloroplasts. The differences between animal cells and plant cells
- The jobs of roots, stems and leaves in plants
- In photosynthesis, light energy is absorbed by chlorophyll and used to convert carbon dioxide and water into sugars
- The equation for photosynthesis: carbon dioxide + water + light energy → glucose + oxygen
- The rate of photosynthesis may be limited by low temperature, shortage of carbon dioxide, shortage of light
- The glucose produced in photosynthesis may be used in respiration or stored as starch
- Carbon dioxide enters a leaf by diffusion through stomata; water is absorbed by root hairs
- Transpiration is the loss of water from a plant
- The effect of temperature, humidity and wind on transpiration
- How plants reduce excess water loss
- Water supports young plants; if short of water they wilt
- Xylem transports water from the roots to the stem and leaves, phloem carries sugars from the leaves to the rest of the plant
- Plants are sensitive to light, moisture and gravity, their shoots grow towards light and against the force of gravity, their roots grow towards moisture and in the direction of the force of gravity
- Plant responses are brought about by unequal distribution of hormones causing unequal growth rates
- How humans use hormones to control growth and reproduction in plants
- The nervous system enables humans to react to their surroundings and co-ordinate their responses
- What type of receptors are found in the eyes, ears, tongue, nose and skin
- The jobs of sclera, cornea, iris, lens, ciliary muscles, suspensory ligaments and retina in the eye
- How waste carbon dioxide and urea are produced and where they leave the body
- What internal conditions are controlled in humans
- Sweating helps to cool the body
- Blood sugar level is controlled by the hormones insulin and glucagon which are produced by the pancreas. The cause and treatment of diabetes
- The effects of solvents, tobacco smoke and alcohol on the body

What you should be able to do:

- Prepare slides of plant cells
- Relate the structure of plant cells to their function
- Examine the external features of a flowering plant
- Investigate the importance of light to plants
- Find if chlorophyll is necessary for starch production
- Find if carbon dioxide is needed for photosynthesis
- Investigate absorption by roots
- Investigate how many stomata there are on a leaf
- Use a potometer to investigate water loss
- Investigate seedling growth
- Investigate if rooting hormone helps cuttings to grow
- Investigate tasting. Find the blind spot
- Label on a drawing the sclera, cornea, iris, pupil, lens, ciliary muscle, suspensory ligaments, retina and optic nerve
- Use a smoking 'machine'

EXTENSION

What you should know:

- How energy from respiration is used to make other compounds
- The functions of nitrate, potassium and phosphate in plants; the symptoms of their deficiency
- What osmosis is and its importance in plants
- How information is transmitted in the parts of a reflex action
- The mechanism of the kidney and the role of ADH
- Mechanisms involved in the control of body temperature
- How insulin and glucagon control blood sugar level

What you should be able to do:

- Investigate the effect of mineral shortage
- Investigate osmosis in cells
- Analyse a reflex action in terms of stimulus → receptor → co-ordinator → effector → response
- Investigate the cooling effect of alcohol

Unit A
Looking at cells

Internal structure

Plant cells have a number of special features (Figure A.1) which affect their biology. The cellulose cell walls give the cells strength and a definite shape. Many cells in parts of the plant exposed to light become green. This is due to the green pigment **chlorophyll** contained in **chloroplasts** in the cytoplasm.

During a plant's development, cells become specialised for different tasks. Figures A.2 and A.3 show the arrangement of tissues in different parts of a plant, and give examples of cells that make up the tissues. For instance, the surfaces of the young stem and leaves are covered by a single layer of cells with a waterproof **cuticle** on the outsides (Figure A.3). Conducting cells are grouped together into veins, or **vascular bundles**, consisting of tissues called **xylem** and **phloem**. In stems, veins carry water, mineral salts and dissolved foods, and also act as supporting columns. In young roots, the conducting material runs through the centre, strengthening the roots against the tug of the wind (Figure A.2).

Most leaves are flat structures, lying so they receive an optimum amount of sunlight. The network of veins carries materials through the leaf and provides strength. Most of the cells in the waterproof outer layer of the leaf do not contain chlorophyll. There are some pores or **stomata**, particularly in the lower surface. Each stoma is surrounded by a pair of green **guard cells**, which alter the size of the pore according to the conditions. For example, if conditions are dry, the guard cells will close the stomata. This allows the plant to prevent loss of water by evaporation.

Inside the leaves (Figure A.3), the cells are green, containing chloroplasts. The upper green cells are arranged in a regular manner and receive much of the sunlight which falls on the leaf. The lower cells are arranged irregularly. Air channels run from the stomata between the lower cells.

Compare a leaf cell with the *Amoeba* and the human cheek cells shown on pages 4 and 5. List the similarities and differences between a plant cell and an animal (human) cell.

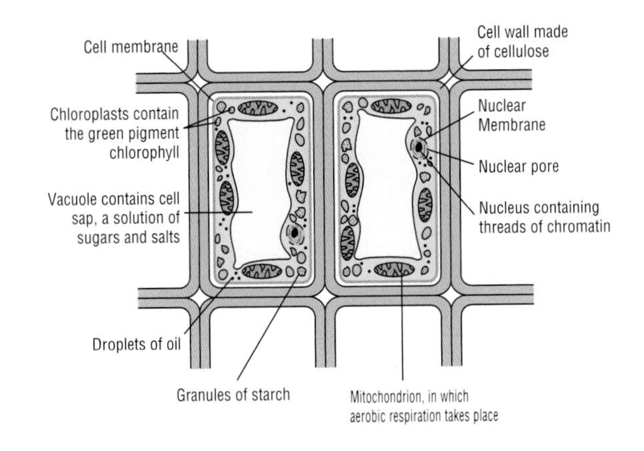

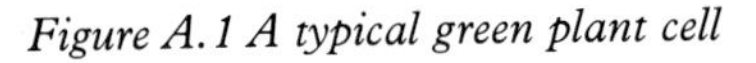

Figure A.1 A typical green plant cell

> **Activity A.1**
>
> **Plant cells**
>
> 1 Prepare a slide of a strip of cells from the inner surface layer of a fleshy leaf of an onion. Draw a few cells, using a microscope to show them in detail.
>
> 2 Compare the pattern of cells seen in cross-sections of a young buttercup stem and root.

Figure A.2 The positions of the vascular bundles in the stem and root of a young buttercup plant

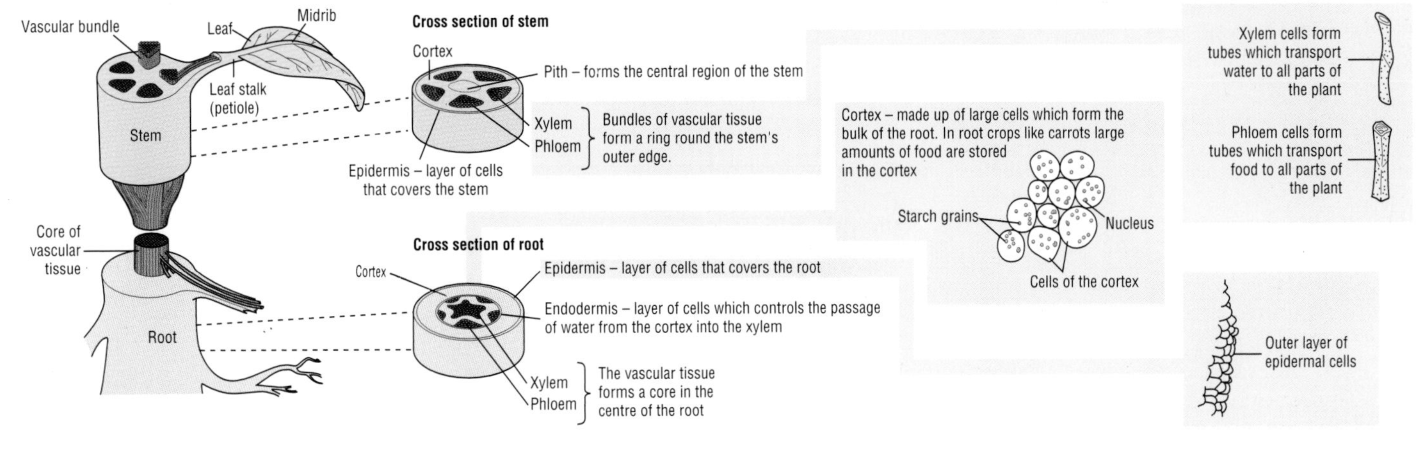

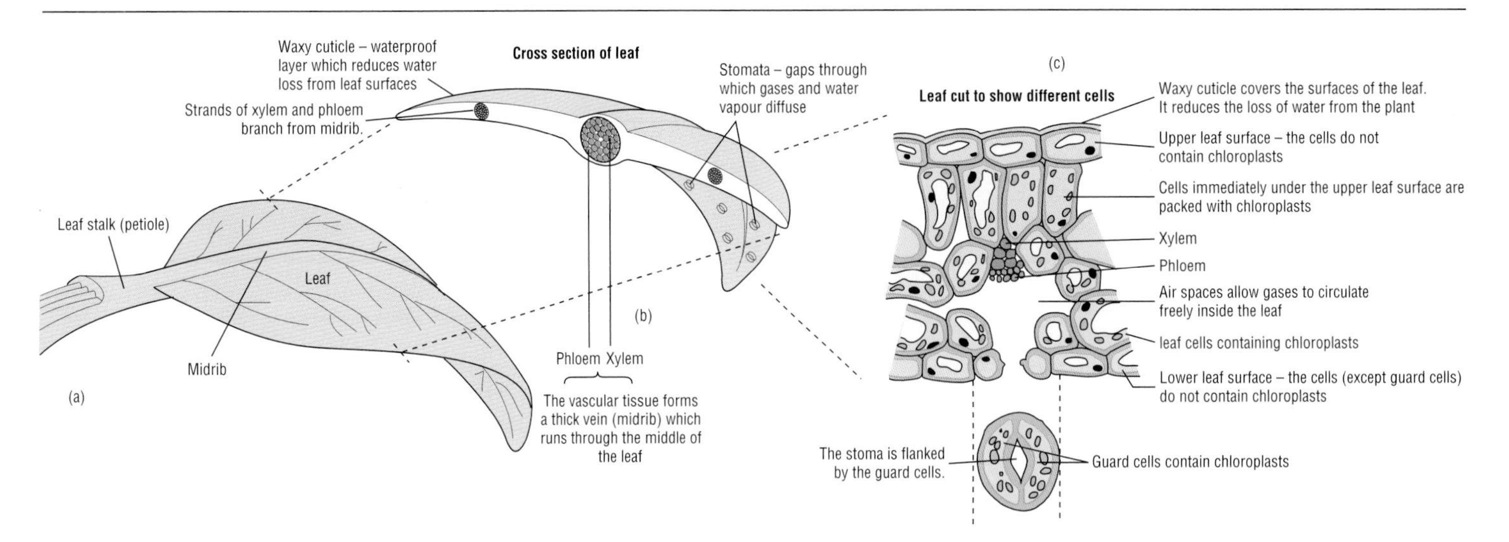

Figure A.3 Leaf structure at increasing magnification: (a) the whole leaf, (b) a section across part of the leaf, (c) the arrangement of cells within the leaf

Summary

- All plants are made up of cells. Plant cells are similar to animal cells in some ways, but different in other ways.
- Like animal cells, most plant cells have a nucleus, cytoplasm and a cell membrane.
- Plant cells have cell walls, and often also have chloroplasts for photosynthesis, and a vacuole filled with cell sap.
- Cells are specialised and organised as tissues within a plant to carry out different functions.
- Xylem tissue transports water, and helps support the plant.
- Phloem tissue transports dissolved food materials around the plant.

Questions

1. In what ways is a root different internally from a stem?
2. In which cells of the leaf are there the most chloroplasts? Why do you think there are more chloroplasts here than in other parts of the leaf?
3. Compare the structure of the leaf cell on page 55 with the structure of *Amoeba* or the human cheek cells shown on pages 4 and 5. Then copy and complete the table below. Put a tick in each space if the structure is present. Add any comments needed to distinguish between a plant cell and an animal cell.

Structure in cell	Plant cell	Amoeba/human cheek cell
Cell membrane		
Cell wall		
Cytoplasm		
Nucleus		
Chloroplast		
Vacuole		

Unit B
What do plants look like?

Buttercups grow in meadows. They are non-woody annual plants, which means that they grow from seed to maturity and produce new seeds all within one growing season. They then die with the onset of the first frosts, leaving the seed to lie dormant through the winter ready to grow when conditions improve the following spring. Non-woody plants are called **herbaceous** plants.

Every kind of plant is different, but buttercups show the features typical of most plants. Figure B.1 shows the features of a buttercup plant. Notice that the part of the plant below ground forms a root system which anchors the plant firmly into the ground. Above ground, the stem holds the leaves and flowers upright. It also supports the tissues which transport food and water between the various parts of the plant.

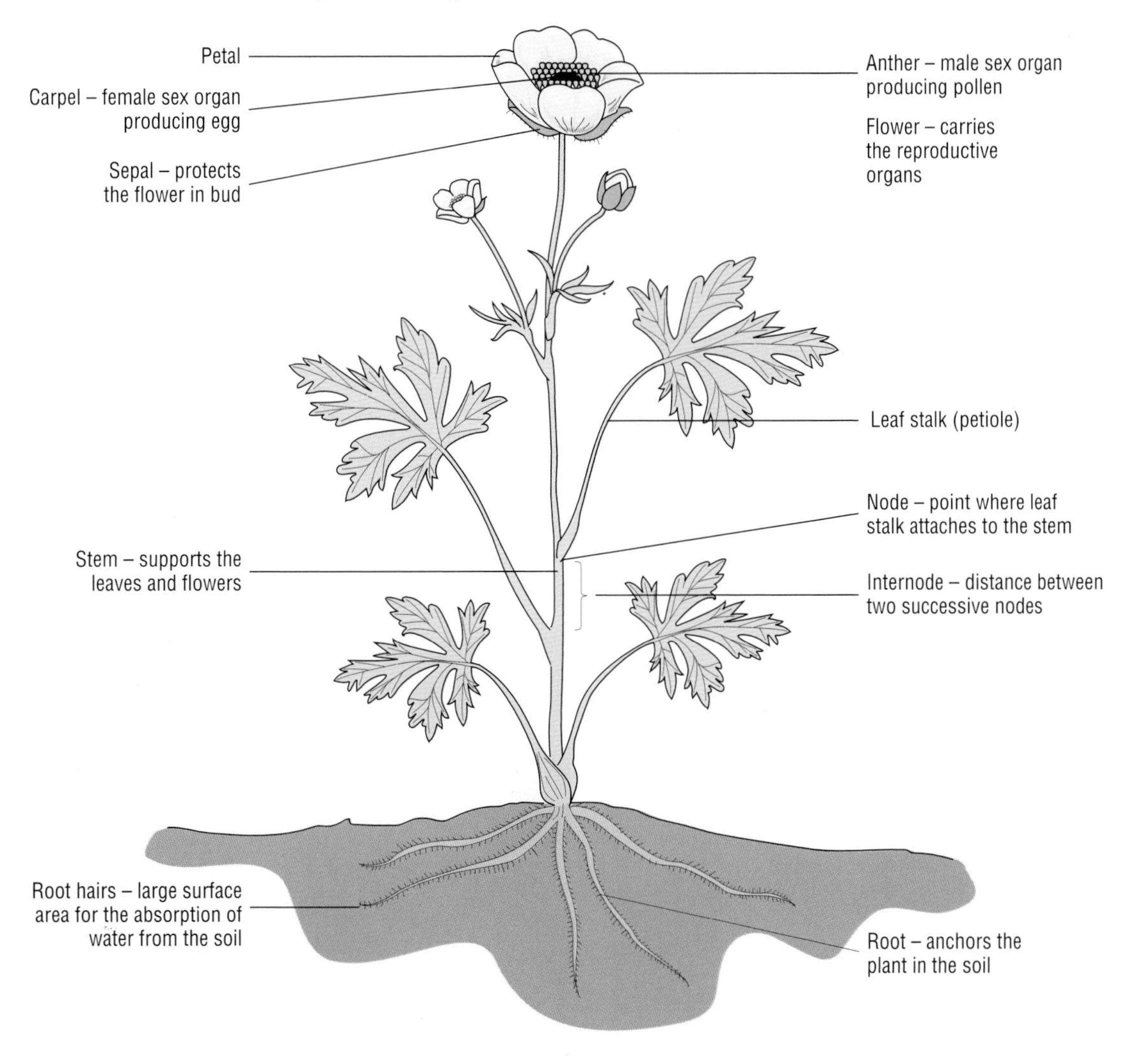

Figure B.1 A buttercup shows features typical of many plants

Activity B.1

Examining the external features of a flowering plant

You will need: flowering plant with root; hand lens.

1. Find as many as possible of the plant features shown in Figure B.1.
2. **The root system:** Feel the surface of the root. Examine the tip of a root with a hand lens. The tip of the root is protected by a layer of cells. Behind this is the region where the cells are dividing (Figure B.2). Just behind the tip of an undisturbed growing root the surface of the root is covered by fine threads or root hairs (see Figure E.4, page 69).
3. **The stem:** Follow the stems branching from the top of the root. Are there any leaves or buds on the shoot? Each bud encloses a growing region which will gradually produce a new stem or a flower. Are the buds and leaves arranged on the stem in a particular pattern?
4. Fruits develop from flowers. How many flowers and fruits are there? How are they arranged?
5. Examine a leaf and leaf stalk. Feel the upper and lower surface of the leaf and look at each surface with a hand lens. Are there any differences between the two? Where the base of each leaf or leaf stalk comes off the stem, there is a bud, but it may be too small to find.
6. Write a description or make labelled sketches to show the roots, stem and other parts of the plant.

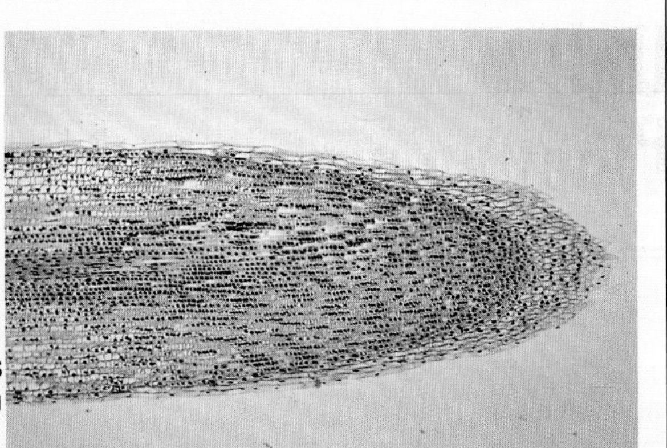

Figure B.2 The dividing cells are near the root tip

Summary

- A flowering plant has a root system and a shoot system.
- There are two patterns of root system – a tap root or fibrous roots.
- The shoot is composed of stem, leaves, buds, flowers and fruits. At the base of each leaf is a bud.

Questions

1. In what ways is a root different externally from a stem?
2. Dandelions (Figure B.3) and their relatives are found on bare ground and verges at most times of year. Examine a dandelion plant carefully and complete a table like the one below to compare it with a buttercup plant.

Feature of plant	Buttercup	Dandelion
Root Shoot Leaf shape and appearance Arrangement of leaves on stem Flower Arrangement of flowers Appearance of fruit Number of seeds per plant (estimate)		

Figure B.3 A dandelion plant

Unit C
Photosynthesis and respiration

Activity C.1

The importance of light

Is there the same amount of starch in a leaf which has been in the dark as one which has been in the light?

You will need: geranium plant which has been kept in dark for 48 hours, ethanol, beaker, boiling tube, forceps, goggles, Bunsen burner, tripod, heat-proof mat, gauze, watch glass, iodine solution.

1 Move your plant into the light. Immediately, remove one leaf of the geranium and test for starch as follows. **Ethanol is flammable: take care.**

Test for starch in green leaves
Place the leaf in a beaker of boiling water until it becomes limp. **Turn off all Bunsen burners in the room**. Using forceps, transfer the leaf to a boiling tube. Cover it with ethanol. Stand the boiling tube in the beaker of hot water. (Why does the ethanol boil when you do this?) When the ethanol has removed all the green chlorophyll from the leaf, wash the leaf gently with tap water. Place the leaf in a watch glass of iodine diluted to straw colour. Remove the leaf, wash it and hold it up to the light. If there is any starch present it will show up as black speckles.

2 After the geranium plant has been in the light for several hours, remove another leaf and test it in the same way. **Remember the safety precautions**.

3 Describe and give an explanation of your results.

Investigation C.3

How does light affect photosynthesis?

Where does the food come from that all green plants and all animals depend upon? Food is manufactured in green plants by the process of **photosynthesis**. (*Photo* = light, *synthesis* = building up.) Under natural conditions, the light energy used during photosynthesis is sunlight. It starts a series of chemical reactions involving carbon dioxide and water. The reactions can be summarised as:

carbon dioxide + water —**light energy**→ carbon compounds + oxygen
(carbohydrates)

If carbon dioxide and water are mixed together they will not react in this way unless sunlight and chlorophyll are present. The carbohydrates produced are soluble sugars which are often built into more complicated substances or used in respiration (see page 36). Sugars are usually changed to insoluble starch for storage.

Activity C.2

Is chlorophyll necessary for starch production?

Devise an experiment to demonstrate to a friend that starch is produced only where chlorophyll is present in a leaf. You can use a geranium plant with variegated leaves (that is, leaves with both green and white areas). What safety precautions must you take?

Figure C.1 Variegated leaf of a geranium

Chlorophyll, the green pigment (or coloured substance), is essential for photosynthesis. It is found in **chloroplasts** in all green cells.

Activity C.4

Will photosynthesis occur if all the carbon dioxide is removed?

Set up a geranium plant which has been in the dark for 48 hours as shown in Figure C.2. Why must the plant have been kept in the dark? What is the function of the sodium (or potassium) hydroxide? Place the plant in the light. After several hours, test both the leaf inside the conical flask and another leaf on the plant for starch as in Activity C.1. **Remember the safety precautions when using ethanol.**

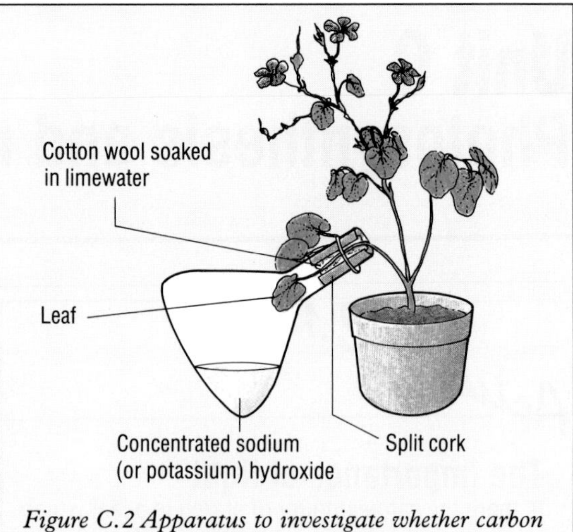

Figure C.2 Apparatus to investigate whether carbon dioxide is necessary for starch production

Photosynthesis in a flask

Simple green organisms, such as the alga *Chlorella* found in pondwater (Figure C.3), also photosynthesise. Each *Chlorella* is a single cell. *Chlorella* can be grown in a culture solution in a large flask (Figure C.5). The effect of conditions on their rate of photosynthesis can be monitored. *Chlorella* has become a very important experimental organism and it can be used to show where the oxygen produced in photosynthesis comes from. Nowadays, the conditions in the flasks can be measured and controlled using a computer interface. A micro-fermenter could be adapted to study photosynthesis in *Chlorella*. What conditions would be needed for *Chlorella* to photosynthesise and so produce oxygen?

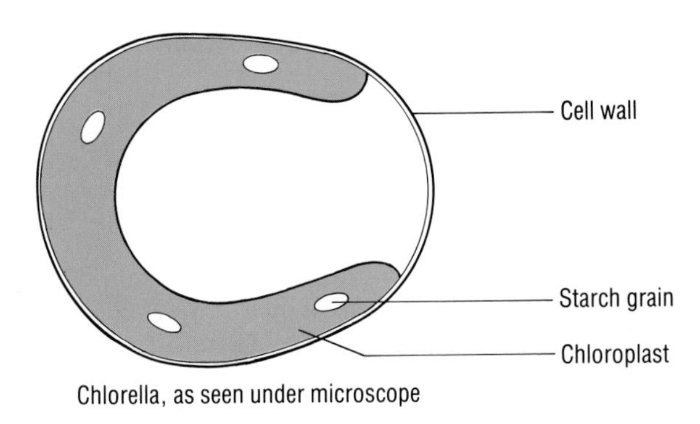

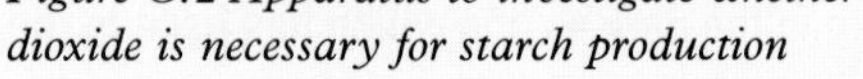

Figure C.3 Chlorella, *magnified*

Activity C.5

Effect of carbon dioxide and light

1. Examine the lower graph in Figure C.4. Describe what happens to the rate of photosynthesis as the amount of carbon dioxide increases up to 500 parts per million (ppm). Can you suggest an explanation? What happens at higher concentrations of carbon dioxide?
2. What do you think would have happened in Activity C.4 if there had been very little carbon dioxide present?
3. If fuel is burnt to heat a greenhouse, carbon dioxide is produced. How might this help the growth of greenhouse crops?
4. How does the upper curve of the graph change at a carbon dioxide concentration above 500 ppm? Try to explain why there is this difference.

At low concentrations of carbon dioxide, the amount of carbon dioxide available limits the rate of photosynthesis. It is the **limiting factor**, whatever the light level is. At higher concentrations of carbon dioxide, the rate of photosynthesis will only increase if the light is bright enough. The light is then the limiting factor.

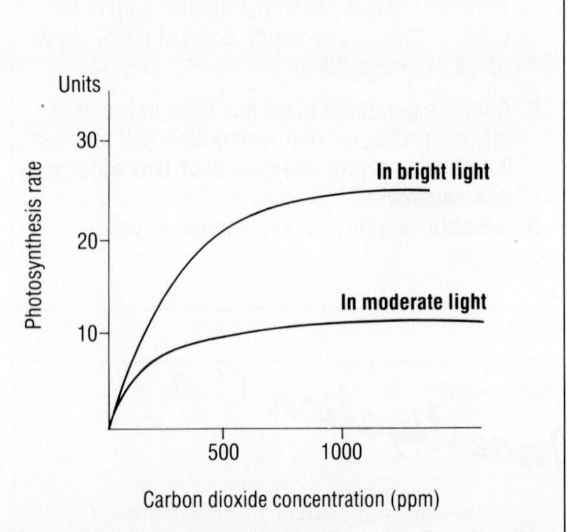

Figure C.4 Graph showing the effect of carbon dioxide concentration on the rate of photosynthesis

Figure C.5 A sample of a culture of Chlorella

An early investigation of plant nutrition

Today, we all know of the existence of the gases carbon dioxide and oxygen. It was not until 1754 that the Scottish chemist Joseph Black first showed how to produce carbon dioxide, which he named 'fixed air'. In 1774, Joseph Priestley produced oxygen, which he called 'dephlogisticated air'.

Some small steps in understanding plant nutrition or photosynthesis had already been taken by this time. In the fifteenth century, Cardinal Cusa wrote about placing weighed earth in a pot and then growing a plant in it. Even after collecting about 50 kg of plant and seeds from the plant, the weight of the pot of soil was scarcely changed. He thought that most of the material of the plant did not come from the soil but 'that the plants which were picked have their weight rather from water.' If Cusa were doing his experiments today, what other explanation might there be of where the plant material came from?

He continued: 'Therefore waters thickened in the earth have taken on the properties of earth and by the work of the sun they have been condensed into the plants.' Notice that he is suggesting that sunlight is important.

Cusa then suggested that if the plants were to be burned, the ash would show what weight of the plant had come from the water. Cusa could not know that much of the material of the plant came from the air, and would be converted back to carbon dioxide and water vapour by burning. All that would be left behind would be the ash containing the small weight of minerals which the plant had taken from the soil.

- What were the two important steps forward in Cusa's suggestions?
- In what two ways do Cusa's suggestions not agree with today's understanding?

Activity C.6

Jean-Baptiste Van Helmont's experiment

In 1648, Van Helmont did an experiment which is more widely known than Cusa's work. It is shown by Figure C.6

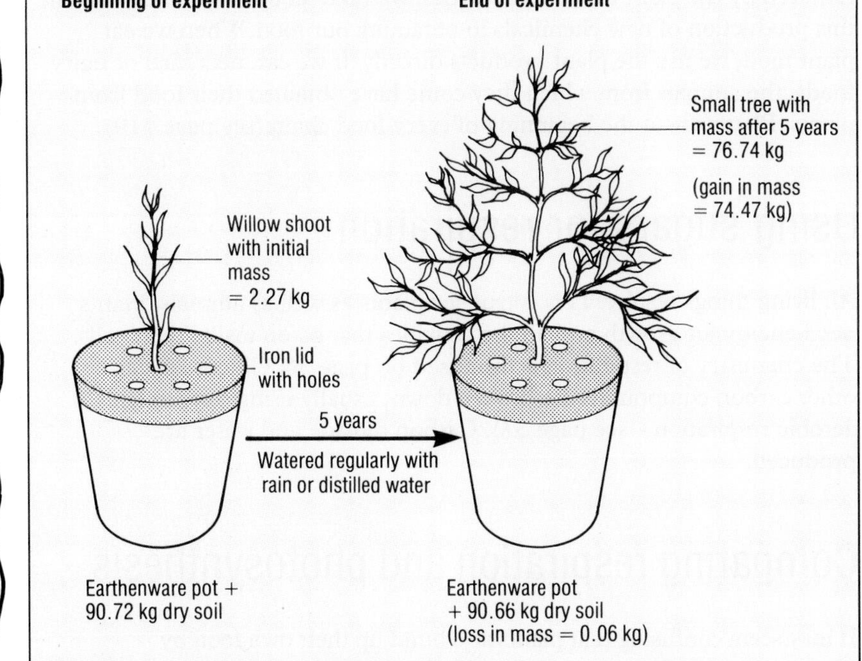

Figure C.6 Van Helmont's experiment

- Why did he use rain or distilled water?
- What conclusions can be made from Van Helmont's experiment?

For more than a century after this, it was generally believed that plants gained all their nourishment through their roots by absorption of water and other materials from the soil. People who have not learned any biology still believe this today.

- Why is this description of plant nutrition incorrect?

Using water containing a heavy form or **isotope** of oxygen, ^{18}O, it is possible to show that the oxygen produced during photosynthesis comes from water, not from carbon dioxide. Light is essential for this stage. Carbon dioxide then combines with the hydrogen produced, to make carbohydrates. Light is not necessary for this second stage, but if there is enough carbon dioxide and hydrogen, more carbohydrates will be produced at higher temperatures. Carbon dioxide that is 'labelled' with a radioactive isotope of carbon, ^{14}C, can be used to trace what carbon compounds are formed during photosynthesis.

The products of photosynthesis

During photosynthesis, carbon dioxide is converted to sugar molecules. Sugars are soluble and can be transported around the plant. They can be used for respiration (see page 36) to provide the plant with the energy it needs for growth and other activities within its cells. Sugar molecules can also join together into long chains, or **polymers**, such as starch and cellulose. Starch molecules are stored in starch grains inside the cells (see Figure A.2, page 55). A different arrangement of sugar molecules is found in the chain of cellulose, which is needed for the formation of plant cell walls. If sugars react with nitrates, they form amino acids, which then join together to form proteins (see Figure B.6, page 12).

Photosynthesis is therefore essential for producing the chemicals required by the plant for all its activities. We (and all other animals) exploit this production of new chemicals in obtaining our food. When we eat plant food, we use the plant products directly. If we eat meat, fish or dairy foods, the animals from which they come have obtained their food from plants. Plants are at the beginning of every food chain (see page 119).

Using sugars for respiration

All living things respire. This is true for plants as well as animals. Plants need energy for growth and all the activities that go on inside their cells. The chemistry of respiration is discussed on page 36. Glucose sugar or other carbon compounds are broken down, usually using oxygen (this is aerobic respiration – see page 37). Carbon dioxide and water are produced.

Comparing respiration and photosynthesis

It may seem confusing that plants also build up their own food by photosynthesis *using* water, carbon dioxide and light energy. This food is later used in all the plant's cells for respiration to release energy where the plant needs it.

The processes of respiration and photosynthesis can be compared using a table. Copy Table C.1 and fill in the gaps with the missing information.

Discovering the importance of light

One of Joseph Priestley's many experiments showed that sometimes plants 'restored air' which had been made 'bad' because animals had been breathing in it. Explain this result. Can you suggest why Priestley did not always get the same results with his plants?

In fact, Priestley did not realise that light was important and so he did not control the light in his experiments. He sometimes tried them when it was dark. What results would you expect in the dark?

Shortly after, John Ingen-Housz, a doctor to Austrian royalty, described the 500 experiments he carried out during a few months of leave in 1778 – 'Experiments on Vegetables, Discovering their Great Power of Purifying the Common Air in the Sunshine and of Injuring it in the Shade and at Night'. He not only showed the importance of light for the process we now call photosynthesis, but he also realised that it took place in leaves but not in fruits.

Slowly, evidence was built up about how plants obtained their food. The Chemist Justus von Liebig was interested in the chemistry of plants. In the 1840s, he showed that the carbon compounds of plants (the carbohydrates and oils) were built up from the carbon dioxide in the atmosphere.

In the twentieth century, photosynthesis has been investigated in much more detail. For instance, Robin Hill in 1939 took chloroplasts on their own and showed that in light they produced oxygen. In this first stage of photosynthesis, the Hill reaction, water is split in the presence of light. The light energy is transformed to chemical energy. The second stage of photosynthesis does not use light, but is affected by temperature. The hydrogen from the water plus carbon dioxide are built into carbohydrates. The chemical reactions use energy.

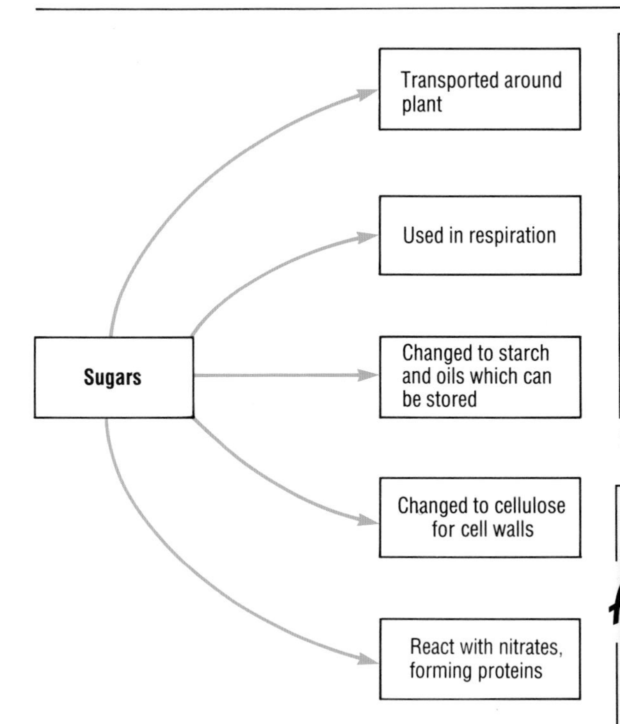

Figure C.7 How the plant uses sugars

WORLD TIMES

WE MUST NOT LOSE THE RAINFORESTS

The world's tropical rainforests contain a vast range of plant species. Many of these have never been studied. Some are beautiful; others produce unusual, but very useful chemicals. The plants producing rubber latex and quinine (used to treat malaria and in tonic water) can both be found in rainforest areas. If we allow the forests to be destroyed, we shall also lose many useful chemicals, as yet unknown to us.

The rainforest plants are a great chemical factory, drawing carbon dioxide from the atmosphere to build their own carbon-containing compounds. This chemical manufacture rejoices in the name of photosynthesis. As the plants photosynthesise, they release oxygen into the air. Who can tell what would happen to the balance of gases in the atmosphere if we allow the world's rainforests to disappear? Scientists are unable to agree on the effect of such destruction.

Biological process	Organisms involved	What happens to energy?	Substances used	Substances produced
Respiration	all plants and animals	energy is released		
Photosynthesis	green plants	light energy is used		

Table C.1 A comparison of respiration and photosynthesis

Activity C.7

Plant productivity

Some of the chemicals produced by plants are very specialised and many have very useful properties. Rubber latex comes from rubber trees, the digitalis drug used for treating certain heart ailments comes from foxgloves, and so on. We are still learning about the chemicals produced by plants, particularly those in rainforests where many of the plants have been very little studied. This is just one reason why we should protect our rainforests (see page 183).

Throughout the world, sixty-five billion (65 000 000 000) tonnes of carbon from carbon dioxide is built into plant chemicals by the process of photosynthesis each year (see page 59). Some kinds of vegetation are more important than others in building up carbon compounds. Table C.2 gives the areas of land occupied by certain kinds of vegetation (Column 1), the area (Column 2) and the weight of new carbon compounds produced in a square metre each year (Column 3). This information has been used to estimate the total amount of new material produced each year (Column 4).

Region of vegetation	Area (10^6 km^2)	Production of new material (g/m^2/year)	New material produced (billion tonnes/year)
Tropical rainforest	17	1000–3500	37.4
Temperate deciduous woodland	7	600–2500	8.4
Temperate evergreen woodland	5	600–2500	6.5
Savanna grassland	15	200–2000	6.0
Tundra and alpine	8	10–400	1.1
Desert and semi-desert	18	10–250	1.6
Cultivated land	14	100–3500	9.1
Open ocean	332	2–400	41.5

Table C.2 Production of new material from different areas of the world

1 In Table C.2, which region has the highest rate of production of new material per square metre?

2 Why is the total production of new material so great in open ocean?

If photosynthesis is more rapid than respiration, it hides the fact that respiration is occurring. Plants build up more food by photosynthesis than they use for respiration. Some of the surplus food may be used by animals. The chemical energy in the food is passed along the food chain.

Summary

- Photosynthesis is the manufacture of carbon compounds by the green cells of plants in the presence of light, using carbon dioxide and water. Light energy splits the water molecules, releasing oxygen gas. In summary, photosynthesis is:

$$\text{carbon dioxide} + \text{water} \xrightarrow[\text{green chlorophyll}]{\text{light energy}} \text{carbon compounds} + \text{oxygen}$$

- Respiration occurs in all living cells. Energy is released for the organism's activities. The process can be summarised as:

glucose + oxygen → carbon dioxide + water (+ energy)

Questions

1 Estimate the total leaf surface of a tree near your school. In what way is the total amount of leaf surface significant for the plant's photosynthesis?

2 What four conditions are necessary for photosynthesis? What is produced by the process of photosynthesis?

3 Explain how you would investigate each of the following:

a Whether the concentration of carbon dioxide in the water around Canadian water weed changes when it is illuminated. (**Hint:** place the water weed in a tube of water to which red hydrogen carbonate indicator has been added. Hydrogen carbonate indicator turns from red to purple as carbon dioxide is used up; it changes from red to yellow if more carbon dioxide is added. What result would you expect?)

b Whether the rate of oxygen production by Canadian water weed is affected by the colour of light. Use coloured plastic filters to alter the light falling on the weed. How would you measure the oxygen production?

c Whether the green colour called chlorophyll obtained from leaves using ethanol is a single pigment. (**Hint:** Paper chromatography can be used to separate coloured substances.)

d Whether there are more chloroplasts containing chlorophyll in the cells near the upper or the lower surface of a leaf. (**Hint:** examine a section cut through a leaf under a microscope.)

4 Give two reasons why it is important to protect tropical rainforest.

5 Make a drawing of a green plant and use arrows to show the net or overall movement of carbon dioxide and of water:

a in daylight,

b in darkness.

Unit D
Plants need minerals

Plants make their own food by the process of photosynthesis. For this process, they need light and use carbon dioxide from the atmosphere and water absorbed from the soil.

Plants also need very small quantities of minerals for healthy growth. Mineral **ions** are absorbed through the roots from the dissolved chemical compounds in the soil. When garden centres sell bottles of 'plant food' they are selling solutions of some of the important minerals; these can be added to the soil in which the plants are growing. When plants are unable to absorb enough of an important mineral they show signs of the deficiency.

If too little nitrate, phosphate, potassium, iron, magnesium, sulphate or calcium is absorbed by the plant, its appearance and growth will be affected in recognisable ways (Figure D.2). For instance, potassium helps photosynthesis and magnesium is needed for the plant to manufacture chlorophyll. If magnesium is in short supply, the leaves, particularly the older ones, become mottled or pale. Iron is part of the chlorophyll molecule and shortage of this mineral will also make leaves pale. A few substances, known as trace elements, are also required in very small amounts. At higher concentrations, trace elements may even be toxic (poisonous) to the plant. Copper is one of the trace elements.

Activity D.1

What is the effect of mineral shortage?

Set up a series of eight similar green *Tradescantia* cuttings, with their roots standing in the following eight culture solutions (see Figure D.1):

- distilled water
- minus nitrogen
- minus potassium
- minus phosphorus
- minus iron
- minus magnesium
- minus calcium
- full culture.

These solutions are made up according to the recipes of the nineteenth-century German plant biologist, Julius von Sachs. Each solution lacks one mineral. Which of the cultures listed are controls?

Leave the plants in a good light, but not full sun, for at least six weeks. If necessary, top up the solution with distilled water. Record your results in a table, showing these characteristics:

- height of each cutting
- colour of leaves
- number of leaves
- size of leaves
- feel of leaves
- branching of roots
- length of longest root.

How could you improve this investigation?

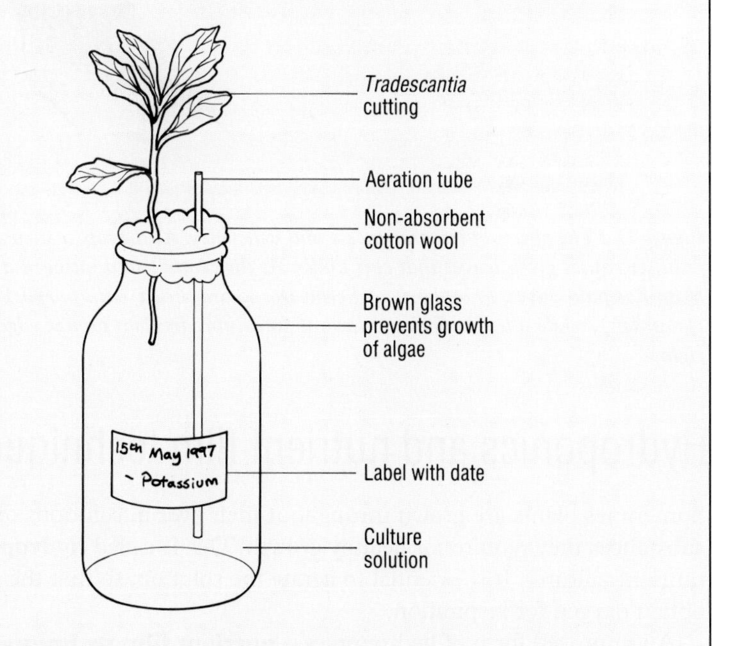

Figure D.1 A Tradescantia *cutting growing in a culture solution*

Nitrate, phosphorus and potassium are the substances which are most frequently in short supply in soils. Nitrogen is the most important because it combines with the sugars produced during photosynthesis to form amino acids (see page 62). These amino acids join together to form large protein molecules. Plants which lack nitrogen grow very poorly.

sugar (from photosynthesis) + nitrate (from soil) → amino acids → proteins

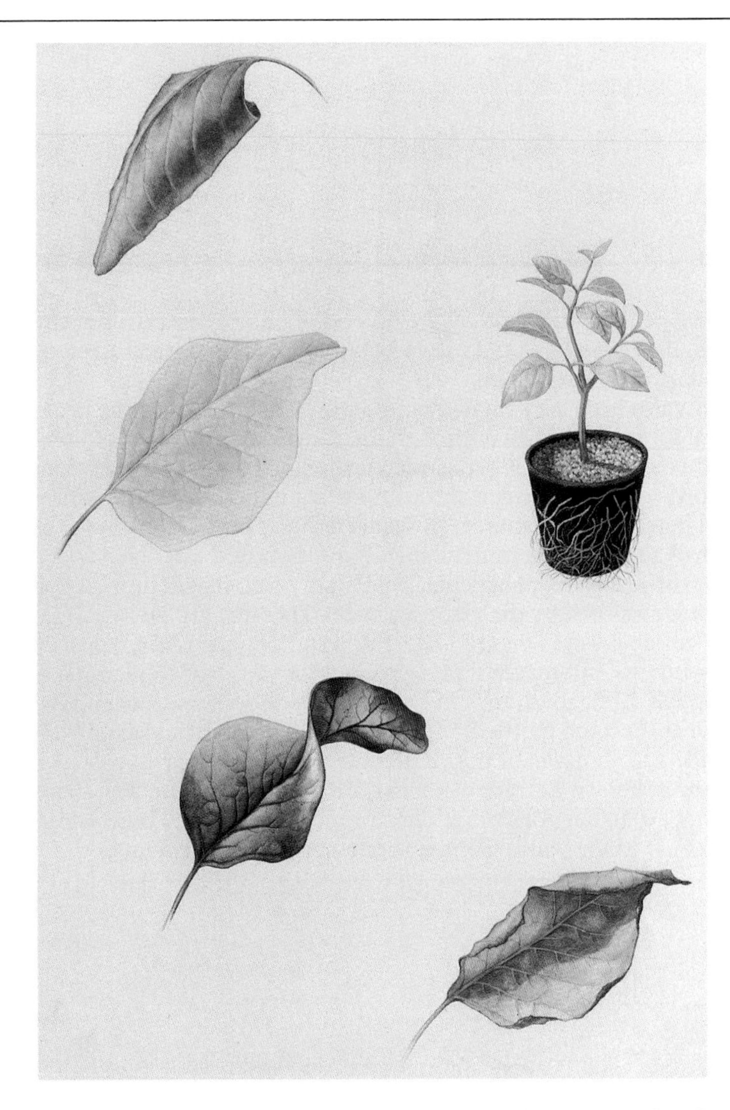

Figure D.2 The effects of mineral excess and deficiency. At the top, a surplus of fertiliser causes green leaves that curl under. At the centre left, a nitrogen deficient plant has pale leaves. Phosphorus deficient leaves are curled, with purplish undersides (lower left), while a leaf suffering a lack of potassium develops bronze edges (bottom right)

Hydroponics and nutrient film technique

Sometimes plants are grown throughout their lives in solutions of the substances they require for healthy growth. This is called **hydroponics** (or nutrient culture). It is essential to aerate the solution, so that the roots can obtain oxygen for respiration.

An improved form of hydroponics – **nutrient film technique** – involves circulating the solution through the containers in which the plants are grown (Figure D.3). The solution or nutrient film contains the correct nutrients at the best pH or acidity for growth. The solution is aerated and all the conditions are maintained by computer control. Nutrient film technique is being tried out at the Desert Development Centre in Egypt. Solar energy is used to pump the solution through the system. This is one way of growing crops in desert areas; but it may not be a realistic method of making the desert bloom. The technical facilities and maintenance may be too expensive.

Figure D.3 Tomatoes being grown by nutrient film technique. The solution circulates to the plants through the tubes

Summary

- Different mineral salts – particularly nitrates, phosphates and potassium salts – are needed for healthy plant growth.
- If an important mineral is in short supply, plants show signs of deficiency.

Extension questions

1 A farmer brings you plants taken from different fields. From their appearance you suspect that the plants are suffering from shortage of different minerals.

Field A: the plants were stunted in growth and the older leaves on each plant were yellow.
Field B: the plants showed poor root growth and the younger leaves on each plant were purple.
Field C: the plants had yellow leaves with dead spots.

You are aware that soil often suffers from shortages of nitrate, phosphorus and potassium. For each field find out:
a which mineral is in short supply
b why the deficiency is affecting the plants.

2 Figures D.4 and D.5 show how the yields of winter wheat and grass increase as more nitrogenous fertiliser is used. After studying the graphs, say what mass of nitrogen you would apply to a 100 hectare field to avoid waste and to give a maximum yield of:
a winter wheat
b grass.

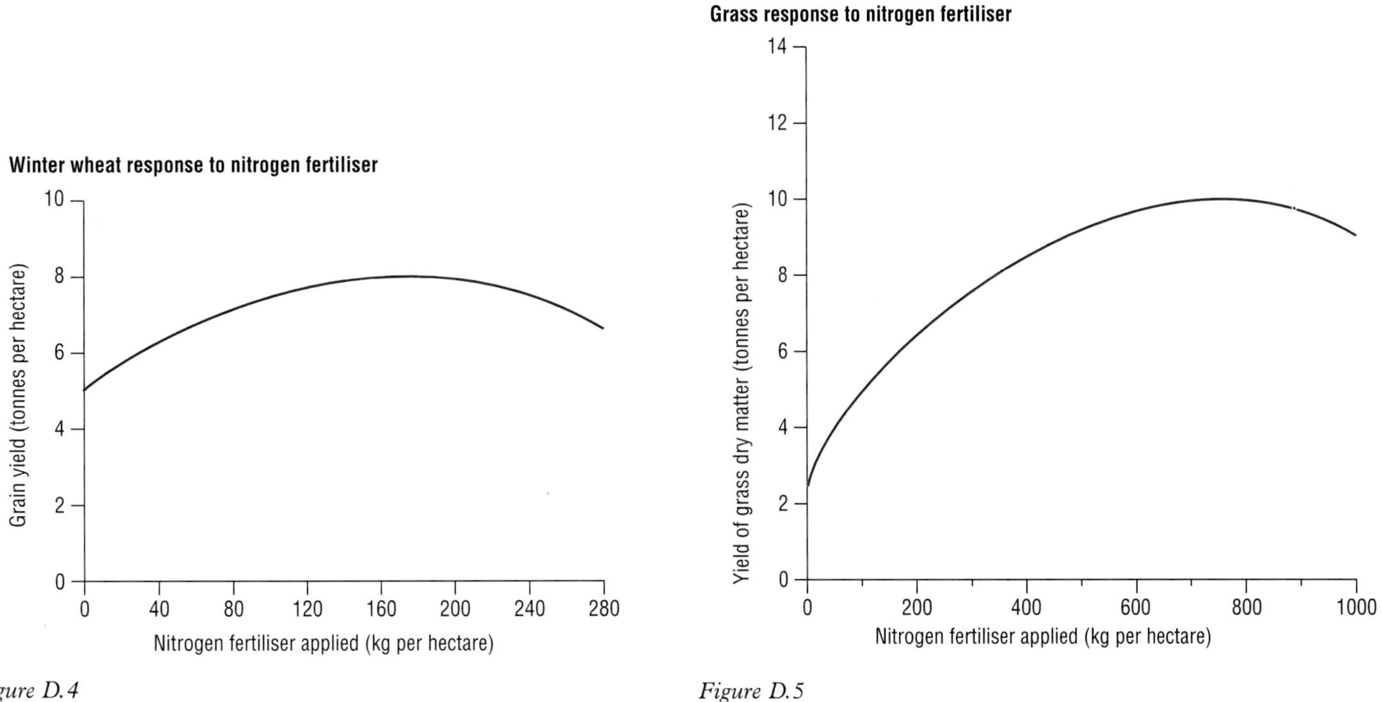

Figure D.4

Figure D.5

Unit E
Water and carbon dioxide

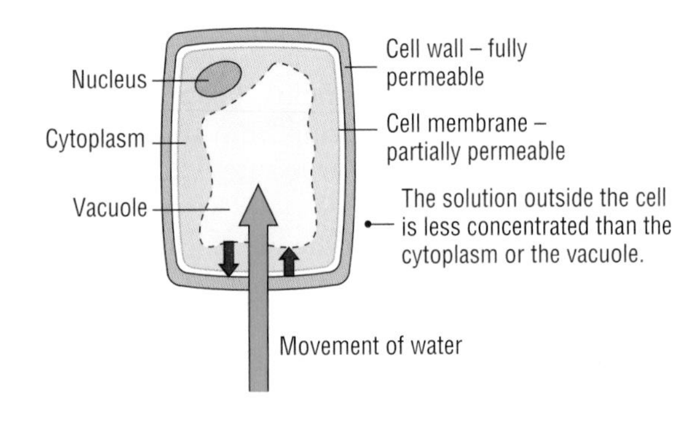

*Figure E.1 Water moves into the cell by osmosis. The increased pressure of water in the vacuole presses the cytoplasm against the strong cell wall. The pressure keeps the cell firm (**turgid**). Non-woody (herbaceous) plants and young plants are supported by the firmness of their cells. The turgid cells press against each other and hold the plant upright. If the cells lose water the plant wilts*

All living organisms need water. Water is transparent and is usually liquid at the temperatures found on earth. It is a good solvent and permits many chemical reactions to occur. If a plant lacks water it wilts. Every plant needs water because:

- it keeps cells firm (turgid)
- it allows transport of dissolved chemicals
- it allows many metabolic activities to occur
- it is required for photosynthesis and other chemical reactions.

Diffusion

Diffusion is the way in which molecules of a gas or a dissolved substance spread out (see page 47). They move from regions of high concentration to regions of low concentration. The molecules move down a **concentration gradient**. Perfume molecules, for example, spread through the air by diffusion.

In plants, both water and mineral salts and ions can diffuse through cell walls, which are **permeable**. Water is also able to pass through the cell membrane, into the cytoplasm of the cell (see Figure A.1, on page 55). However, this membrane is only **partially permeable**. Although molecules of water can diffuse through the membrane, the molecules of many dissolved substances cannot pass through it.

Osmosis

The vacuole of a plant cell contains dissolved chemicals. If you place a plant cell in water, some water will enter the cell. It diffuses from the water outside, through the partially permeable cell membrane, through the cytoplasm, to the solution in the vacuole (see Figure E.1). This is an example of **osmosis**.

Osmosis is a special case of diffusion. There is more water outside the membrane than inside, so there is a concentration gradient of water. The water moves down the gradient through a partially permeable membrane, into the cell.

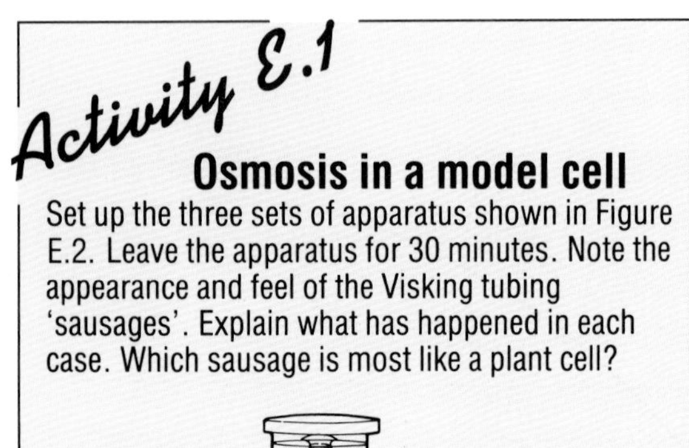

Osmosis in a model cell

Set up the three sets of apparatus shown in Figure E.2. Leave the apparatus for 30 minutes. Note the appearance and feel of the Visking tubing 'sausages'. Explain what has happened in each case. Which sausage is most like a plant cell?

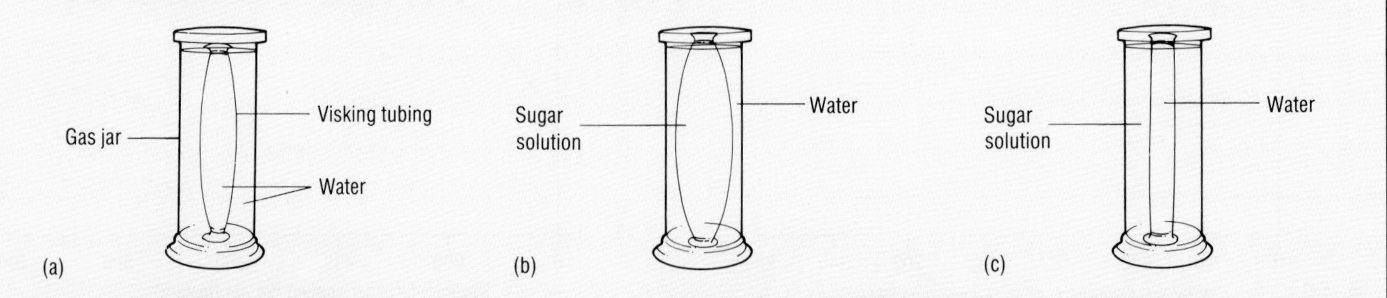

Figure E.2 Three Visking tubing 'sausages': (a) filled with water and standing in water; (b) filled with sugar solution and standing in water; (c) filled with water and standing in sugar solution

Activity E.2

Osmosis in cells

You will need: microscope, lamp, two slides, coverslips, pipette, mounted needle, scalpel, forceps, rhubarb (or other red plant cells), 10 per cent sugar solution, water, labels.

1. Label the slides **A** and **B**.
2. Strip a piece of epidermis (surface layer) off the rhubarb stalk. Place a small piece of this epidermis on each slide.
3. Place a drop of sugar solution on the epidermis on slide **A**. Place a drop of water on slide **B**. Lower a coverslip onto each drop.
4. After 15 minutes, examine each slide under the microscope to see whether the pink colour fills the whole cell. Make coloured drawings of a few cells from each slide.

   When water enters the cell, the colour spreads throughout the cell but is paler. If the cell is placed in sugar solution, water leaves the cell. **Plasmolysis** occurs – that is, the cytoplasm shrinks away from the cell wall. As the water moves out by osmosis, the red cell sap appears as smaller patches. The colour is more concentrated and looks darker. Use arrows on your diagrams to show the direction of water movement and labels to explain what has happened to the cells on each slide.
5. If possible, use the computer program *Osmosis* (from Audio-Visual Productions).

Activity E.4

Absorption by the roots

You will need: two groundsel plants (or similar), two beakers, water and dilute red ink (or aqueous eosin), scalpel, watch glasses.

1. Place one plant in a beaker, so that its roots are standing in red ink; place the other in a beaker with its roots in water.
2. At half-hourly intervals compare the stems, leaves and flower buds, looking for signs of redness in the plant standing in red ink. When red colour appears, cut thin slices from the main root and stem of both plants. Compare them to find the path of the red stain. If red colour appears in the plant standing in ink it must have been absorbed through the roots. It travels through the xylem vessels in the veins which lie in the centre of the root and in small bundles around the stem (see Figure A.2 on page 55).
3. Try standing celery stalks or a white-flowered plant in red ink. Where does the red ink appear in each case?

Absorption by the roots

Roots anchor the plant in the soil and absorb water from the soil. The root hair's near the tip of each root increase the root's surface area. Water is absorbed into them by osmosis. It then travels inwards, from cell to cell, passing by osmosis along the concentration gradient. As long as the cells are alive, water will be absorbed. No energy from respiration is required.

Activity E.3

How does osmosis make cells turgid?

Figure E.3 What will happen when the screw clip is opened?

When a cell becomes firm because it is swollen with water it is described as **turgid**. If you leave the model shown in Figure E.3 for 30 minutes or longer, it will become very turgid. **Turgor pressure** has built up inside the model cell. Hold the jet over a large sink. Open the screw clip. Explain what happens.

Activity E.5

Examining roots

1. Dig up a dandelion plant and a grass plant. Wash off the soil and draw each plant. The dandelion has a main or **tap root**; the grass has **fibrous roots**.
2. Make a drawing of the root hairs on some seedlings. Plants grown on a damp surface produce particularly long root hairs.
3. Using a microscope, examine a cross-section of a young root showing root hairs. Draw a few cells to show the root hair cells (see Figure E.5).

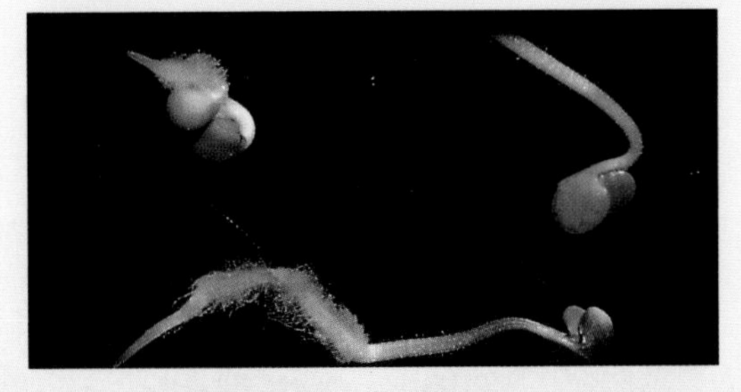

Figure E.4 Root hairs on a mustard seedling

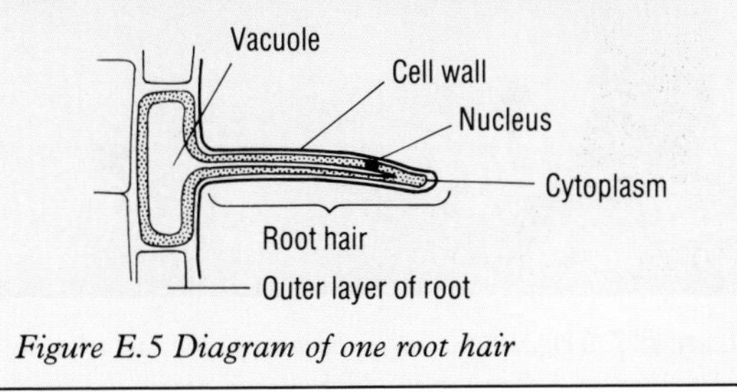

Figure E.5 Diagram of one root hair

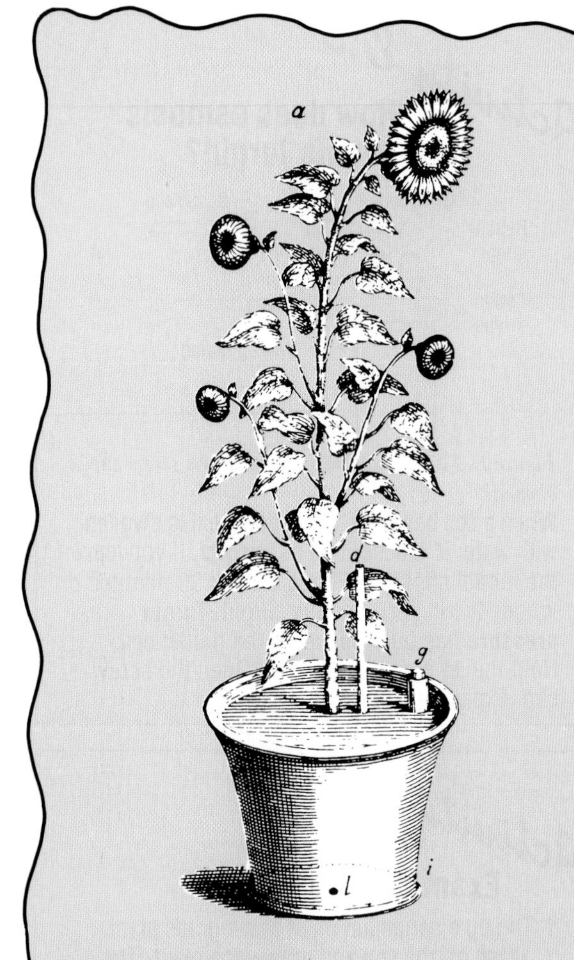

Figure E.6 Stephen Hales' apparatus for showing water uptake: **a** is a sunflower plant; **d** is a tube for entry of air; **g** is a tube (with a cork) for watering the plant; **i** is the pot with a waterproof cover over the soil

Stephen Hales' study of water in plants

Stephen Hales (1677–1761) went to Cambridge University just as the great physicist Isaac Newton left Cambridge. Although he was studying to be a clergyman, he was interested in chemistry, animals and plants. Hales became a great experimenter, following Newton's example by taking more measurements ('staticks') than most earlier scientists. He tried to find out answers to questions about plants. His investigations were carefully planned. The techniques and apparatus he devised made experiments with gases easier for other scientists.

Hales' experiments on plants were very important. He investigated growth, water in plants and the importance of air. At that time, the different gases in air were not known. In 1727 he wrote a book, *Vegetable Staticks*, describing his experiments. *Vegetable Staticks* begins with an experiment on sunflower plants 'shewing the quantities imbibed and perspired' (or transpired).

He sealed a pot containing a sunflower plant, and found the average mass of water lost each day. He made an allowance for moisture lost by evaporation through the pot. After his investigation he removed the leaves and measured their surface area on a grid. He also estimated the surface area of the roots and measured the area of the cross-section of the stem. He worked out the ratios of the areas and the rate of movement ('velocity') of water.

In 1727, Hales wrote, 'We have also many proofs of the great force with which plants and their several branches and leaves imbibe moisture, up their capillary sap vessels ... If therefore these Experiments and Observations give us any further insight into the nature of plants, they will then doubtless be of some use in Agriculture and Gardening ... But as it requires a long series and great variety of frequently repeated Experiments and Observations, to make a very small advance in knowledge of the nature of vegetables, so proportionably we are from thence only to expect some gradual improvements in the culture of them.' Today, we have much better techniques and far more understanding of how plants function, but the study of the uptake and movement of water through crops is still an important field of research.

Minerals (dissolved chemical compounds) enter the roots from the soil by active transport. Energy from respiration is used to absorb minerals.

Both water and minerals pass into the xylem vessels in the vein or vascular bundle in the centre of the root. The solution in the veins flows up the stem towards the leaves. The xylem vessels are very fine tubes, formed from many cells lying end to end. Xylem is specialised for transporting solutions up the plant (see Figure A.2 on page 55).

Figure E.7 Irrigating a sugar beet crop; the reel of hose is attached to a spray gun which moves across the field

Water shortage and irrigation

Plants cannot survive without water. In our gardens we may occasionally need to use watering cans and hoses. In dry summers, farmers may need to irrigate their crops. For instance, sugar beet is a crop which often needs extra water. Farmers can buy the extra water from the water authority. One method of irrigation uses equipment which automatically moves across the field spraying out water (Figure E.7).

In many parts of the world, there is a water shortage most of the time. In Northern Africa, dry conditions are spreading south from the Sahara desert, across a region known as the Sahel. Many hours each day are spent carrying water from rivers, pumps and wells to the crops. Simple, easily-maintained watering systems can make a dramatic difference to the people's lives, as well as to the quality of the crops.

Figure E.8 Simple pump irrigation in the Gambia

The functions of leaves

a Photosynthesis. As you know, green leaves carry out photosynthesis. Their large flat area makes them ideal for absorbing light. The surface cells of the leaf are covered by a cuticle which is fairly waterproof. There are pores or stomata in the leaf epidermis, often on the lower side only. The opening of the pores is affected by a pair of guard cells (see Figures E.9 and E.10). They are usually open during daylight and close at night. In sunlight, carbon dioxide for photosynthesis enters the leaf through the stomata. Oxygen leaves through the same pores.

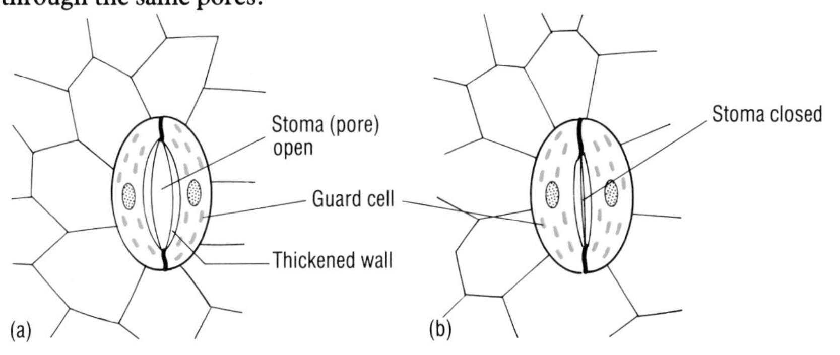

Figure E.9 Stoma with guard cells: (a) open; (b) closed

Inside the leaf, the gases diffuse through the spaces between the cells. These spaces increase the area of the leaf's internal surfaces for exchange of gases into or out of the cells, depending on the concentration gradient (see page 47).

The leaf's shape is ideal for photosynthesis but the leaf structure also enables the plant to control its loss of water to the atmosphere.

b Transpiration. Water travels up to the leaves in the xylem vessels, in the veins (see Figure A.2 on page 55). It then passes by osmosis into the surrounding cells. Some water evaporates from the surface of these cells, into the spaces between the cells. The water vapour diffuses out through the stomata when they are open. This loss of water from the leaves is called **transpiration.**

Guard cells are the only cells in the leaf epidermis which contain chlorophyll. Their walls have thick regions and contain fibres. In the light, guard cells photosynthesise. The sugar produced makes them take in water and swell. Their peculiar walls cause them to change shape as they swell. As a result, the stoma opens, allowing water vapour to escape into the air. In most plants, stomata close at night.

Stomata also close if the plant wilts, so that further loss of water is prevented. The closing of the stomata in this situation helps to keep conditions inside the plant cells unchanged so that they can carry out essential activities. If much more water is lost than enters the plant through the roots, the plant may not recover.

This is an example of **homeostasis** (see page 90) – the process of maintaining a steady internal environment in the cells. Homeostasis is important in both plants and animals.

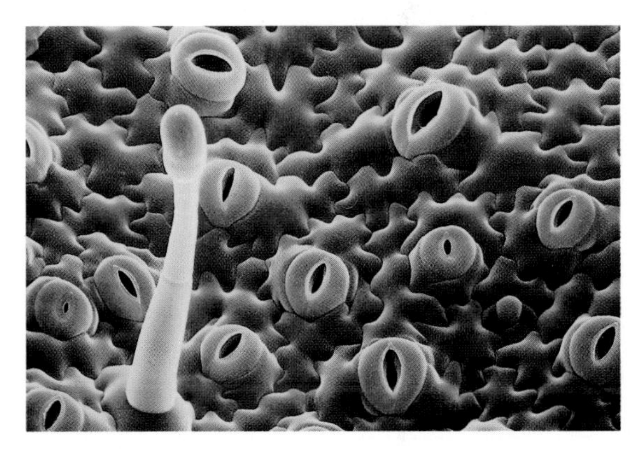

Figure E.10 Scanning electron micrograph of open stomata

Activity E.6

How many stomata are there on a leaf?

1. Using forceps, pull a small piece of epidermis off a leaf (iris leaves work well). Mount it in water on a slide. Examine it with a microscope and draw a stoma with its guard cells.
2. **a** Brush a little clear nail varnish (preferably coloured slightly pink) onto the lower surface of a smooth leaf. Let the nail varnish set firm.
   b Peel the nail varnish off with forceps. Mount it on a slide and look at the impression of the leaf surface.
3. Use the method in **2** to compare the number of stomata on a vertical iris (or grass) leaf and the upper and lower surfaces of a privet leaf.

Activity E.7

Measuring water loss

Investigation E.8

Investigate how plants lose water.

Activity E.9

A computer simulation of transpiration

If there is one available, use a computer software package that simulates transpiration, to investigate how transpiration changes in:

a hot, dry desert conditions
b hot, humid conditions, such as those in a tropical rainforest
c cool, windy conditions

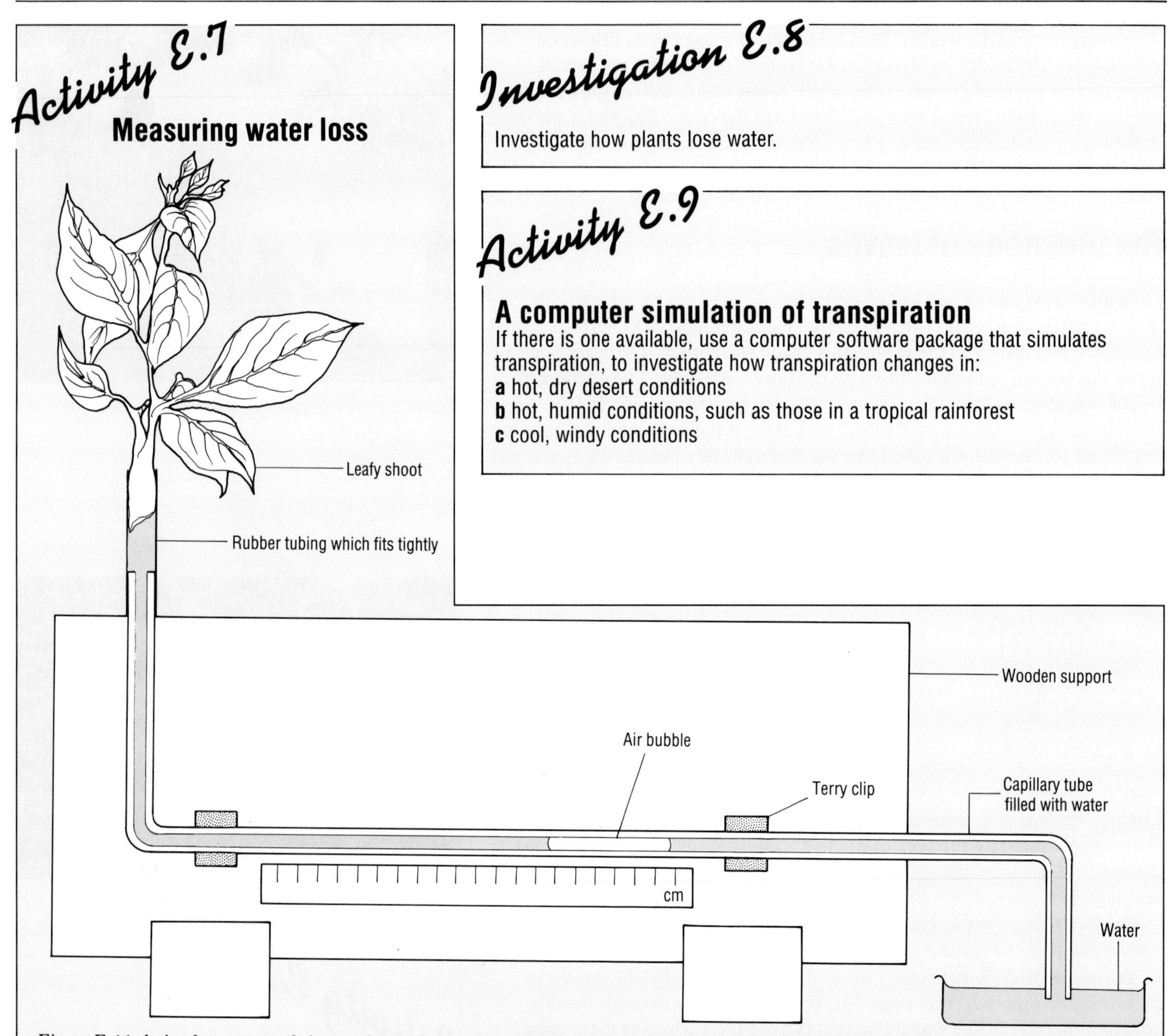

Figure E.11 A simple potometer being used to measure water uptake

It is difficult to measure transpiration in a whole plant accurately. One method uses a small shoot or twig in a **potometer** (see Figure E.11). Keep the cut end of the shoot under water. Fill the potometer with water and insert the cut end of the shoot into the upper end of the glass tube. The shoot must fit tightly. Allow an air bubble into the other end of the tube. Measure the distance the bubble moves along the tube in five minutes.

If you want to take a second reading, place the lower end of the tube in water and squeeze the rubber tubing gently to expel the bubble. Then remove the tube from the water to allow another bubble in.

Try to use the apparatus to compare the movement of water along the tube, with the shoot under different conditions – warm, cool, windy, light, dark. Allow the plant to settle in each set of conditions, with the lower end of the tube in water, before you begin to take readings. How many readings should you take in each situation?

Describe your experiment and results. What do the readings measure? Do you think they are a satisfactory way of comparing transpiration in different conditions?

Summary

- Diffusion is movement of molecules from regions of high concentration to regions of low concentration down a concentration gradient.
- Osmosis is movement of a solvent (water) from a dilute solution through a partially permeable membrane to a more concentrated solution. It is important for the movement of water into and out of plant cells.
- Minerals enter a plant by active transport.
- Leaves are important for maintaining the homeostasis of a plant.

Questions

1 Make a list of all the leaf structures. In a second column, give the functions of each structure.
2 Make labelled sketches of a leaf to show each structure. Use arrows to show the path of water as it moves through a leaf and is transpired.

Extension questions

3 Explain what is meant by a partially permeable membrane.
4 Look at Figure E.12. Explain fully with the aid of a diagram what has happened in the plant cells to produce the change from photograph (a) to photograph (b).

(a)

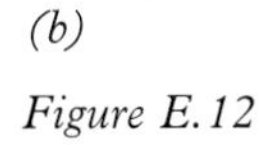

(b)

Figure E.12

Unit F
Moving materials

How does water travel in the xylem?

In all flowering plants, including trees, water travels upwards in the xylem vessels of the veins. The tallest trees in the world are over 110 metres high. How does sap or solution in the xylem reach the top of the tree? The mechanism is still not fully understood.

The most satisfactory explanation depends on transpiration. When the stomata are open during daylight, transpiration will take place. As long as there is a good enough supply of water in the soil, more water will travel up the xylem vessels in the veins to replace that which is lost from the leaves. The loss of water by transpiration keeps water moving through the plant.

The continuous movement or mass flow of water upwards in the xylem vessels is sometimes called the **transpiration stream**. The water also carries dissolved minerals from the soil to all parts of the plant. As water evaporates during transpiration it cools the leaves.

Figure F.1 How does water reach the top of tall trees, such as 'General Sherman' in California, 85 m tall?

Movement of substances in plants

Sometimes the movement of solutes (dissolved substances) around plants is by means of physical processes such as diffusion. In other situations in the plant, it is an active process requiring energy from respiration.

Solutes are absorbed into the root from the soil. The concentration of a solute which has accumulated in the cells of the root may be greater than the concentration of the same substance in the soil. Movement through the cell membrane against the concentration gradient is carried out by a process known as active transport (see page 47). This uses respiration energy. The solutes travel across the root and up the xylem vessels in the veins, by means of a combination of diffusion, active transport and flow in the transpiration stream. The mechanisms involved are not all fully understood.

Sugars and other chemicals produced within the plant also move about the plant in solution. All solutes may be carried upwards through the xylem in the transpiration stream.

Both upward and downward movement of solutes is possible in the **phloem** (see Figure A.2 on page 55). Phloem consists of living cells, lying close to the xylem in the veins. Respiration energy is used for phloem transport. The sugars produced by photosynthesis in the leaves are moved downwards to the rest of the plant through the phloem. In some plants, such as carrots and beets, surplus food is stored in the roots (Figure F.2). Most of the stored food is in the form of insoluble starch. Water therefore does not accumulate in storage cells as a result of osmosis.

Figure F.2 In root crops like carrots large amounts of food are stored in the cells of the cortex

Surviving dry conditions

Many plants living in dry situations have evolved ways of overcoming or surviving water loss. For example, plants may lose less water through their leaves because

- they are able to close the stomata
- they have leaves with fewer stomata
- the stomata are sunk in pits or surrounded by hairs which reduces transpiration
- their leaves have a more waterproof surface
- the leaves are nothing more than spines, as in cacti.

The North American creosote bush has a network of roots which absorb dew that collects just below the dry surface soil. The soil around the bush is too dry for other plants. Gradually, new stems grow out in a circle. The older parts of the plant dry out and die.

It has been calculated from the size of the circle that one creosote plant is more than 11 000 years old. It has certainly survived dry conditions successfully.

The mesquite is a tree with a root system which is much larger than the shoot. It extends many metres deep to reach what little water there is. The tree has been grown to provide shade and firewood in desert conditions.

Some plants survive water shortage by storing water. In cacti, water is stored in the green fleshy stems; in stonecrops, the leaves store water.

Summary

- The path of water in a plant is:
  a from the soil into the root hair cells by osmosis
  b inwards across the cells of the root, by osmosis down the concentration gradient
  c into the xylem vessels in the veins
  d up the xylem vessels through the stem to the leaves
  e into the cells around the veins of the leaves, by osmosis
  f into the leaf spaces by evaporation
  g out to the air through the stomata in the leaf surfaces by transpiration.
- Transpiration helps to move water up a plant.
- Solutes move around a plant by diffusion and active transport.

Questions

1 Distinguish between transpiration and the transpiration stream.
2 Prepare a display of the ways in which plants survive dry conditions.

Extension questions

3 What makes the cells of a dried pea absorb water when the pea is placed in water? Describe what happens inside the cells. Use scientific words to describe the difference between the cells in the dried pea and the cells in the soaked pea.
4 Explain why water does not accumulate in cells that store starch.

Unit G
Growth and sensitivity

First things first! 'Responses' and 'stimuli' are everyday words but what do we mean when talking about living things responding to stimuli? Think of it like this. The environment is changing all the time. Some changes are long-term, others short-term. Because these changes cause plants and animals to take action, the changes are called **stimuli**. The actions which plants and animals take are called **responses**. Being able to respond to stimuli means that living things can alter their activities according to what is going on around them.

Figure G.1 Responding to light. A time-lapse camera was focused on these sunflower seedlings and a series of pictures taken at 6 hour intervals. The plants took just 24 hours to respond as shown to intense light from the left

Plant growth movements

Do you keep plants in the house? If you do you may have noticed that they bend towards the window. Intense light from a particular direction (in this case the window) is a stimulus to which plants respond by growing towards it (Figure G.1).

The benefit to the plant of this response is clear: the leaves receive as much light as possible for photosynthesis.

Figure G.1 shows that plants are never still. The time-lapse camera helps us to see what happens. Stems twist and turn and flowers and leaves move in daily rhythm. Plant movements are **growth movements** in response to stimuli. They take longer than the quick nervous system responses of animals, which will be discussed in the next Unit.

There are two kinds of growth movement:

- A **nastic** movement is a response to a stimulus which comes from all directions. For example, temperature change is the stimulus for tulip and crocus flowers to open and close. They open when the temperature rises and close when it falls.
- A **tropic** movement (or **tropism**) is a response to a stimulus which comes from one direction. Tropisms are positive if the plant grows towards the stimulus, negative if it grows away (Figure G.2).

Figure G.2 Different tropisms

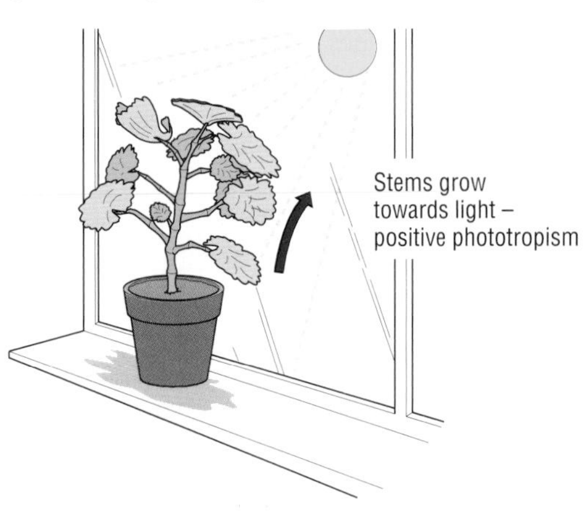

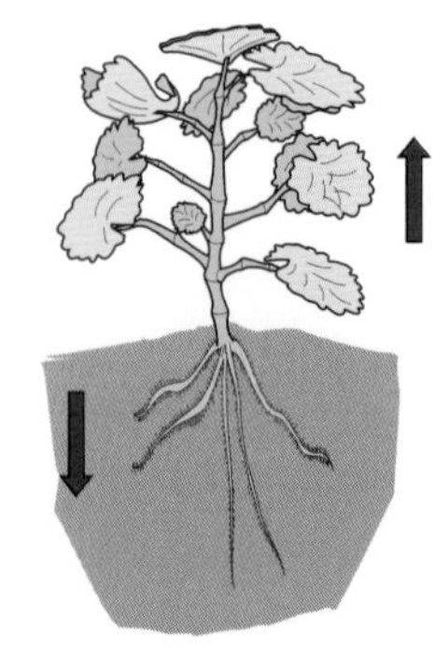

Activity G.1

Anne's Amaryllis

Anne was fascinated by the rapid growth of an Amaryllis bulb in her living room. She measured the height of the shoot each day. Her measurements were made to the nearest 0.5 of a centimetre and continued until the flower began to open. Her results are shown in Table G.1

Figure G.3

Day	Height (cm)	Day	Height (cm)
0	bulb planted		
7	7.5	18	46.0
8	8.5	19	51.5
9	10.5	20	54.0
10	13.0	21	58.5
11	16.5	22	59.0
12	21.0	23	61.0
13	24.5	24	61.0
14	28.0	25	61.5
15	34.0	26	64.5
16	38.5	27	64.5
17	42.5	28	64.5

Table G.1

1. Make a table showing how much the plant grew each week.
2. Plot a graph of Anne's results, drawing the best curve possible to fit the points. Try to explain the shape of the curve.
3. How could Anne have improved her investigation of the growth of Amaryllis plants?

Activity G.2

Investigations into seedling growth

You will need: a tray and several pots of wheat seedlings.

a If you are provided with a tray of wheat (or maize) seedlings, how would you measure their growth? Discuss your ideas with other members of the class and, if possible, try some of your ideas.

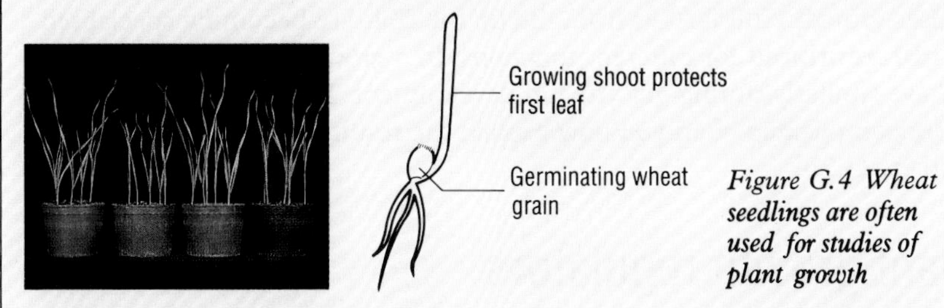

Figure G.4 Wheat seedlings are often used for studies of plant growth

b Mark a wheat seedling with waterproof ink at 1 mm intervals, starting from the tip of the growing shoot (see Figure G.4). Allow the plant to grow. After two days, which region of the shoot has grown the most?

c How would you use the apparatus shown in Figure G.5 to measure growth of a wheat seedling? In what ways will the use of the apparatus be an improvement on other methods of studying plant growth?

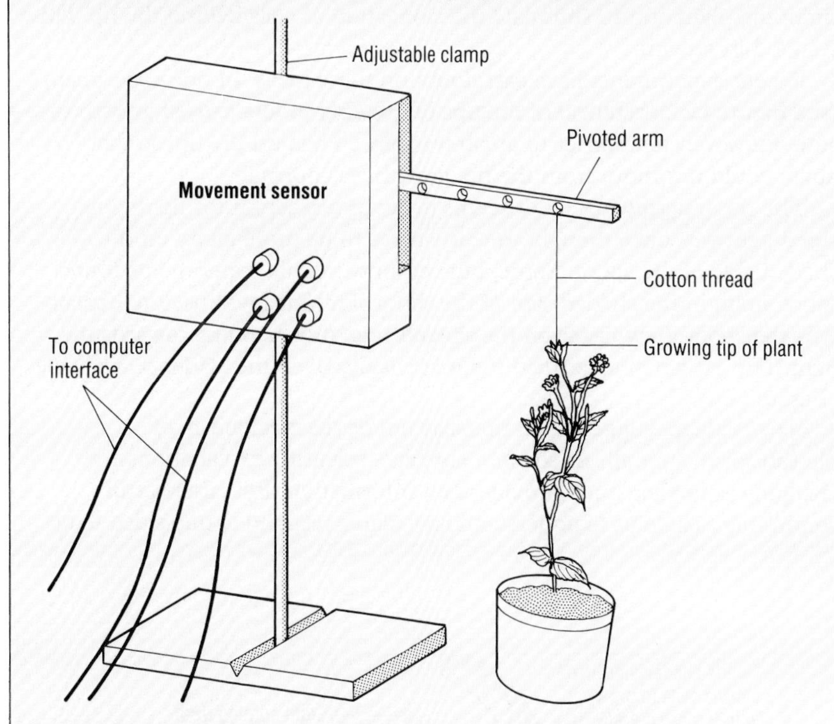

Figure G.5 A movement sensor being used to measure plant growth

d Place one pot of watered wheat seedlings in a dark cupboard. Keep another pot in similar conditions but in the light. After a week compare the growth of the two sets of seedlings.

A plant growing under a stone is taller but more straggly than normal. It is yellow rather than green. Such a plant is described as **etiolated**. What features of etiolation do the plants grown in the dark show?

What is growth?

Growth is often defined as **an irreversible increase** in size. Try explaining what this means to a friend. Why is the definition not completely satisfactory?

If you examine the cells at the tip of a root (see Figure B.2, page 58) you will see small, newly divided cells at the point or apex. Farther away from the tip, the new cells are elongating (becoming longer). **Elongation** causes growth. Still farther from the tip the cells are becoming specialised or **differentiated** for different functions. In a growing shoot, the cells behave similarly during growth and development. In a bud, it is a little more complicated. Can you now explain the results you obtained in Activity G.2**b**?

Plant growth regulators

We now know that the growth of plants is affected by chemicals known as **growth regulators** or plant hormones. In the 1920s, experiments by Frits Went on growing shoots of wheat seedlings suggested that there was a growth-regulating chemical called **auxin**. This auxin was later identified as indolyl-3-acetic acid or IAA. IAA was thought to travel from the growing tip of the plant and to stimulate the elongation of cells behind the tip. This resulted in growth.

Recent experiments have cast doubt on these ideas. In one experiment (see Figure G.7), removal of the tips of growing shoots caused growth to stop. However, if a barrier to auxin was placed behind the tip, so that auxin could not move from the tip, growth was normal.

The tip of a growing shoot seems to act as a receptor for light. One-sided light was once thought to cause there to be more auxin on the shaded side of the shoot. Some, but certainly not all, experiments found more auxin in the shaded side of the shoot. This idea was used to explain why the shoot grew faster on the shaded side, thus bending towards the light. The explanation seemed the most likely, given the evidence available (Figure G.6).

New evidence suggests that this may not be the mechanism of phototropism after all. As so often happens, scientific explanations are changing as further study provides new information. Research is still continuing to help us to understand how plants respond to one-sided light.

Light and growth

Charles Darwin is well-known for his work on evolution. He also carried out some interesting experiments on the growth of plants. In 1880, he and one of his sons, Francis, wrote *The Power of Movement in Plants*. They noticed that a growing grass bent 'towards the light of a small lamp.' You may have seen plants growing on a window-sill bending over in a similar way. A growth response like this is called a **tropism**. If the response is to light, it is called **phototropism**.

The Darwins thought that the 'uppermost part (of the plant) determined the curvature of the lower part.' They measured the region which bent and made drawings. 'When little caps of tinfoil ... were placed on the summits of the cotyledons, though this must have added considerably to their weight, the rate of the amount of bending was not thus increased.' They also tried the effect of other treatments such as cutting off the very tip of the growing shoot, covering the tips with clear and with blackened glass, and restricting bending by attaching 'splints' to the growing shoot. Careful experiments with the seedlings led Charles and Francis Darwin to conclude 'that the exclusion of light to the upper part ... prevents the lower part, though fully exposed to a lateral light, from becoming curved.' The Darwins had shown that the tip of the shoot was particularly sensitive to light. The effect of uneven lighting was to cause the region behind the tip to bend as it grew. As an explanation of their findings, they suggested 'that when seedlings are freely exposed to a lateral light some influence is transmitted from the upper part to the lower part, causing the latter to bend.'

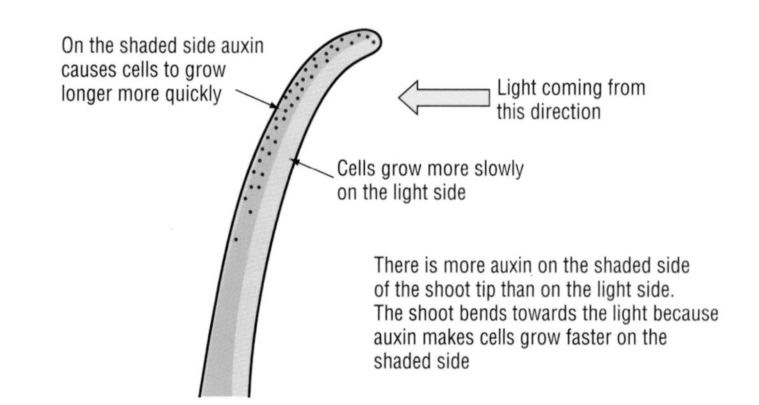

Figure G.6 A widely accepted explanation as to why plants grow towards intense light

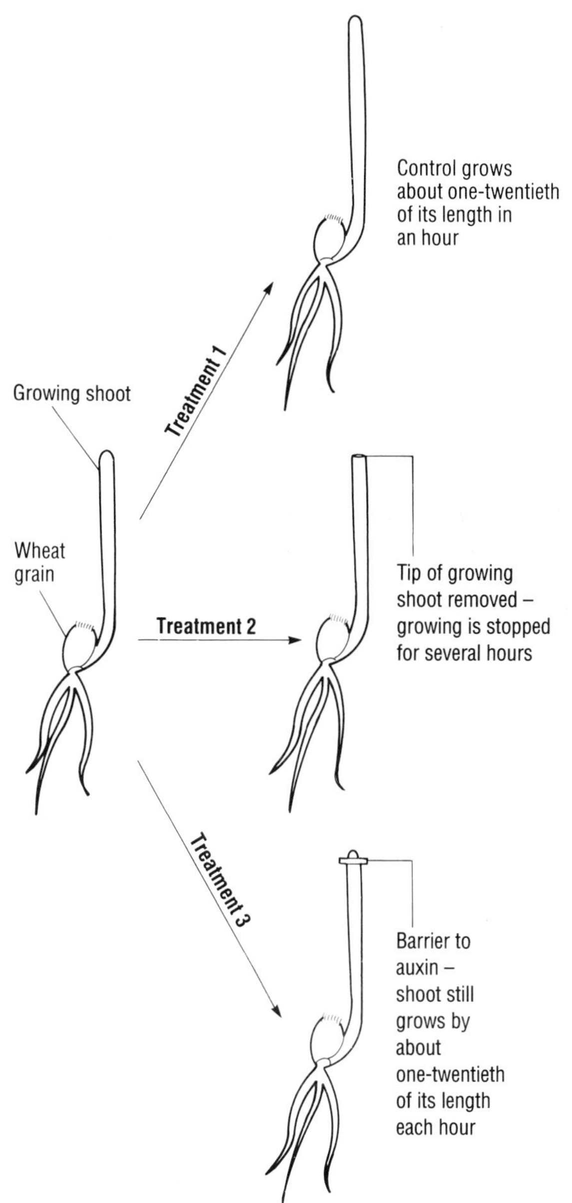

Figure G.7 This simple experiment shows that auxin from the tip of the growing shoot is not necessary for growth

Figure G.8 Mesembryanthemums open in response to sunlight

Other plant responses

Some flowers turn towards light and close when a cloud covers the sun (Figure G.8). Leaves of the sensitive plant *Mimosa pudica* (Figure G.9) fold up if the plant is touched or blown by the wind. More leaves fold if the stimulus is violent. A Venus flytrap leaf (Figure G.10) responds if an insect crawls across it; three projecting hairs on the inside of the leaf are stimulated. The leaf suddenly folds, trapping the insect, which the plant then digests.

In each of these examples, scientific research is helping to show how the plant receives the stimulus and then produces a response.

Figure G.9 The sensitive plant Mimosa pudica *responds to touch by folding its leaves*

Activity G.3

Does rooting hormone help cuttings to grow?

You will need: a potted plant with many side shoots (such as geranium, fuchsia, or begonia rex), two tubs of clear rooting gel or two pots of potting compost, new rooting hormone powder, old rooting hormone powder.

Some people believe cuttings will root more easily if the root is first dipped in water and then in rooting hormone compound. Is the rooting hormone helpful? Does it matter how long you keep the rooting hormone on the shelf in your shed? Devise an experiment to find the answers to these questions. For cuttings, use shoots with no flowers and with three or four pairs of leaves. Cut below a node and remove all but the top two pairs of leaves.

Figure G.10 A Venus flytrap leaf folds to trap insects

Using plant hormones

After the discovery of auxin the search was on for other plant growth substances. Today we know that a range of substances control plant growth. For example, growers control ripening fruit by keeping fruit in sheds in an atmosphere that contains ethene. As little as one part of ethene per million parts of air is enough to speed up ripening. Figure G.11 shows some of the uses of plant hormones to affect root, stem and fruit development.

The substance 2,4-D (2,4-dichlorophenoxyacetic acid) is a synthetic auxin. It kills plants by making them grow too fast. The plants become tall and spindly. However, 'grassy' plants, including the food crops barley, wheat and oats, are not affected by 2,4-D at concentrations which destroy broad-leaved plants like docks, daisies and dandelions. Gardeners and farmers put this selective effect of 2,4-D to good use – as Figure G.12 shows.

Substances that kill unwanted plants (weeds) are called **herbicides**. Dicamba is another herbicide which behaves like auxin. It can be used in the form of a 'weed pencil' (see Figure G.13) to dab a spot on single deep-rooted plants such as dandelion and bindweed.

Figure G.11 (a) The cabbage plants on the left have been treated with a plant growth substance called gibberelin. The rapid growth of the stem normally occurs only as a result of an environmental stimulus like the onset of winter frosts

Figure G.11 (b) The cutting on the right was placed in a dilute solution of auxin (10 mg/dm^3) 14 days before the photograph was taken. At the same time the cutting on the left was placed in water. Gardeners use rooting preparations that contain auxin to encourage root growth in cuttings. Too much auxin prevents root growth

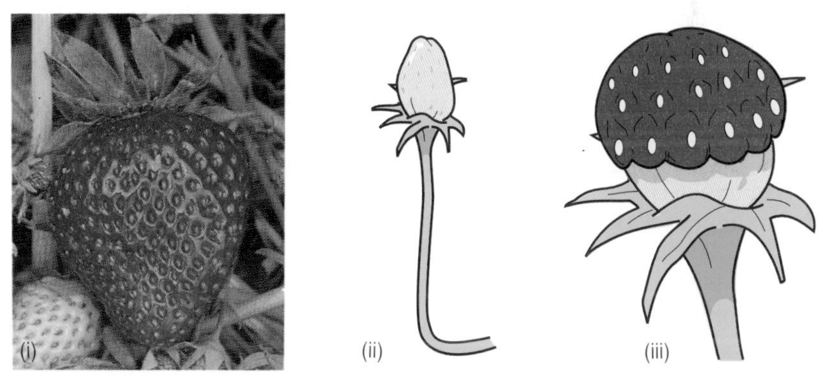

Figure G.11 (c) Seeds produce auxin: (i) shows a normal strawberry; in (ii) the strawberry remains undeveloped after all of its seeds were removed; (iii) shows a strawberry that had three rows of seeds removed. This part of the fruit remains undeveloped. The fruit develops normally where the seeds remain. Market gardeners treat the carpels of some crop species with auxin paste to produce fruits without fertilisation. Examples are seedless cucumbers and tomatoes

Figure G.12 Spraying wheat with 2,4-D kills the broad-leaved weeds which would otherwise compete with the wheat crop for growing space, nutrients and water. The yield of wheat increases and the extra money the farmer receives more than offsets the cost of buying and using the chemical

Figure G.13 Using a weed pencil

Summary

- Growth can be defined as an irreversible increase in size. It is measured in various ways.
- Plants are sensitive to light and other environmental factors.
- Growth, development and plant responses are controlled by chemicals known as growth regulators (hormones) which are transported around the plant.
- We use plant hormones to control the processes of growth and reproduction in plants.

Questions

1 Explain how the early view of how auxin affected growth could have been used to explain why plants grown in the dark are taller than usual.
2 Comment critically on the evidence that shoot tips respond to light.
3 Plan an experiment to show whether gravity has an effect on the direction of growth of the roots of pea seedlings.
4 Gardeners sometimes pinch out the main growing tip of plants to make side shoots grow. Explain how trimming a privet or yew hedge improves the hedge. How would you investigate whether removing the main growing tip of a geranium plant affects the way it grows?
5 You have been told that plants grow faster in the dark than in the light. You don't believe it, but decide to test the idea. How would you investigate this suggestion using the apparatus in Figure G.5? How would you test which of these conditions give the best growth?
a continuous light
b continuous darkness
c short, alternating light and dark periods.
6 Explain what is meant by:
a a herbicide
b a selective herbicide.

Forest recovers from Agent Orange damage

Agent Orange is a herbicide which causes trees to lose their leaves. Most of the trees die. The herbicide persists for a long time in the environment. It contains 2,4-D and dioxin which damages animals, including humans. During the Vietnam war in the early 1970s, Agent Orange was sprayed on large areas of forest to make it easier to flush out hiding North Vietnamese soldiers. The result was an environmental and human disaster. The area was left a dusty, barren landscape, covered only with grass, and liable to soil erosion.

After the war ended in 1975, Vietnamese scientists tried planting young trees in a trial area. In the hot, dry season grass fires destroyed the saplings. The scientists then tried planting foreign trees. These grew rapidly, providing shade from the sun. Later, native species were planted under the cover provided by the foreign trees. After 12 years of trials, the native species had become established. Now, larger areas can be restored in the same way. Even more encouraging is that replanting of some of the damaged mangrove swamps is already beginning to lead to recovery. The Vietnamese people have shown resourcefulness, determination and scientific ability in tackling the environmental problems left by the war.

Agent Orange had another harmful effect. It caused damage to the fetuses (the unborn children) of Vietnamese mothers. Some of these children had Vietnamese fathers; others were the children of American soldiers. Sadly, many were born with deformities.

Used wisely, herbicides can be beneficial. However, like many useful discoveries they can be put to undesirable uses. The use of Agent Orange in Vietnam is an extreme example. Often, those misusing a scientific discovery do not fully realise the consequences of their actions. Science brings us new possibilities - it is we who have to decide whether to use them for good or evil.

Unit H
Reacting to changes in our surroundings

A B C D E F G H I

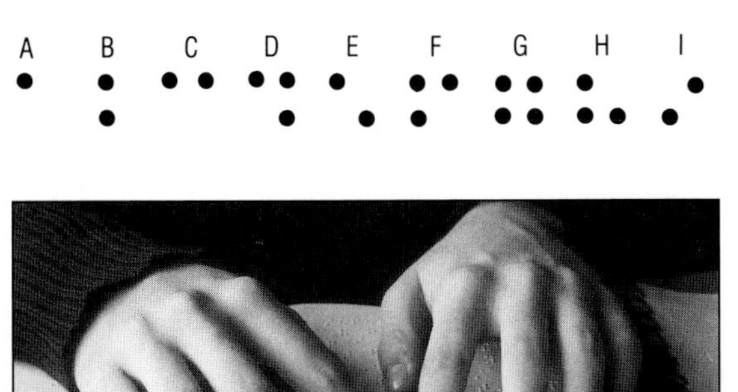

Figure H.1 Braille

The senses

We learn about our surroundings by using our **sense organs**. The sense organ is a **receptor** (that is, it 'receives' the stimulus). Most of us can use eyes, ears, taste, smell and touch to pick up information about our environment. Someone who is blind or deaf has to obtain more information through their other senses. This can be very difficult, because most things in our surroundings are designed for sighted and hearing people. Blind people use the sense of touch in their fingertips to read Braille, a system of raised dots (Figure H.1).

Our sense organs, especially our ears and eyes, are very complex. In addition to hearing, our ears detect our position in relation to gravity and changes in the movement of our head. The nose can smell many different chemicals and the tongue can distinguish salt, sweet, sour and bitter. Your skin has several different kinds of receptors, so that you can sense pain, hot and cold, touch and pressure (see Figure I.2, page 91).

Activity H.1 How good are you at detecting hot and cold?

You could try this activity at home.

You need: 3 containers, ice and hot water.

1. Fill the containers as follows:
   - **a** with ice cold water
   - **b** with hot water – just hot enough to allow you to put your finger into it (get a helper to judge this)
   - **c** with warm water
2. Place the first finger of your left hand in the cold water and the first finger of your right hand in the hot water, both at the same time. Leave them in for one minute.
3. Take out both fingers and dip them repeatedly but alternately (that is one at a time) in the warm water, for about a second each time. Does the warm water feel the same temperature to both fingers?
4. Describe what you have done and what you felt.
5. Discuss the results of your experiment. What did you find out? Why do you think there is a difference in sensation between the two fingers? Does the result mean that your fingers cannot judge whether an object is hot or cold? How should you test the temperature of a baby's bath water?
6. Write up your conclusions and the main points of your discussion.

Reacting to stimuli

What happens if you touch something hot? You probably cry out and jerk your hand away. This is just one way in which you react or respond to a change in your surroundings. Another example might be the way you produce more saliva if you smell food. The gland or muscle which carries out the response is the **effector**. Information has travelled through your nervous system from a sense organ or receptor to the effector.

Stimulus → sense organ (receptor) → nervous system → effector → response

The nervous system

The nervous system consists of the **brain** inside the skull, the **spinal cord** protected by the vertebrae, and **nerves** (Figure H.2). The brain and spinal cord are called the **central nervous system**. The nerves form the **peripheral nervous system**.

Each nerve consists of a long fibre made up of nerve cells or **neurons**, lying side by side. The brain and spinal cord are also made up of neurons. Each nerve cell, or neuron, has a cell body containing the nucleus. Thin extensions of the cell body form either dendrites or axons. A dendrite can be very long, running all the way from a receptor to the spinal cord; an axon can also be very long, sometimes running all the way from the spinal

Activity H.2 Examining a section through the spinal cord

You need: slices of neck of lamb, microscope slide of a transverse (cross) section of spinal cord, microscope,

1 Examine the neck of lamb (use Figure H.3 to guide you). Look for the whitish spinal cord surrounded by the bone of a neck vertebra. See if you can find a pair of nerves leaving the spinal cord between two vertebrae.
2 Examine the slide under low magnification and try to find all the parts shown in Figure H.4.

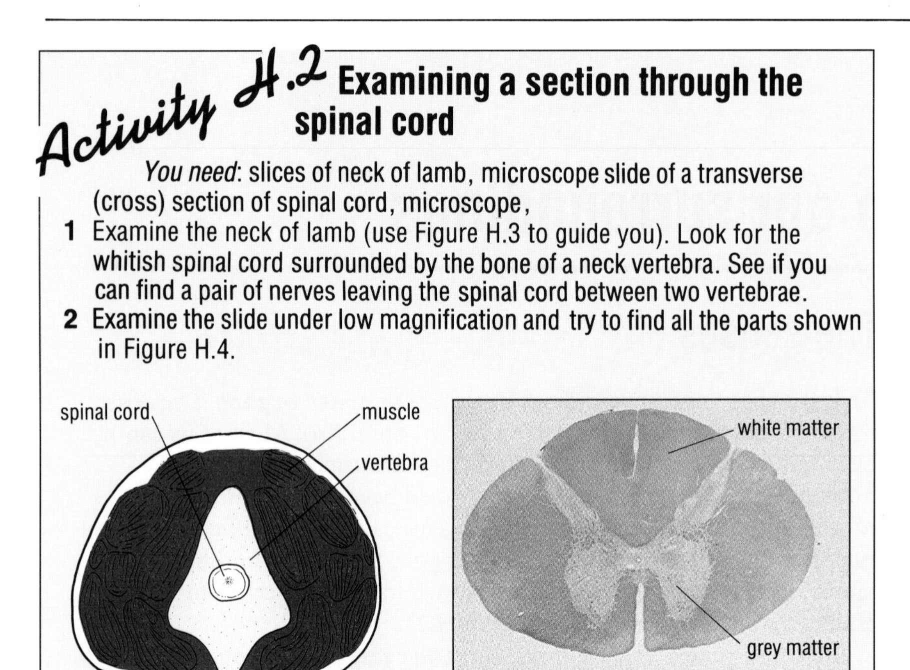

Figure H.3 The cut surface of the neck of lamb

Figure H.4 Photomicrograph of a section through the spinal cord. How much detail can you find under the microscope?

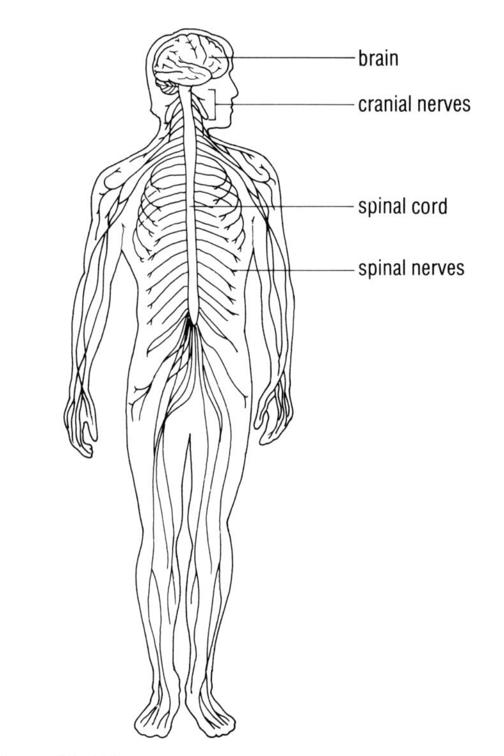

Figure H.2 The nervous system

axon can also be very long, sometimes running all the way from the spinal cord to the hand or foot. There are two main types of neurons: sensory neurons run from the sense organs to the central nervous system; motor neurons run from the central nervous system to the muscles. Figure H.5 shows the main differences between the two types of neurons. Parts of the neuron are covered by a sheath of white, fatty material called myelin. Uncovered parts appear grey.

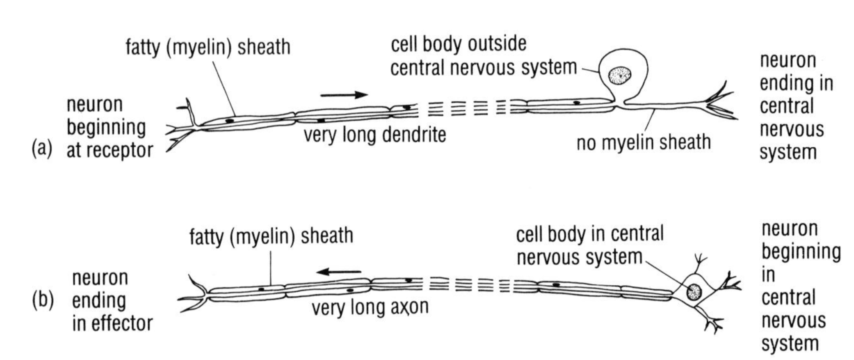

Figure H.5 (a) Sensory neuron. (b) Motor neuron. The arrows show the direction in which information travels in the form of nerve impulses

The nerve impulse

Information travels along neurons in the form of electrical impulses. When one neuron meets another there is a small gap or **synapse** between them. A chemical is released which transmits the impulse across the synapse. Similarly, when a motor neuron reaches a muscle fibre a chemical is released. This transmits the impulse to the muscle fibre, causing it to contract.

Activity H.3

Making model neurons

You need: two 20 cm lengths of multi-strand single core wire, insulated with white plastic; 2 small white glass beads; wire strippers; sellotape.

1 Making a motor neuron model
Strip 4 mm of plastic from the end of the first length of wire. Place a bead on the exposed wire so it fits tightly against the plastic. The bead is the cell body. Spread out the strands of exposed wire. Remove 1 mm of plastic from the other end and spread out the strands of the wire you have exposed. Use the sellotape to fasten this model in your notebook.
2 Making a sensory neuron model
Strip 12 mm of plastic from the second length of wire. Place a bead over this bare end so that it is touching the plastic. Spread out 1 mm at the tip of the strands of exposed wire. Remove 1 mm of plastic from the other end and spread out the strands of the wire you have exposed. Fasten this model in your notebook.
3 Label the two models. What criticisms can you make of these models? How could you improve them?

Activity H.4

Observing the knee jerk reflex and the action of stretch receptors (in pairs)

1 One person sits on a high stool with legs hanging down and relaxed. The other person taps below the knee with the side of the hand. What happens?
2 One person stands upright with eyes closed. The other watches from the side and notices any swaying backwards and forwards.
3 Write down what you observed. How would you explain it?

Activity H.5

Making a model of the spinal cord and reflex arc

You need: plasticine (2 colours); 3 pieces of white coated wire as in Activity H.3; 3 white beads; wire strippers.

1 Make a plasticine model of a piece of spinal cord (Figure H.7).
2 Make a sensory neuron and a motor neuron as in Activity H.3 (see Figure H.7).
3 Make an intermediate neuron as in the drawing. You will need a shorter piece of wire.
4 Place the neurons correctly on the end surface of the plasticine. This surface represents a cross-section of the spinal cord. In reality there are many very fine neurons lying side by side, making up the nerves.

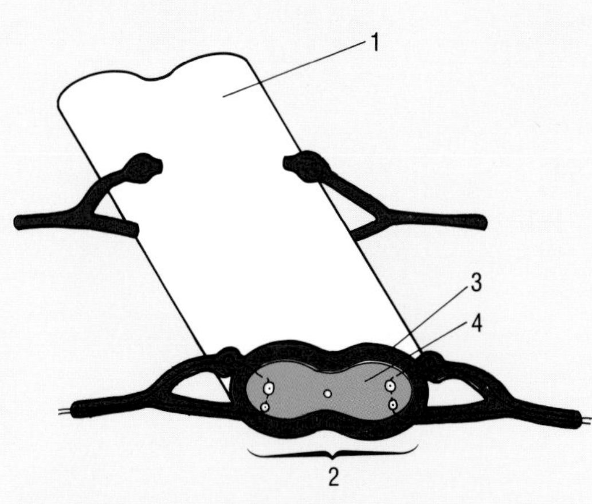

Figure H.7 A plasticine model of a spinal cord and reflex arc. Can you label it?

Simple behaviour: a reflex action

If something pricks your hand, you move it away without thinking. This is a **reflex action** – a quick, automatic response. You could not control the reaction. It is **involuntary**. We carry out many reflex actions. Each one involves:

a) stimulus of a receptor or sense organ;
b) transmission of a nerve impulse inwards along a sensory neuron to the central nervous system (brain or spinal cord);
c) transmission of the impulse through an **intermediary neuron** (or **relay neuron**) in the central nervous system;
d) transmission of a nerve impulse outwards along a motor neuron from the central nervous system;
e) response of the effector organ (muscle contracts or gland secretes – see page 92).

When a doctor needs to test how well a patient's nervous system is working, he uses the knee jerk reflex. The patient sits relaxed, with the lower leg dangling. The doctor taps gently, immediately below the knee. This pulls on a tendon which in turn pulls the muscle in the thigh. There are special **stretch receptors** in the thigh muscle. When they are stretched they are stimulated and set off the steps of a reflex action ((*a*)-(*e*)). Finally, the thigh muscle (the effector) contracts, causing the lower leg to kick forwards. The **reflex arc** or pathway of an impulse in this reflex action is shown in Figure H.6.

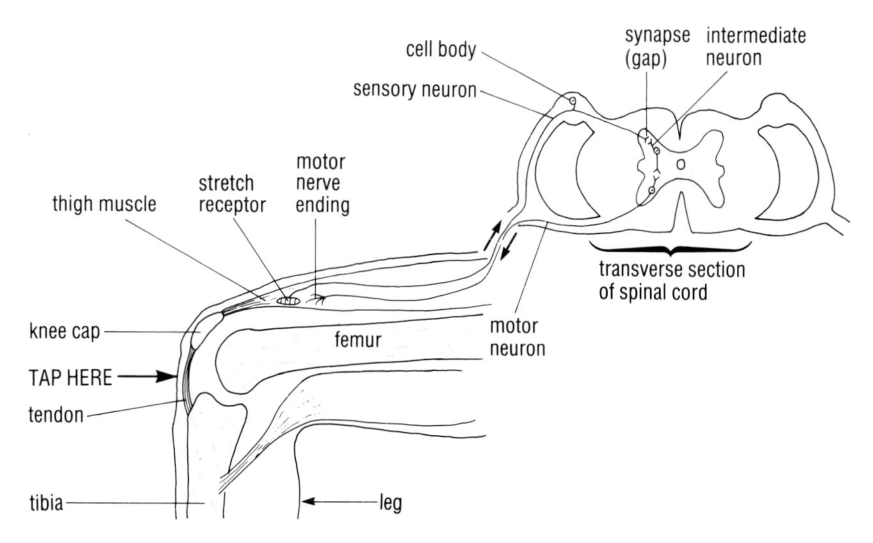

Figure H.6 The knee jerk reflex arc – the cross-section of the spinal cord is on a much larger scale than the thigh

The knee jerk reflex does not seem to be of much use to the body. But the stretch receptors are very important. They are stimulated by any stretching of the muscle in the front of the thigh. If you are standing up for a long time you sway gently. When you move backwards, the stretching of the receptors causes a reflex action. The muscle contracts and pulls you back into an upright position. Stretch receptors and reflex actions in other muscles work in a similar way to keep your body in the correct position.

Reflex actions help you to react quickly; steps *a-e* take place in a fraction of a second. An impulse also passes up the spinal cord from the intermediary neuron to the brain so that the brain is aware of what has happened.

Normally, if you pick up something hot you drop it quickly. But, if it is your mother's best plate you do not do so. Why? Before you pick up the plate you note its value. An impulse travels to a synapse in the reflex arc. This prevents transmission of the impulse along the motor neuron of the reflex arc. It is as if the impulse said, 'Whatever happens, you must not drop this plate.' It prevents or **inhibits** the impulse travelling so the reflex action does not occur.

More complex behaviour

Compared with everything else your nervous system does, reflex actions are very simple. Most of your behaviour involves more complicated nerve pathways. Memory and learning are still only partly understood. Some of your actions are involuntary, but many are **voluntary**, that is, they are under the conscious control of your brain. The brain is the thinking and control centre for all your body's activities.

If you tried out the knee jerk reflex you will have seen how quickly nerve impulses must travel. The impulse passes to the spinal cord and back incredibly fast. The speed of the reaction is affected by our state of health, or how tired we are. Sometimes a quick response can be very important. Can you think of any examples?

Eyes for vision

Sight is a very important sense. Figure H.9 is a diagram of a section through the eye. It shows the complex structure which enables us to focus and detect the light falling on the eye.

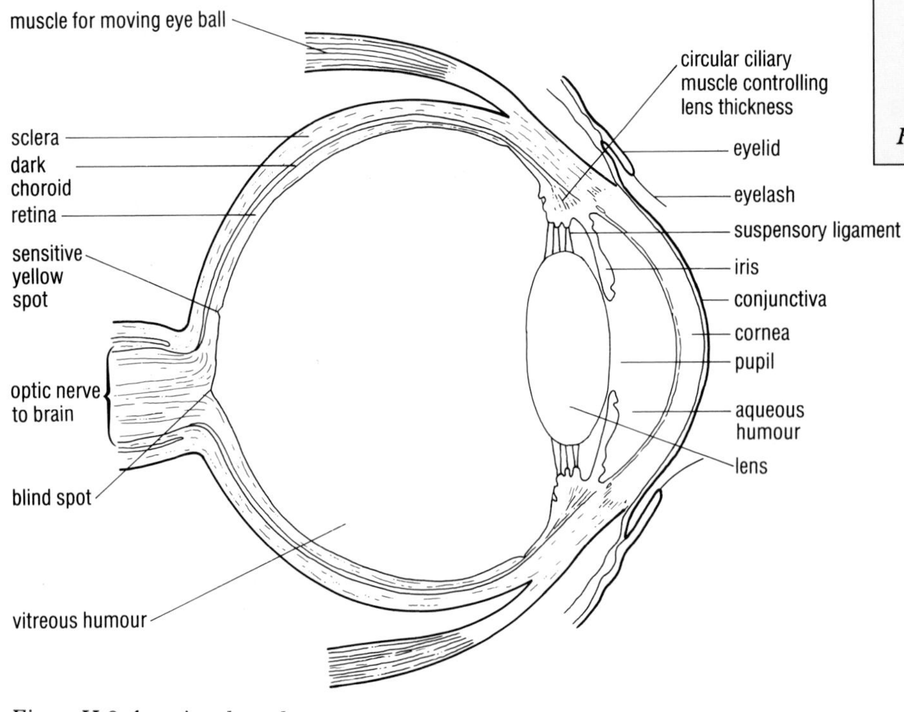

Figure H.9 A section through an eye

Activity H.6

Test your own reaction time

You need: a strip of card about 3 cm wide and 40 cm long.

1. Making the reaction timer.
   Mark the card off in centimetres. Draw an extra-thick line 2 centimetres from one end and label it 'Start here'. Number the centimetre marks from this line, 1, 2, 3, and so on.
2. Get a friend to hold the strip so that it hangs with the dark line between, but not touching, your thumb and forefinger. Ask your friend to let go of the strip.
3. When your friend lets go, grasp the strip between your thumb and forefinger as quickly as you can. Read off the nearest number on the strip.
4. Repeat until you have three similar readings. Work out the average reading.
5. Obtain your reaction time from the graph (Figure H.8).

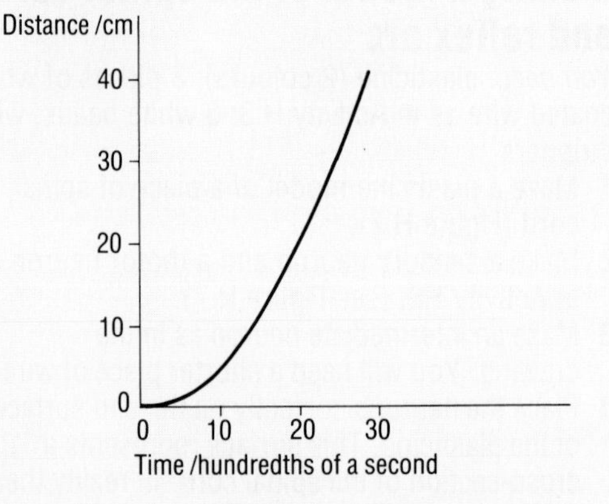

Figure H.8 Graph for finding your reaction time

Activity H.7

Looking at eyes

You need: sheep's eye (keep in refrigerator), instruments, dish, mammal's skull, model eye, small piece of newspaper.

WORK IN PAIRS.

1 Draw one of your partner's eyes, showing the iris, pupil, lids, and the opening of a tear duct towards the inner corner of the lower lid. (This drains to the nose.)
2 Examine the outside of the sheep's eye (or model). Find the six muscles which move the eyeball. Find the optic nerve at the back of the eye. Find the hole for the optic nerve at the back of the socket in the skull.
3 Use sharp scissors to cut around the eye and separate front and back halves. Place these in a dish of water.
4 Examine the lens in the front half. Carefully remove and examine it. If the lens is still transparent, place it on the newspaper. Can you read through it? Compare the sheep's lens with the model lens. Describe the lens and its effect.
5 Look carefully at the other half of the eye and note the black colour. Can you see where the nerve leaves the eye?
6 Examine the eye and the model to find the other structures shown in Figure H.9.
7 WASH YOUR HANDS after clearing away.

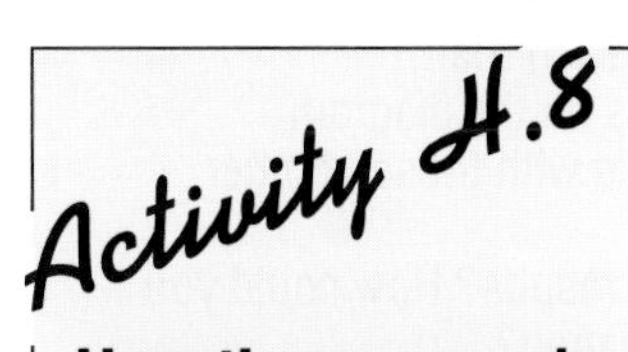

How the eye works

1 Read the passage on the functions of the parts of the eye.
2 Write a list headed 'Light passes through these layers of the eye'.
3 Make a table with columns headed 'Parts of the eye' and 'Functions'.
4 Write your own explanation of how the lens thickness is altered.

Functions of the parts of the eye

The **eye ball** is protected by a bony socket, the eyebrows, eyelids and eyelashes. It is moved around in the socket by six short muscles. The front of the eye is covered by a thin transparent **conjunctiva** which is kept clean by liquid from the **tear glands** in the eyelids. Behind this is the tough, transparent **cornea** which joins the tough **sclera** covering the back of the eye.

Light enters the eye through the conjunctiva and cornea and passes through the **pupil,** a hole in the centre of the coloured **iris.** It then passes through the flexible **lens** and across the back of the eye to fall on the **retina.** The retina contains receptor cells of two kinds – **Cones** to detect Colours when the light is bright, and **rods** which pick up light even when it is quite dark, but which cannot detect colour. Outside the retina is the dark **choroid** layer, which prevents reflection of light, and then the sclera.

In front of the lens the eye is filled with a liquid (**aqueous humour**). Behind the lens is the jelly-like **vitreous humour**.

When the light reaches the eye from an object it is refracted (bent) by the substances it passes through in reaching the retina, where an image of the object forms. If the light is dim the iris muscles cause the pupil to get bigger, if the light is bright the pupil will be smaller. The lens is held in place by **suspensory ligaments** around its edge. The ligaments are attached to a ring of **ciliary muscle.** When the ring of muscle contracts the lens becomes rounder from front to back. The lens is able to focus the light from near objects when this happens (Figure H.10).

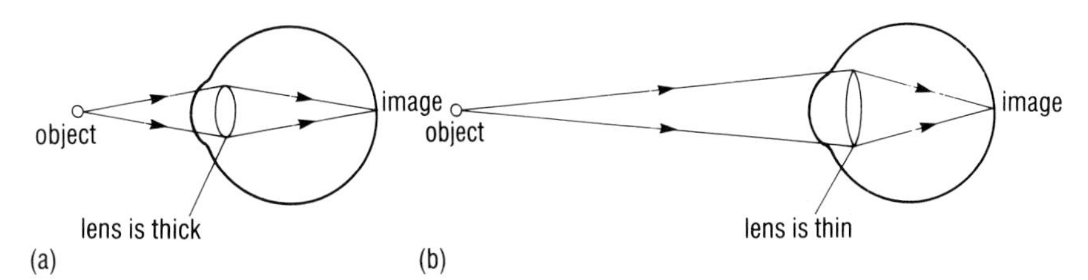

Figure H.10 Focusing on objects at different distances is called accommodation. What difference is there between the lens for (a) a near object? (b) a distant object?

Someone who is **short-sighted** (who has **myopia**) is unable to focus the light from a distant object sharply on the retina. **In long sight,** near objects are difficult to focus. As people get older the lens becomes less flexible and focusing becomes more difficult. Find out how the focusing occurs in each case and what kind of spectacle lenses are needed to correct these faults.

When light falls on the rods or cones of the retina, these pass impulses to the neurons forming the **optic nerve** which goes to the brain. Where the optic nerve leaves the eye there are no cones or rods and there is a **blind spot.** Because we have two eyes we do not usually notice the blind spot. Having two eyes also helps us to interpret the shape and distance of objects. Can you explain why?

Ears for hearing

Ears are specialised to detect the vibrations of sounds and also to obtain information about the position and movement of the head, helping us to keep our balance. The **ear lobe** (or pinna) collects sound vibrations which pass along the outer passage to the thin **eardrum** (see Figure H.11). This moves and in turn sets three little ear bones moving. The bones transmit the vibration across the **middle ear** to the **oval window**.The **eustachian tube** which leads to the throat allows the pressure inside the ear to be adjusted. When you have a cold or climb mountains your ears may 'pop' because the pressure inside your ear hasn't adjusted to the change in atmospheric pressure.

The vibration passes through the oval window to the **inner ear,** which is filled with fluid. The **cochlea** in the inner ear is coiled like a snail. It contains sensitive receptor cells which can detect the vibrations caused by sounds received by the eardrum. An impulse passes along the auditory nerve to the brain. Different parts of the cochlea detect different pitches. The pressure in the inner ear is adjusted by a **round window** which moves as the oval window vibrates.

The middle and inner ear are surrounded by bone to protect them and to prevent vibrations affecting the functions of the ear.

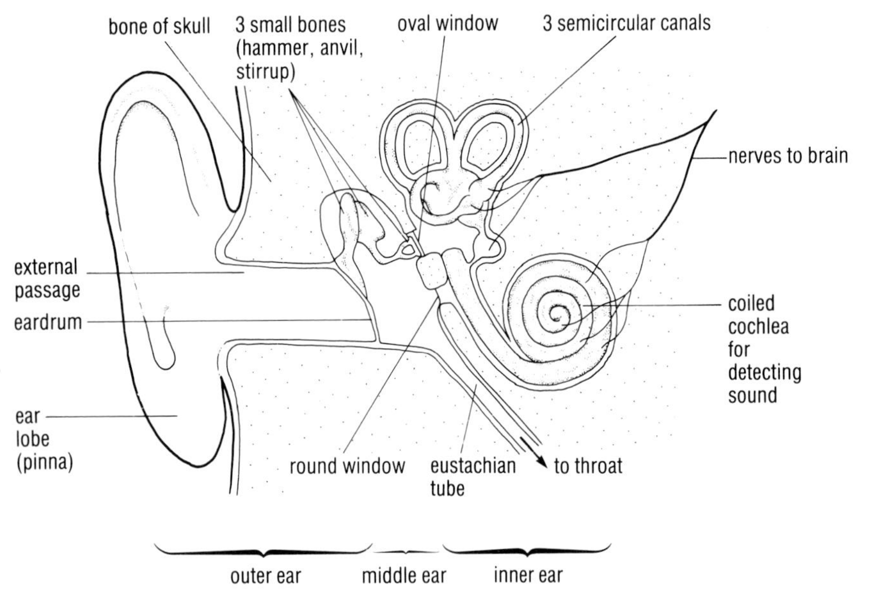

Figure H.11 The structure of an ear, shown diagrammatically

Balance

Three tiny semicircular canals (see Figure H.11) in the inner ear detect head movements. They are at right-angles to each other, so the fluid inside at least one of them will move as the head moves. The movement of the fluid stimulates receptors at the ends of the canals.

Other small structures in the inner ear can detect the position of the head in relation to gravity.

Chemical senses

Smell and taste are very important in helping us to choose our food. Smell can also warn us of danger. When we smell things chemical receptors in the moist area inside the nose detect and distinguish between very small amounts of chemicals in the air.

The tongue also has chemical receptors – **taste buds.** It may surprise you to know that the tongue is only capable of detecting sweet, salt, bitter and sour. All the other sensations which help us distinguish foods are picked up by our noses. This is why your sense of taste is so poor when you have a cold.

Activity H.9

Finding your blind spot

1. Place your thumbs, nail uppermost, side by side on the edge of the desk.
2. Close your left eye. With your right eye look at your left thumb-nail.
3. Now, slowly move your right thumb to the right. What happens to it?
4. When it disappears you have found the blind spot for your right eye. Can you find the blind spot for your left eye?

Activity H.10

Tasting

This is an unusual science activity because you will be allowed to taste in the laboratory!

You need: a tissue for each person, 4 short straws per person, clean containers of water, sugar solution, salt solution, vinegar (sour), and black coffee (bitter).

WORK IN PAIRS.

1. Copy Figure H.12.
2. Investigate which region of the tongue detects each of the flavours. The person being tested will need to have closed eyes. Rinse and wipe the tongue between each trial.
3. Record your findings on the diagram.
4. Compare your results with those of other groups in the class.
5. How reliable are the results? How could you improve the investigation?

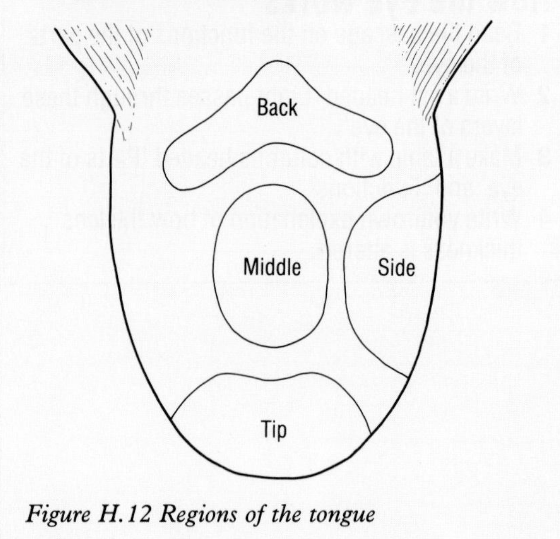

Figure H.12 Regions of the tongue

Summary

- Humans respond to stimuli in their surroundings.
- Information is detected by sense organs (receptors), nerves transmit impulses around the body and effector organs (muscles and glands) carry out the reactions.
- The sequence reads:

stimulus → receptor → nervous system → effector → reactions

- The brain is the control centre of the nervous system.

Questions

1 We respond to stimuli all the time. When a doorbell rings (stimulus) we answer the door (response); when we feel hot (stimulus) we take jumpers and coats off (response).
a List at least seven stimuli to which you have responded today, and state your responses.
b Name the sensory receptors responsible for detecting the stimuli.
2 Why is it easier to study an object in detail if you look straight at it?
3 One of the first hearing aids was the 'ear trumpet', which was a large hollow horn held to the ear. Why do you think this device improves hearing ability?

Extension questions

4 Make a list of the labels for Figure H.7 on page 85.
5 Copy the table below and complete the gaps.

Examples of reflex actions

Reaction involved	Stimulus	Receptor (Sense organ)	Effector	Does it go through brain or spinal cord?
Eye blink	Movement across eye	Eye	Eyelid muscles	Brain
Pupils of eye get smaller		Eye	Muscles of iris in eye	
Sneeze				
Producing saliva				
Swallowing				Brain
Dropping hot object				Spinal cord
Knee jerk				
	Drawing pin prick	Sense organ in skin of foot		

6 Arrange the following parts of the reflex arc in the correct order: effector, sensory neuron, receptor, motor neuron, intermediary neuron.
7 Explain the differences between:
a voluntary and involuntary behaviour
b stimulus and response
c receptor and effector
d reflex action and reflex arc.
8 Find out about the experiments of the Russian scientist Pavlov, who studied the reactions of dogs.

Unit I
Keeping the internal environment constant

The internal environment

The cells of the body behave normally when in the right surroundings. Some of the conditions they need are:

- the correct temperature
- the correct mixture and concentration of chemicals
- sufficient water
- the correct acidity or pH

The conditions inside the body are known as the **internal environment**. Claude Bernard, a French professor of medicine in the nineteenth century, wrote: 'The unchanging nature of the internal environment is necessary for free life.'

Homeostasis is the name given to the way in which the body keeps the internal environment constant. It is an essential feature of living things. In order to continue living, changes within the body must be controlled.

One example of homeostasis is **thermoregulation** or the way in which the body normally keeps its internal body temperature steady at 37 °C. If illness or very cold surroundings cause the body temperature to drop below 35 °C, the body cannot any longer correct its internal temperature. A person with such a low temperature (**hypothermia**) may die. In cold winters, old people risk suffering from hypothermia. It is also possible to suffer from exposure on mountains, moors and even the seashore if it is very cold. In severe cases of exposure a person may become hypothermic; it is important to get medical help quickly.

The skin keeps body temperature steady

Figure I.2 shows a magnified view of what a small piece of skin would look like if you cut through it. Usually the skin is only about one or two millimetres thick, but its structure is quite complicated. Not all the things you can see in Figure I.2 are present in all parts of the skin.

The skin has an outer part called the **epidermis** and an inner **dermis**. The hairs start in **follicles** which are folds of the epidermis. On the side of each follicle is a **sebaceous gland** which produces an oily substance. A muscle stretches from the follicle to the inside of the epidermis. When this muscle shortens or **contracts** it pulls the hair and its follicle upright. Imagine all the hairs doing this together – this is what gives us goose pimples when we are cold or frightened.

In dogs or other mammals with more hair than humans, the hairs have an important function. When the animal is cold, the hairs stand upright and trap air between them. Air is a good insulator, so this layer of air stops the animal's body heat escaping. Our hair is not much good for this purpose, so *we* have to wear clothes to keep warm.

Activity I.1

a Use a hand lens to examine the skin in the following regions: underside of forearm; back of forearm; palm of hand; finger-tip; back of hand and fingers.
b Make a table to describe the features which are present in each region. Mention hairs, wrinkles, folds, ridges for gripping objects, tiny holes or **sweat pores** (see Figure I.1).

Figure I.1 A fingertip with sweat pores between the ridges of the fingerprint

c Why do you think the skin on some parts of the body has wrinkles and folds?

Activity I.2

Look at Figure I.2 and work out which structures would be in each region of the skin you examined in Activity I.1.

Activity I.3

a Use a hand lens to examine a microscope slide (provided by your teacher) of a section cut through the skin. Can you see any hairs?
b Now, examine the same slide with the low magnification of the microscope and see how many of the things shown in the diagram you can see. How does your slide section differ from the diagram?

Activity I.4

Place a drop of ethanol on the back of your hand. Allow the ethanol to evaporate away. Do you feel any cooling effect? Where has the heat energy needed to convert the alcohol to a vapour come from?

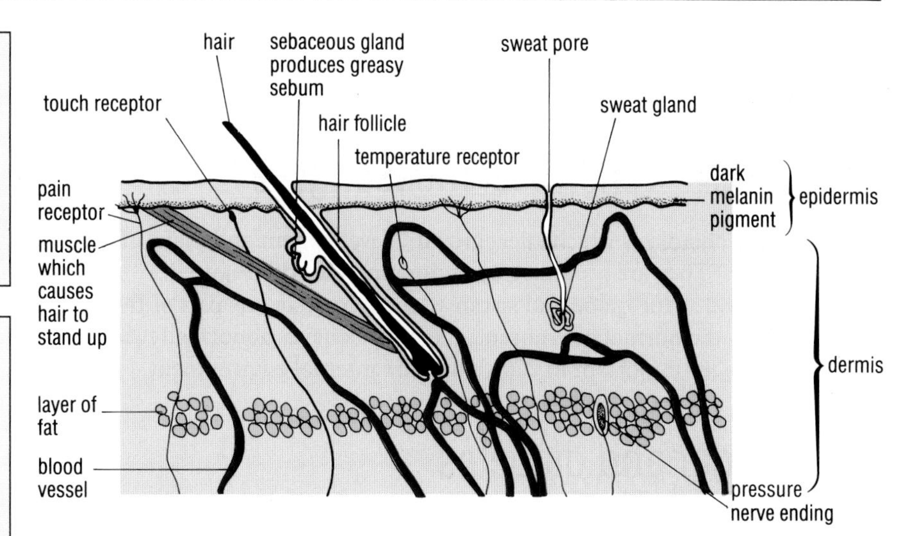

Figure I.2 A section through the skin, magnified

The hairs in a human do not have much value for keeping the body warm. However, the skin has other ways of helping to keep the body's temperature steady. Deep in the dermis are cells filled with fat. These act as a good insulator to keep you warm on cold days.

Body temperature is monitored and controlled by the thermoregulatory centre in the brain. The centre has receptors (see page 83) sensitive to the temperature of blood flowing through the brain. Also, temperature receptors in the skin (Figure I.2) send nerve impulses to the centre, giving information about skin temperature. In cold weather, the blood vessels running to the surface of the skin between the fat cells constrict, so less blood flows to the skin surface and less heat is lost to the outside. If you become hot, the blood vessels expand, or dilate. When this happens, more blood flows to the surface of the skin and loses heat to the outside. This is why fair-skinned people look flushed when they get hot.

Sweat glands produce or **secrete** a liquid, **sweat**, when a person is hot. This moisture spreads out over the surface of the skin and evaporates. Activity I.4 will help you to understand how the evaporation of sweat cools the body. Sweat evaporates more slowly than ethanol but it works in the same way; it takes heat from the blood flowing through the dermis of the skin and so cools the body.

Shivering also helps us keep warm! Small muscles in the skin contract and relax repeatedly if the body is cold. As the muscles contract, they produce some heat.

Homeostasis

The body regulates change in the same way that a thermostat in a water bath or central heating system monitors and controls temperature. A thermostat detects changes in temperature. If the temperature rises it switches off the heat. If the temperature drops it switches on the heat.

In the brain there are receptors to detect changes. Most of them are in the brain, as we have seen. These receptors continuously test the internal environment. If they detect a change, a **feedback** process leads to correction of the internal conditions. The control mechanisms may be triggered off by nerve impulses from the receptors or by chemical messengers (**hormones**) which travel around the body in the blood.

Hormones – chemical messengers

Hormones are chemicals which affect growth, development and many of the body's activities. They are produced by **endocrine glands** (Figure I.3). Hormones travel in the blood and so can affect more than one part of the body. If one of the hormones is absent, the body behaves abnormally. Some hormones are very important in controlling the internal environment.

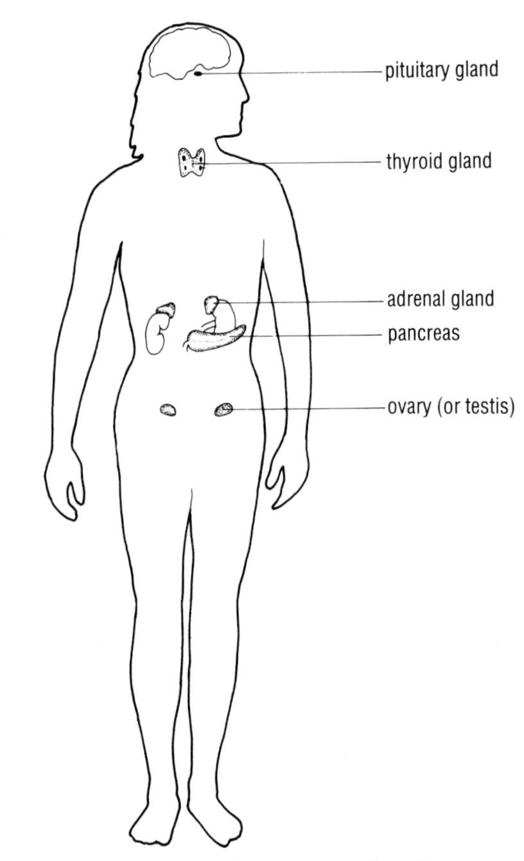

Figure I.3 These endocrine organs produce hormones

Blood sugar and diabetes

The concentration of glucose sugar in the blood must remain steady if cells are to work correctly. When a person eats carbohydrate, more glucose enters the body. During exercise, glucose is used for respiration. In most people the body can regulate the amount of sugar in the blood. But some people are unable to control their blood sugar. They suffer from sugar diabetes.

In the early 1920s Nicolas Paulescu from Romania and Frederick Banting and Charles Best in Canada found that a juice or extract from the pancreas (see Figure I.3) helped sufferers from diabetes. The important chemical in the extract is a hormone called **insulin**. Banting was awarded the Nobel Prize for his part in discovering insulin. Soon scientists were able to purify insulin so that it could be used to treat diabetes. In the 1950s Frederick Sanger of the Medical Research Council in Cambridge worked out the chemical structure of insulin. It was found that the hormone insulin is a protein – this was the first protein for which scientists discovered the arrangement of amino acids in the molecule. Since then insulin has been made artificially.

The discovery of insulin helped scientists and doctors understand what happens in the body. When the glucose concentration in the blood is *high* the pancreas secretes insulin into the blood. This causes glucose to be stored as glycogen (in the liver) and as fat. If the glucose level is *low* the pancreas secretes **glucagon**. This hormone increases the level of glucose in the blood by promoting the conversion of glycogen in the liver into glucose.

Diabetics produce too little insulin. If a diabetic eats very much carbohydrate, the body does not store the extra glucose. Instead, the level of glucose in the blood rises sharply. While the glucose level in the blood is too high the person will be unwell. Some of the extra glucose is excreted through the kidneys and the glucose in the blood gradually falls again. But, when the diabetic is active, there will be no stored glucose to use for respiration to give energy (see page 36). The blood sugar will then become too low, and the person may go into a coma (become unconscious).

Fortunately, the discovery of insulin made it possible to treat diabetics. They inject themselves with insulin regularly. This regulates the glucose concentration so that they can eat the carbohydrate foods they need to provide them with energy. Diabetics who follow a doctor's advice can expect to live normal, healthy lives.

Control of water in the body

About two-thirds of your body is water. Water enters and leaves the body in various ways (Table I.1), but it is important that the amount in the body stays the same.

Gains of water (cm^3/day)		**Losses of water (cm^3/day)**	
Drinking	1400	Excretion of urine	1500
In food	800	From the lungs	350
Produced by chemical reactions in the body	300	As sweat through the skin	550
		In faeces (through anus)	100
Total	**2500**	**Total**	**2500**

Table I.1 The body's water budget

Activity I.5

Examining kidneys

You need: model of human or other mammal; kidney in fat; slide of section of injected kidney (set up and with sketch to aid interpretation); dish; instruments; microscope.

1. Find the kidneys, kidney blood vessels, ureter, bladder and urethra on the model.
2. In a group, look at a kidney. Find the blood vessels and ureter. Discuss how to tell these three tubes apart. Carefully remove the protective fat. Feel the texture of the kidney. Cut the kidney lengthwise into two equal halves. Find the parts shown in Figure I.5.
3. NOW WASH YOUR HANDS
4. Make a labelled sketch.
5. Examine the slide under the microscope. Look for the structures shown in Figure I.4. Make labelled sketches.

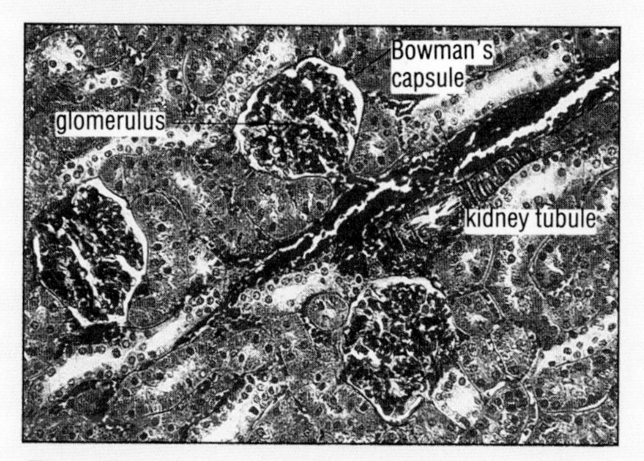

Figure I.4 Glomeruli in kidney cortex

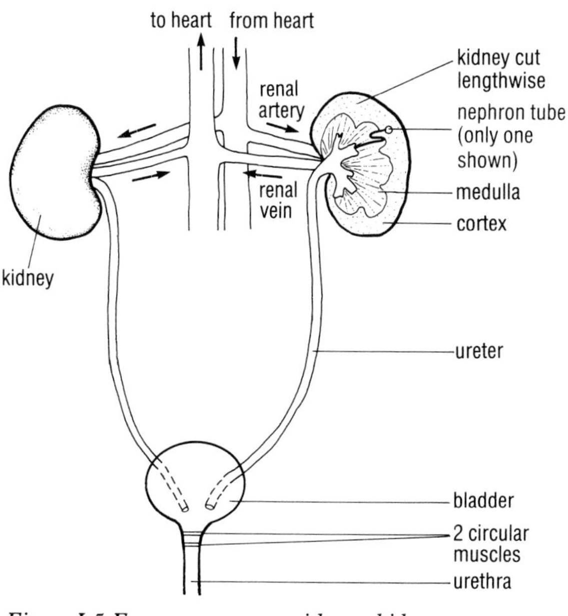

Figure I.5 Excretory system, with one kidney cut lengthwise

When you have a drink of water, it passes through your digestive system until it reaches the large intestine. There the water is absorbed into the blood.

Receptors in the brain detect how much water is in the blood. If there is too much water, most of the surplus is excreted through the kidneys in the urine. A dilute urine is produced until the level of water in the blood returns to normal.

If the brain detects that there is too little water in the blood, then the kidneys retain more water. The pituitary gland secretes anti-diuretic hormone (ADH), which makes the walls of the collecting duct of each nephron (Figure I.6) 'leaky'. Water is absorbed back into the body from the urine and a more concentrated urine is produced. The person also has the sensation of being thirsty. Drinking restores the correct water level in the blood. If a person cannot get enough water to drink, or cannot retain water in the body because of diarrhoea or excessive sweating, the person could suffer dehydration. This is a serious condition, and must be treated quickly, because the cells of the body only function properly if there is the right amount of water present.

Uses of water in the body

Water is important for the body in many ways. It is a very good solvent and many chemicals in the body are carried by the blood in solution. Unwanted substances can be diluted by water and excreted from the body in solution.

Water is also necessary for chemicals to pass through cell membranes and for many of the body's chemical reactions.

Water in the body helps to keep the body temperature steady at 37 °C. As water evaporates from the skin it helps cool the body.

Looking at the kidneys

The two kidneys are at the back of the body at about waist level. They can be easily damaged in accidents, by some drugs and poisons, and by illness including diabetes and high blood pressure. They are protected by fat and each has an artery and vein carrying blood to and from it. A tube, the **ureter**, leads from the kidney to the **bladder** (Figure I.5). Liquid from the bladder travels to the outside along the **urethra**.

The work of the kidneys

The kidneys are important for **excretion** (removal of unwanted products from the body) and for controlling the amount of water, glucose and other substances in the blood. Unwanted amino acids are changed in the liver to energy-giving molecules plus urea containing nitrogen. This process is called **deamination**. If urea becomes very concentrated in the blood it is poisonous. It is removed by the kidneys.

Each kidney has a **cortex** surrounding a **medulla**. In the cortex many substances pass from the blood into the ends of thousands of tiny tubes (**kidney tubules**). The liquid in the tubules is carried across the medulla, through the pelvis into the ureter. By the time the liquid reaches the ureter,

ureter, some dissolved substances have been removed and the liquid is called **urine**. The blood **reabsorbs** substances which are useful. The urine contains waste chemicals which the body does not need.

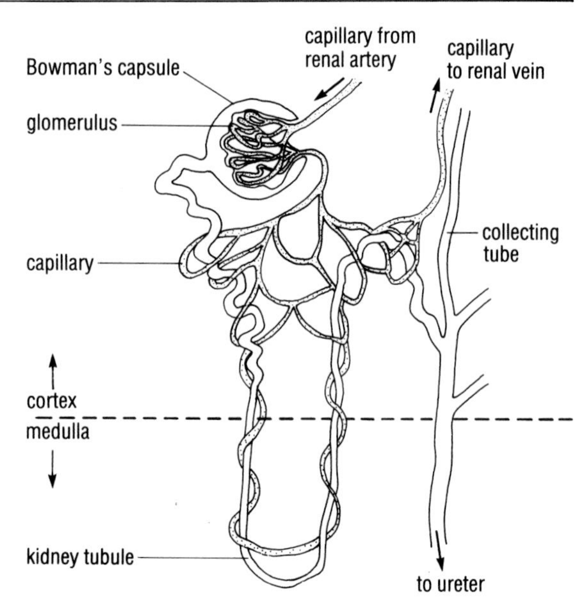

Figure I.6 A single kidney nephron

Where do wastes come from?

The thousands of chemical reactions in cells break down compounds and form new ones. The waste products formed could be harmful if they were allowed to accumulate (Table I.2). Excretion removes them from the body.

Waste substances	How waste is made	Where waste is made	Where waste is excreted
Carbon dioxide	Cellular respiration (see page 36)	All cells	Lungs
Urea and other compounds containing nitrogen	Deamination of amino acids	Liver cells	Kidney

Table I.2 Waste substances produced by metabolism

Water and mineral salts (as ions) are needed for cells to work properly. However, water and salts in excess of the body's needs are excreted in the urine. Salts are also lost through the skin when we sweat.

Changes in the kidney tubule

The end of each of the million or so kidney tubules in the cortex of the kidney is shaped like a hollow cup. This is called a **Bowman's capsule**. Each capsule surrounds a cluster of twisted capillary blood vessels (a **glomerulus**). One glomerulus, capsule and tubule is called a **nephron**. Figure I.6 shows one nephron and its supply of blood vessels.

Blood containing dissolved wastes reaches the kidney in a renal artery, which splits into branches going to all parts of the cortex. These branches end in capillaries running into the glomeruli. The blood in each capillary is at high pressure and much of the liquid part (plasma), which contains dissolved substances, is forced through the walls of the glomerulus into the Bowman's capsule. This liquid (the **filtrate**) travels along the twisting tubule, which is surrounded by capillaries.

Scientists can use micropipettes to take liquid from various parts of a kidney to find out what is happening. As the liquid passes along the tubule, glucose, salts (as ions) and other useful substances are actively reabsorbed into the blood in the surrounding capillaries. The capillaries join up and the blood leaves the kidney in the renal vein. The composition of the liquid in the tubule changes so that by the time it reaches the end of the tubule it has become urine. Table I.3 compares the composition of blood plasma and urine.

The urine passes into a collecting duct which goes to the ureter (Figure I.5). The urine travels along the ureters and is stored by the bladder until it is released through the urethra. The body produces about 1500 cm^3 of urine every day.

	Blood plasma	Urine
Water	92	95
Proteins	7.5	0
Urea	0.03	2
Ammonia	trace	0.05
Sodium ions (Na^+)	0.34	0.6
Potassium ions (K^+)	0.02	0.15
Chloride ions (Cl^-)	0.37	0.6
Glucose	0.1	0

Table I.3 Approximate composition of blood and urine of a healthy person (g in 100 cm^3)

Activity 9.6

What do the kidneys do?

1 Read the section on the work of the kidneys
a Discuss it with a partner.
b Write a simple explanation of how the kidney works for someone younger than you are. Make sure that you explain all the new words in the passage (the ones in heavier type). Use drawings if it helps to make your explanation clearer.
2 Explain the following:
a Why the human body needs at least one working kidney.
b How urine produced in the kidney reaches the outside of the body.
c Why the volume of urine is often greater in cold weather.
d Why doctors make a chemical test on the urine of diabetics.

Kidney failure

If both kidneys fail, unwanted substances build up in the blood and a person rapidly becomes ill. People whose kidneys have failed can sometimes be treated on a kidney machine, often called an 'artificial kidney'. Materials are exchanged through Visking tubing inside the machine, a process called **dialysis**. But there are drawbacks: kidney machines are very expensive and not all patients can use them. Treatment takes several hours, sometimes up to three times a week, and the patient has to be very careful about what he or she eats and drinks.

Some people are able to have a kidney transplant. This operation has risks but nowadays is usually successful. If it works, the patient (the **recipient**) has to take medicine for some time but life can be much nearer to normal. The transplanted kidney usually comes from the body of someone who has just died (the donor). Permission must first be obtained from the person's relatives. This is always sad for them, but it does mean that use is made of part of the dead person's body. The blood and tissue of donor and recipient have to be similar (or **compatible**) for the transplant to succeed. Kidney transplants are cheaper than using a kidney machine, and enable people to live a more normal life.

Summary

- Homeostasis is the way in which the body keeps the internal environment constant. The body is able to detect changes and to control them. Examples of homeostasis are:
  (a) keeping the internal temperature steady
  (b) keeping the water and glucose levels of the blood constant.
- Hormones (chemical messengers) such as insulin are often important in homeostasis.
- The kidney removes excess water and urea, and regulates the concentration of other substances in the body.

Questions

1 Write down as many functions as you can for the following organs: **a** liver, **b** kidney, **c** pancreas.
2 Explain the differences between: **a** urea and urine, **b** ureter and urethra.
3 Collect information on hypothermia in old people. What advice would you give the elderly on ways of avoiding hypothermia?

Extension questions

4 Distinguish between the roles of the hormones insulin and glucagon in keeping the body's blood glucose level steady.
5 List the different body components that help keep the body temperature steady. Explain how each component does its job.
6 What are the control features of any homeostatic mechanism that keep the internal environment of the body constant? Explain the role of each feature.

Unit J
Drugs and health

Figure J.1 Abusing solvents

Penicillin, and other antibiotic drugs that attack bacteria, have been one of the medical success stories of the twentieth century. Today doctors use an arsenal of drugs to fight disease. Unfortunately, some people become addicted to drugs such as morphine, cocaine and other pain-killers, which affect the central nervous system (see page 83). These drugs produce a sense of well-being and are favourite substances of **abuse** – a term often used to describe the non-medical use of drugs.

Nicotine in tobacco, and alcohol (ethanol) in beers, wines and spirits, are also frequently abused. Although these drugs may be used legally by people over a certain age, smoking and heavy drinking harm the body. Sniffing glue and other substances containing solvents is another common form of drug abuse. Solvent abuse is a growing problem among young people.

Solvents

Glues, paints, nail varnish and cleaning fluids contain volatile solvents like esters and ethanol. Breathing them in (Figure J.1) gives a warm sense of well-being but also produces dangerous disorientation. For example, people might think that they can jump out of high windows without falling.

Solvents slow down bodily functions – affecting, for example, the nerve centres which control breathing and heart rate. Long-term solvent abuse can damage the liver, kidneys, lungs and brain.

Smoking and health

Smoking is a major cause of lung disease. Cigarette smoke is acid and contains nicotine, tar, carbon monoxide and other substances. Smokers take in the smoke through their mouths, by-passing the filters of the nose. Then the smoke is often inhaled into the lungs. Figure J.2 shows some differences between the lungs of smokers and non-smokers.

Do you know anyone with a 'smoker's cough'? Some of the substances in cigarette smoke irritate the lining of the trachea and bronchi, which causes extra mucus to form. Others paralyse the cilia. The mucus, or phlegm, builds up in the trachea and bronchi – and the smoker coughs repeatedly to get rid of it. Smokers also become breathless quickly.

Nicotine affects the nervous system, heart, blood vessels, blood and kidneys. The addictive effect of nicotine is serious. Once the body has become used to it, the person craves for more. This makes it very difficult for people to give up smoking.

Carbon monoxide in smoke can combine with haemoglobin in the blood, so that less oxygen can be carried around the body. People with high levels of carboxyhaemoglobin cannot work well. They can also develop angina pains in the chest.

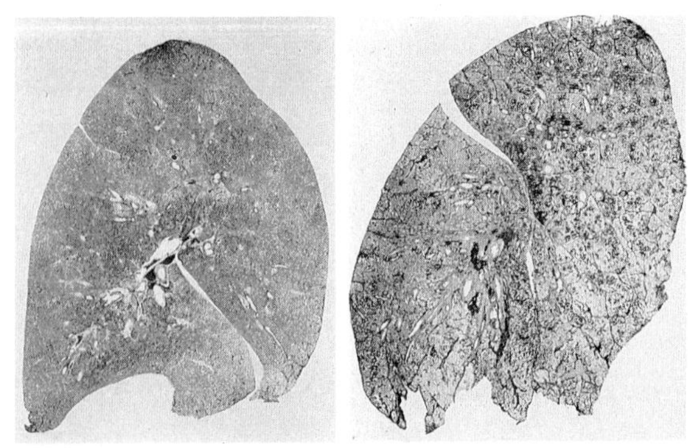

Figure J.2 A normal lung and a smoker's lung. What differences are visible?

Activity J.1

Using a smoking 'machine'

Use the apparatus (Figure J.3) to compare different kinds of cigarettes and a small cigar. Clean the apparatus and replace the cotton wool and indicator each time. Record what you see in the U-tube and any colour change in the indicator.

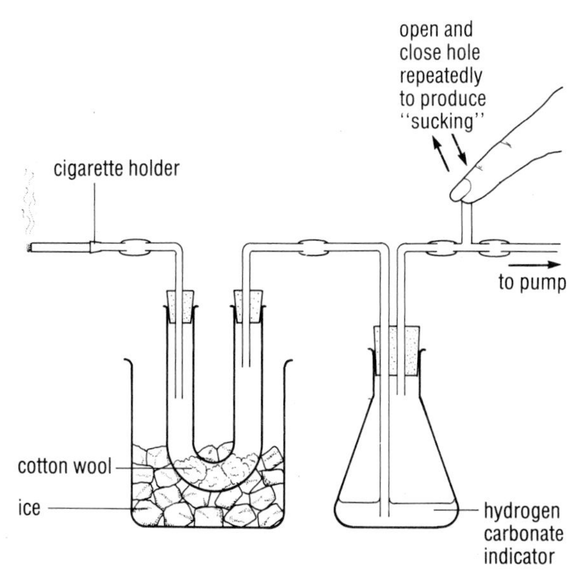

Figure J.3 A smoking 'machine'

The *tar* in smoke is **carcinogenic** (it causes cancer). In cancer, cells reproduce in an unusual way. The extra cells cause a growth or tumour. By the time lung cancer is diagnosed, the growth may be difficult to cure. Cancer cells can also move around in the blood stream and start *secondary* growths elsewhere. Doctors are gradually learning to treat cancer but it is still a very serious illness.

Other illnesses, such as bronchitis (Figure J.4) and coronary heart disease (Figure J.5) are associated with smoking.

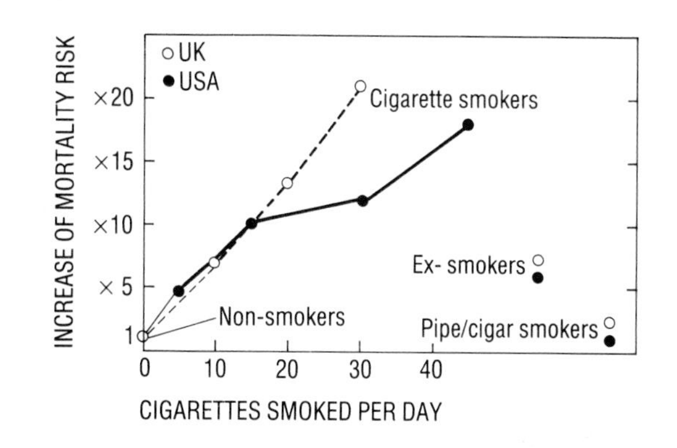

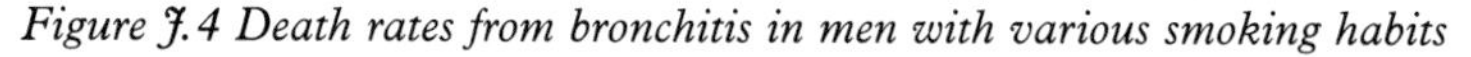

Figure J.4 Death rates from bronchitis in men with various smoking habits

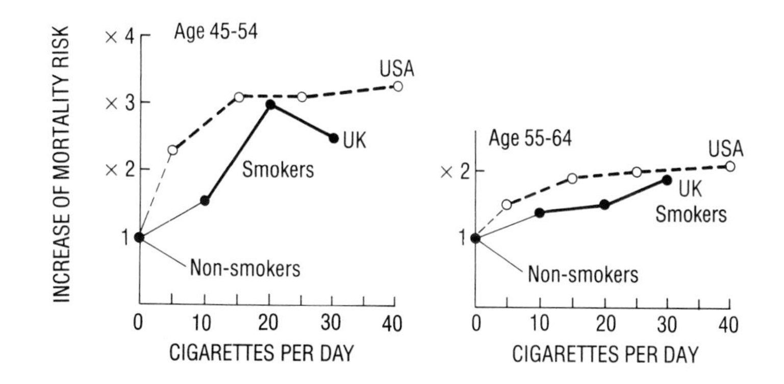

Figure J.5 Death rates from coronary heart disease in men with various smoking habits

Lung cancer, bronchitis and coronary heart disease are all major causes of death in men. They are becoming increasingly serious in women, too. These are all diseases that are more likely to occur in smokers. Table J.1 shows the results of a survey of doctors.

Insurance companies now consider smoking such a health risk that they give cheaper rates for life insurance to non-smokers.

Non-smokers	12
1 to 14 cigarettes each day	about 15
15 to 24 cigarettes each day	nearly 16
25 or more cigarettes each day	nearly 20
Pipe or cigar smokers	just over 12

Table J.1 Annual death rate per 1000 male doctors

There is a strong correlation (relationship) between death rate and smoking cigarettes. One set of data is not proof, however. Some people argue that whatever causes a person to smoke (for example, stress or worry) might also cause the illness. But the evidence against smoking continues to mount up. The Royal College of Physicians said in 1971:

Premature death and disabling illnesses caused by cigarette smoking have now reached epidemic proportions and present the most challenging of all opportunities for preventive medicine...

The campaign to reduce smoking has continued ever since.

120
90
60
30
0
DEATHS PER 100,000
All men
7%
Male doctors
38%
1954-57
1958-61
1962-65

Figure J.6 Death rates from lung cancer in male doctors and all men in England and Wales

Smoking, lung cancer and emphysema

Research into the link between smoking and lung cancer was carried out in the 1950s and '60s. The Royal College of Physicians (a society of doctors) produced a report, *Smoking and Health*, in 1962. This was followed by *Smoking and Health Now* published in 1971. In 1983, the Froggatt Report by an Independent Scientific Committee on Smoking and Health summarised the evidence that smoking is harmful.

Although the early evidence suggested that smoking was harmful, it was not complete proof. When doctors read this evidence, many of them decided to give up smoking. A study of more than 30 000 doctors was made over ten years. The results were impressive. The death rate from lung cancer among doctors (compared with the male population as a whole) began to go down (Figure J.6).

Smoking also weakens the walls of the alveoli (see page 35) and repeated coughing can destroy some of them. This breakdown of the alveoli is called **emphysema**. Figure J.7 shows the result. Destroying the alveoli reduces the surface area for the exchange of gases (see page 36). Why does a person with emphysema become breathless and exhausted easily?

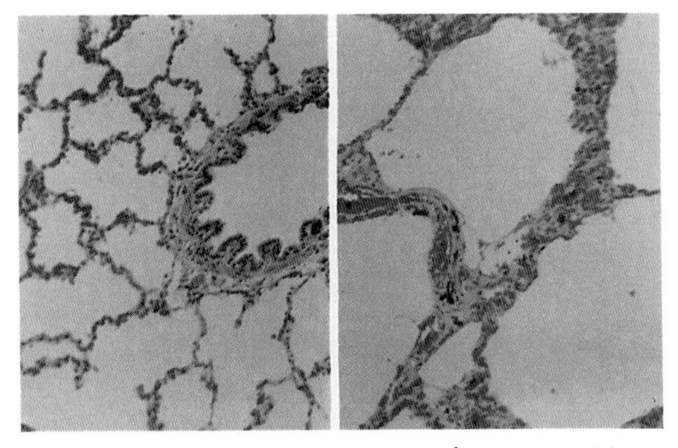

Figure J.7 This photograph shows a lung destroyed by emphysema, caused by smoking

Why do people smoke?

Since there is so much evidence that smoking is harmful, why do so many people smoke?

Once you've started smoking it can be very difficult to stop. But most people manage to give up if they really want to. Even though smoking is harmful, a lot of the damage can be reversed. As soon as a person stops smoking their health begins to improve. Ten years later most of the damage to the body will have been repaired.

Passive smoking

The Froggatt Report, in 1983, collected evidence that other people's smoke can cause illness. This is called 'passive smoking'. In the short term, your eyes, nose and throat may be irritated if you share a room with a smoker. Children of smokers have more lung disease than normal. Working in a smoky atmosphere can increase the risk of heart, circulatory and lung diseases. Fortunately, more and more train compartments and public rooms are now made 'no smoking' areas.

Babies born to mothers who smoke are smaller than normal and may be affected in other ways. Some substances from smoke can pass into the mother's milk. So, it is far better for expectant and nursing mothers not to smoke.

Activity J.2

1. Make a list of reasons why people start to smoke.
2. Make another list of reasons why you think people should not smoke.
3. Make a third list of reasons why people find it hard to stop smoking.
4. Read a pamphlet on stopping smoking. How can people be helped to stop smoking?

Alcohol

Alcohol depresses the activity of the nervous system. It affects different parts of the brain, depending on how much is drunk. Alcohol reduces inhibitions and people feel a sense of well-being. Alcoholic drinks have helped oil the wheels of society since the time thousands of years ago when people first exploited yeast to make wine and beer (and bread!).

A person's behaviour changes as he or she drinks more and more alcohol. A small amount affects the part of the brain which controls judgement. Larger quantities affect the part of the brain which controls movements of the arms and legs. Even more impairs memory. More and more alcohol affects further areas of the brain until vital brain centres that keep us alive are affected. Drinking too much may cause death.

Drinking too much alcohol is one cause of diseases such as **cirrhosis** of the liver (Figure J.8), heart disease and damage to the nervous system.

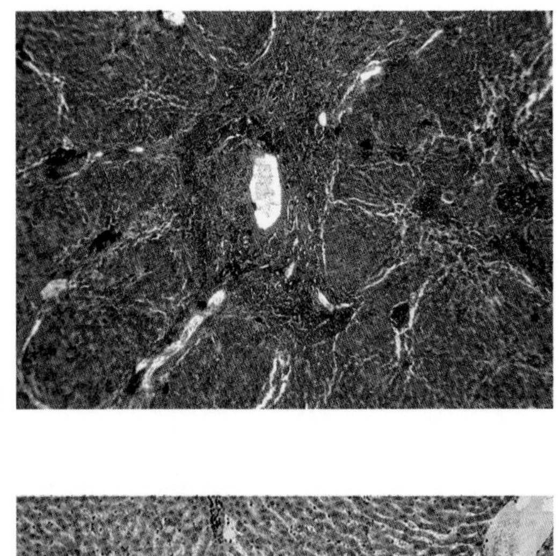

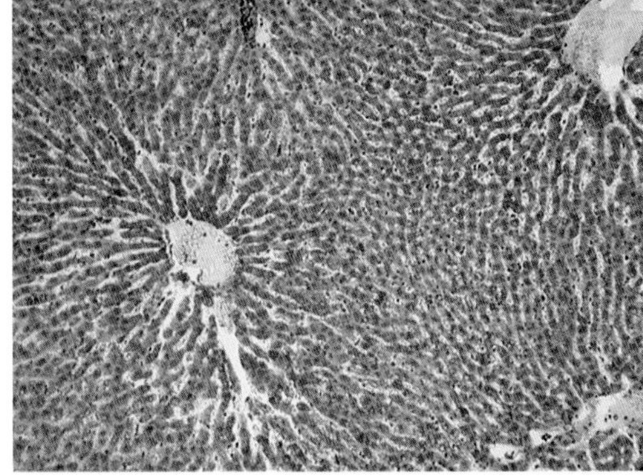

Figure J.8 (a) A section of healthy human liver, seen under a microscope. (b) A section of a liver damaged by cirrhosis, caused by excessive alcohol consumption. The tissue is scattered with fibrous scar tissue, and nodules of regenerating cells

How much alcohol is too much?

The amount a person can drink safely depends on the person's sex, age, size and metabolic rate. For example, the safe level of alcohol for a woman is only about two-thirds as much as that for a man of the same weight. Also different drinks contain different concentrations of alcohol (Figure J.9).

Opinions on 'safe' limits for drinking alcohol vary. Some doctors suggest that a woman should not drink more than two pints of beer a day. Others say that the limit may be a little more. A pregnant woman who regularly drinks beer, wine or spirits increases the risk of the fetus developing abnormally. Fetal growth is also reduced. All doctors agree that drinking alcohol affects your behaviour and that heavy drinking affects your health.

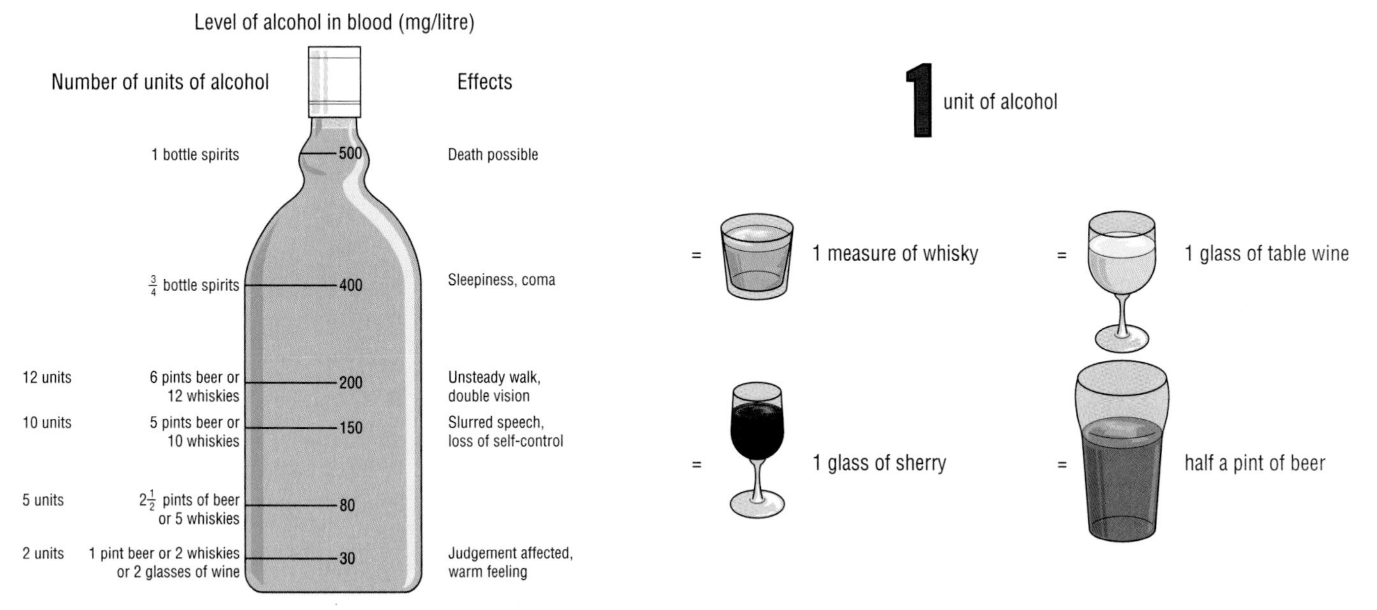

Figure J.10 The effects of alcohol. Drinking alcohol affects your driving ability. NEVER DRINK AND DRIVE

Figure J.9 Units of alcohol

Dependence and withdrawal

People often start to use drugs to experience the sense of well-being ('high') that the drug gives. However, dependence on a drug (**addiction**) can develop quickly. The mechanisms underlying dependence are poorly understood. For example, one idea supposes that dependence on **heroin** (a pain-killer) develops because long-term abuse of the drug reduces production of the body's natural pain-killers (called **endorphins**). The body comes to depend more and more on the drug – so an addiction develops.

Addicts continue to use some drugs not so much for the 'high' but to avoid the symptoms that follow giving up (**withdrawal**). Symptoms vary but are often unpleasant. For example, people who give up smoking often feel irritable and lack concentration through the withdrawal of nicotine.

Summary

- Solvents, nicotine and alcohol (ethanol) are drugs of abuse. They are addictive and damage health.
- Giving up abuse of drugs may result in unpleasant withdrawal symptoms.

Questions

1. Why do you think solvent abuse and drinking too much alcohol have similar effects on behaviour?
2. Find out which parts of the brain are affected as a person drinks more and more alcohol.
3. Decide whether each statement is: definitely true; possibly true; not true.
   a) If you smoke heavily you are more likely to get heart disease.
   b) Heavy smoking causes heart disease.
   c) Heavy smokers always get heart disease.
   d) People who are not smokers do not get heart disease.
   e) Heavy smokers always get lung cancer.
   f) Non-smokers never get lung cancer.
   g) If you smoke heavily you are more likely to get lung cancer.
   h) There is no link between smoking and lung cancer.
   i) Heavy smoking may be one cause of lung cancer.
   j) Heavy smokers always live less long than non-smokers.
   k) Heavy smokers often live less long than non-smokers.
   l) Cigarette smoke may affect non-smokers.
   m) Mothers who smoke have less healthy babies.
   n) Non-smokers sometimes get heart disease or lung cancer.
4. Name three diseases which are more frequent in cigarette smokers. What kind of evidence has been used to discover the effects of smoking? What are the dangers to cigarette smokers of
   a) nicotine; **b)** tar; **c)** carbon monoxide?
5. Prepare a poster called 'Did you know that cigarettes may damage your health?'

End-of-module questions

QUESTION ONE

The table shows some of the jobs of parts of the cell. Choose words from the list for each of the items 1–4 in the table.

nucleus **chloroplast** **cell membrane** **cell wall**

1	controls the passage of materials into the cell
2	strengthens the cell
3	controls the activities of the cell
4	absorbs light energy

(E/F)

QUESTION TWO

The table shows some of the jobs of receptor cells. Choose words from the list for each of the items 1–4 in the table.

receptors in the eye **receptors in the ear**
receptors in the skin **receptors in the tongue**

1	sensitive to taste
2	sensitive to pressure changes
3	sensitive to light
4	sensitive to sound

(E/F)

QUESTION THREE

The table shows the uses plants make of various chemicals. Choose words from the list for each of the items 1–4 in the table.

nitrate **sugars** **potassium** **phosphate**

1	important in photosynthesis reactions
2	important for the synthesis of proteins
3	important as an energy source
4	helps enzymes involved in photosynthesis

(A/B)

QUESTION FOUR

The table is about photosynthesis. Choose words from the list for each of the items 1–4 in the table.

chlorophyll **Sun** **sugar** **oxygen**

1	provides energy for photosynthesis
2	a waste substance from photosynthesis
3	traps energy for photosynthesis
4	a useful substance made by photosynthesis

(E/F)

QUESTION FIVE

Choose from the list **two** factors that will increase the rate of water loss by transpiration from the leaf below.

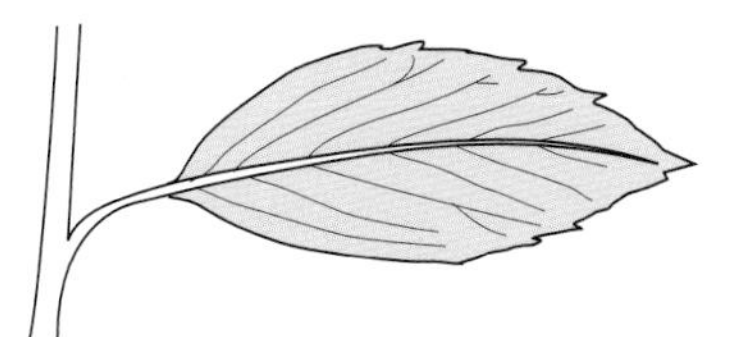

an increase in the wind speed
a decrease in the temperature of the air
an increase in humidity (the amount of water in the air)
a decrease in the amount of oxygen in the air
an increase in the temperature of the air

(C/D)

QUESTION SIX

Drugs affect our bodies. Choose from the list **two** statements that are correct.

alcohol increases our speed of reaction
solvents can result in emphysema
tobacco smoke can result in liver damage
tobacco smoke can result in heart damage
solvents can cause damage to the lungs, liver and brain

(C/D)

QUESTION SEVEN

Cells take in water by osmosis. Choose from the list **two** factors that increase the rate of water uptake by osmosis.

the cells being turgid
more sugar in the cell
more starch in the cell
the pressure in the cell being high
the pressure in the cell being low

(A/B)

QUESTION EIGHT

The drawing shows guard cells on the lower surface of a leaf.

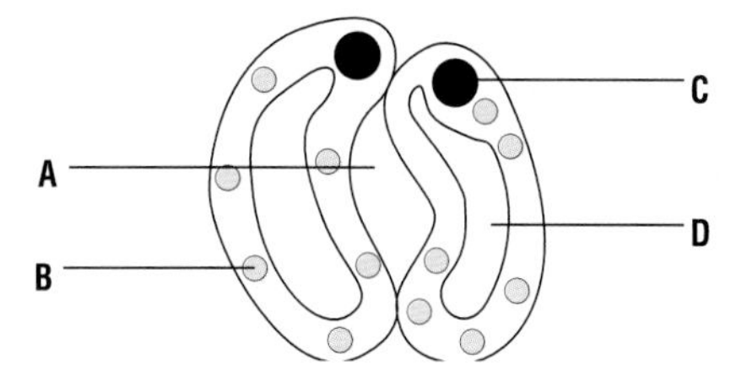

1 Which part shows a nucleus?
A **B** **C** **D**

2 Which part shows the cell sap?
A **B** **C** **D**

3 Which part shows the stoma?
A **B** **C** **D**

4 Which part shows a chloroplast?
A **B** **C** **D**

(C/D)

QUESTION NINE

The diagram shows parts of the eye.

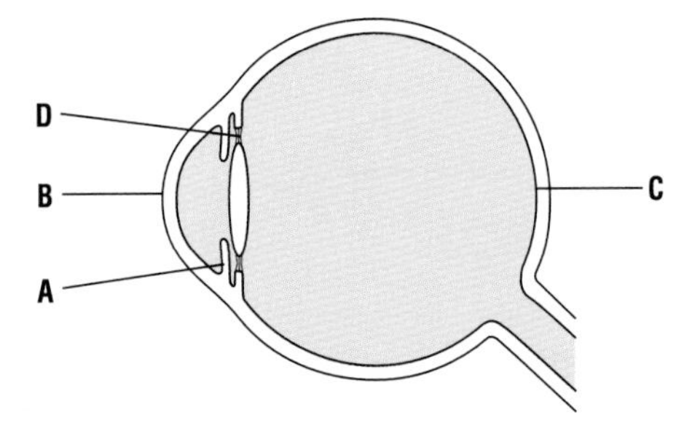

1 Which part allows light into the eye?
A **B** **C** **D**

2 Which part holds the lens in place?
A **B** **C** **D**

3 Which part controls the size of the pupil?
A **B** **C** **D**

4 Which part contains cells sensitive to light?
A **B** **C** **D**

(C/D)

QUESTION TEN

The drawing shows the nerve pathway of a reflex action.

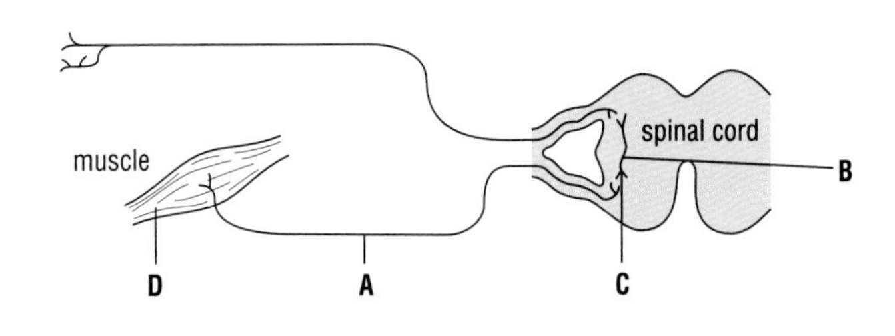

1 Which part is a synapse?
A **B** **C** **D**

2 Which part is a motor neuron?
A **B** **C** **D**

3 Which part is an effector?
A **B** **C** **D**

4 Which part is a relay neuron?
A **B** **C** **D**

(A/B)

QUESTION ELEVEN

1 Urea is a waste product. Which of the following statements is true?
A Urea is produced in the bladder.
B Urea is produced by the kidneys.
C Urea is a breakdown product of sugars.
D Urea is a breakdown product of amino acids.

2 The body controls the amount of water in it. Which of the following statements is true?
A Water is lost only through the skin and kidneys.
B Water is lost through the skin, lungs and kidneys.
C Sugar is dissolved in the sweat lost from the skin.
D The bladder stores water as urea.

3 ADH controls the rate of water loss through the kidneys. Which of the following statements is true?
A The pituitary gland actually releases ADH.
B ADH causes the kidneys to produce more dilute urine.
C ADH causes a reduction in water reabsorption by the kidneys.
D If the water content of the blood is low, less ADH is released.

4 Body temperature must be kept constant. Which of the following statements is true?
A If the core body temperature is too low, blood vessels supplying skin capillaries dilate.
B If the core body temperature is too high, muscles 'shiver'.
C If the core body temperature is too high, sweat glands release more sweat.
D The evaporation of sweat 'warms' the body.

(A/B)

QUESTION TWELVE

Plants lose water from their leaves by transpiration. The graph shows water loss from a plant over a 24 hour period.

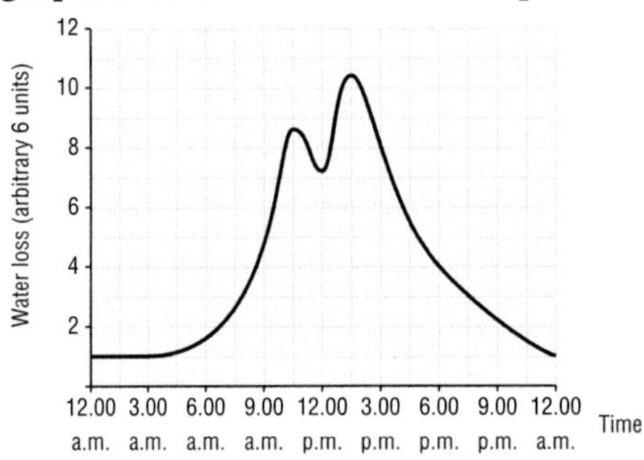

1 Why is transpiration greater at 9.00 a.m. than at 3.00 a.m.?

2 Suggest three possible reasons for the drop in transpiration rate between 10.30 a.m. and 1.30 p.m.

3 Suggest one adaptation that a plant living in dry conditions might have which reduces water loss.

4 Which structures take in water from the soil?

(C/D)

QUESTION THIRTEEN

A farmer was growing wheat in a field. Every 4 hours, for a 24 hour period, a scientist tested several plants to find out how much sugar was in the leaves. The table gives the results.

Time of day	Sugar concentration (% dry mass of leaves)
04.00	0.40
08.00	0.65
12.00	1.85
16.00	2.20
20.00	1.55
24.00	0.60

1 Plot this information on graph paper.

2 Using your graph, estimate:
a the sugar concentration at 10.00.
b when sugar concentration is at its maximum. (C/D)

3 Why does sugar concentration vary in this way? (A/B)

Environment

This module explores how living things are adapted to survive in their environments. It looks at the different physical factors (light, heat, water, availability of oxygen and carbon dioxide) affecting organisms living in a particular location, at food chains and food webs, and at how feeding transfers energy and nutrients between organisms. We consider the role micro-organisms play in the decomposition and recycling of organic material, and also see how human activities can disrupt the natural processes that maintain the stability of the environment.

Environment

Learning Objectives

What you should know:

- Physical factors which affect the growth of organisms include temperature, light, water, oxygen, and carbon dioxide
- Organisms have features which enable them to survive in the conditions in which they normally live
- Organisms compete with each other for space, food and water
- The factors which affect the size of populations
- The relationship between predator and prey populations
- Food chains begin with green plants (producers) which provide food for other organisms (consumers)
- Food chains are often interconnected to form food webs
- The number of organisms at each stage of a food chain can be shown as a pyramid of numbers
- The mass of living material at each stage in a food chain can be represented as a pyramid of biomass
- Materials decay because they are broken down by microbes. Decay is speeded up in moist, warm conditions, with oxygen
- The main stages in the carbon cycle
- How human activities may pollute water, land and air
- How acid rain is produced and its effect on organisms
- The effects of rapid growth in the human population on the environment

What you should be able to do:

- Investigate the oxygen/carbon dioxide cycle in water
- Suggest how particular organisms are adapted to the conditions in which they live
- Suggest the factors for which given organisms are competing
- Investigate population changes in grain weevil
- Interpret population graphs
- Construct and interpret food chains
- Construct and interpret food webs
- Construct and interpret pyramids of numbers
- Construct and interpret pyramids of biomass
- Investigate the activity of soil bacteria
- Make a compost heap

EXTENSION

- The reasons why at each stage in a food chain, less material and less energy are contained in the biomass of the organisms
- Ways in which the efficiency of food chains can be improved
- The main stages in the nitrogen cycle
- All the energy originally captured by green plants is eventually transferred to the environment
- How pollution of water by fertilisers eventually leads to oxygen depletion of the water
- The effects on the environment of large scale deforestation in tropical areas
- Why levels of carbon dioxide and methane in the atmosphere are rising
- The possible effects on the environment of the 'greenhouse effect'

- Evaluate the positive and negative effects of managing food production
- Recognise that practical solutions to human needs may require compromise between competing priorities
- Weigh evidence and form balanced judgements about major environmental issues

Why do organisms live where they do?

Unit A
Physical factors

The word **abiotic** is used to describe the physical factors (non-living part) of the environment. The abiotic part of a pond is the water; its physical and chemical conditions influence where organisms live. The abiotic part of an oakwood is the area covered by trees.

Light

The amount of light is an important abiotic influence on life inside the wood. It affects the rate of photosynthesis and therefore the amount of plant growth under the canopy layer. This in turn affects the animals that depend on plants for food and shelter, and so on along the food chain.

Figure A.1 shows seasonal changes in the amount of light inside an oakwood. As the leaves of the canopy unfold in spring they stop light from entering the wood.

At the beginning of March there are about 11 hours of daylight. This increases to about 16 hours by the end of May. The change in the length of day between March and May makes plants of the **field layer** in the wood (eg bluebells and primroses) flower in May (Figure A.2).

The amount of light needed to cause flowering varies depending on the species of plant. The plants of the field layer take advantage of the spring light shining through the bare branches before the canopy develops and reduces light levels in May and June. Then the trees benefit from the long hours of daylight for photosynthesis (Figure A.3).

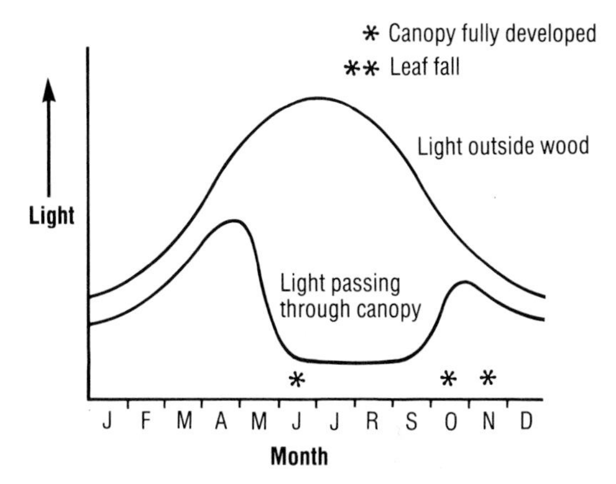

Figure A.1 Seasonal changes in the amount of light inside an oakwood. There may be up to 90% less light inside the wood than outside by the time the canopy is fully developed in summer. Autumn leaf fall allows more light inside the wood once again

Figure A.2 Bluebells flower in spring as day length increases

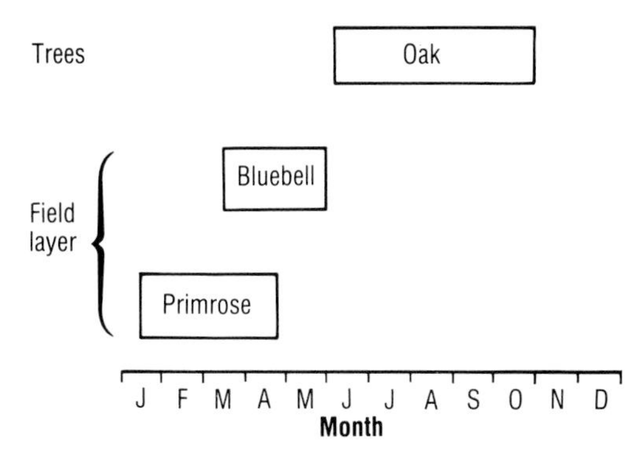

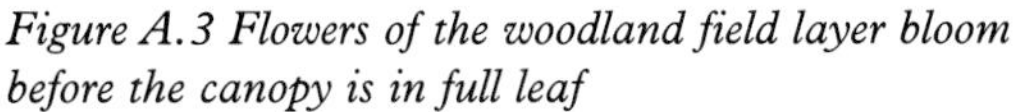

Figure A.3 Flowers of the woodland field layer bloom before the canopy is in full leaf

Figure A.4 Californian desert, USA.
There is little vegetation for most of the year but after the annual rains begin the landscape blooms with flowers for a brief period before drought returns

Rain and temperature

Globally, rain and temperature are probably the most important abiotic influences of all. Without rain, communities perish; for example the droughts of the 1970s and 1980s in the Sahel region of West Africa and in 1976, 1984 and 1995 in Britain damaged both crops and natural vegetation. With rain and warmth, communities flourish (see Figure A.4).

Living things are adapted to seasonal patterns of rainfall. Figure A.5 shows the effect of seasonal rainfall on vegetation in West Africa. Notice that forest forms where there is a lot of rain and little or no dry season. The growth and variety of plant life is prolific in the hot, humid environment, with up to several hundred different species of tree per hectare. Some trees are very large: they are called **emergents**. Also notice the shorter trees and shrubs between the emergents. Some will be sapling emergents but most are low-growing species adapted for living in reduced light.

Moving along the transect A–B, the length of the dry season affects the type of vegetation. Intense heat and lack of water make the development of forests impossible. Instead, the grassland of savanna grows in the rainy season and dries up in between, leaving acacia scrub to stand out against the scorched landscape (Figure A.6).

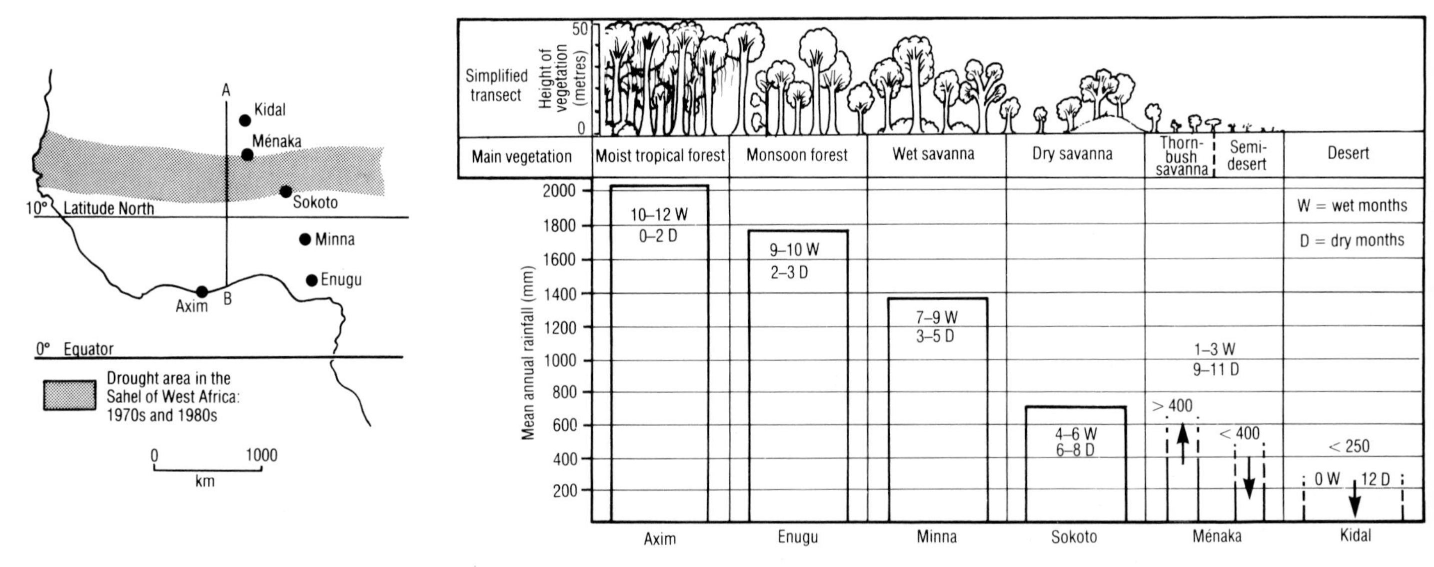

Figure A.5 Weather stations are located at the places named on the map. The mean annual rainfall at each station is plotted on the bar charts and the number of wet and dry months indicated. Changes in the type of vegetation are shown along a transect marked A – B

Plants like the acacia are adapted to cope with life in difficult conditions. They have roots that penetrate up to 6 metres underground to reach the water table. These deep roots supply water to parts of the plant exposed to the dry environment above ground. These plants can survive the drought of the dry season, when shallow-rooted vegetation dies (Figure A.6).

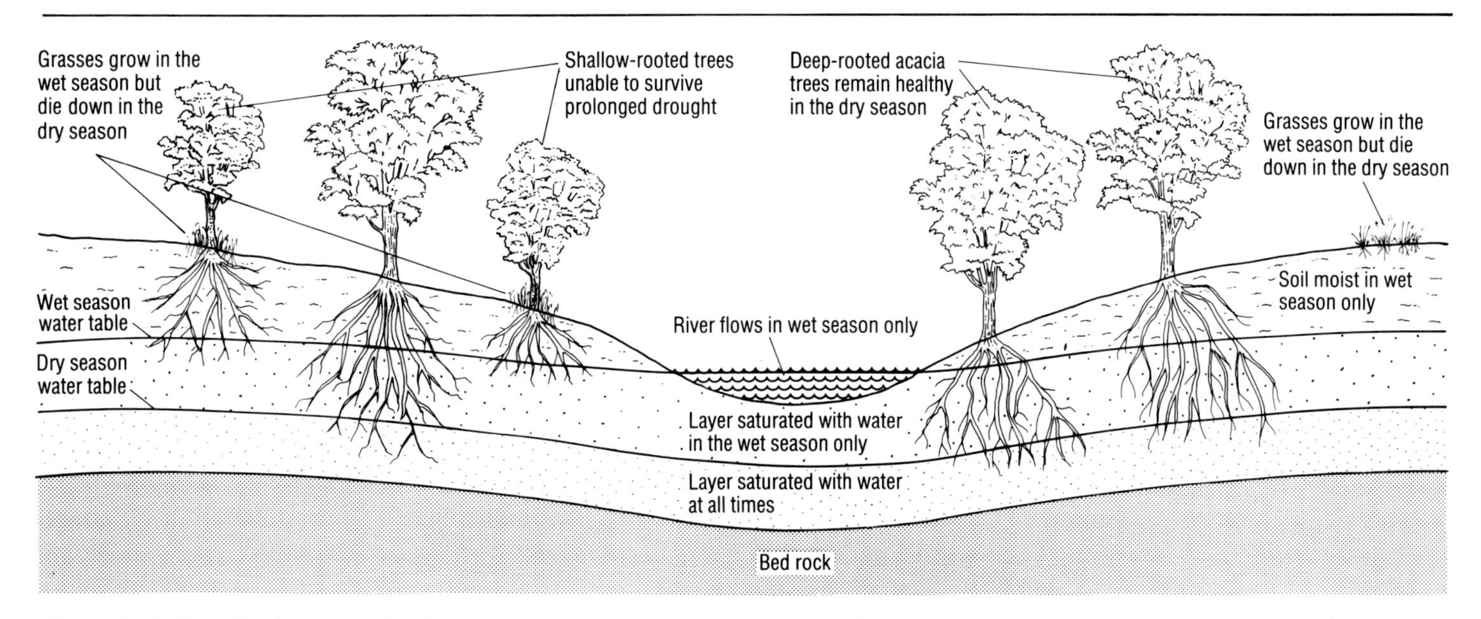

Figure A.6 Soil profile of savanna showing wet and dry season water tables. Depth of root penetration to the water table affects chances of plants surviving drought

Grasses survive drought in a different way. Growing tall and lush in the rain, they flower and produce seeds before the dry season returns. In the dry season the plants wither and die. But the seeds survive, protected by a tough seed coat from the blistering sun. The seeds remain dormant until the next wet season when they sprout and grow rapidly so that they can form seeds before the onset of the next dry period.

The land surface of the Earth is covered by areas of vegetation that roughly correspond to the different temperature zones. These in turn correspond to the distance north and south of the Equator. Figure A.7 shows summer temperatures worldwide and some of the major vegetation areas.

The effects of temperature on types of vegetation can be seen if we climb a mountain. For every 100 metres increase in height the drop in temperature (about 0.5 °C) corresponds to a 1° increase in latitude north or south of the Equator. So, as Figure A.8 shows, plants and animals normally found in colder regions may reach the Equator in the mountain ranges that run from north to south.

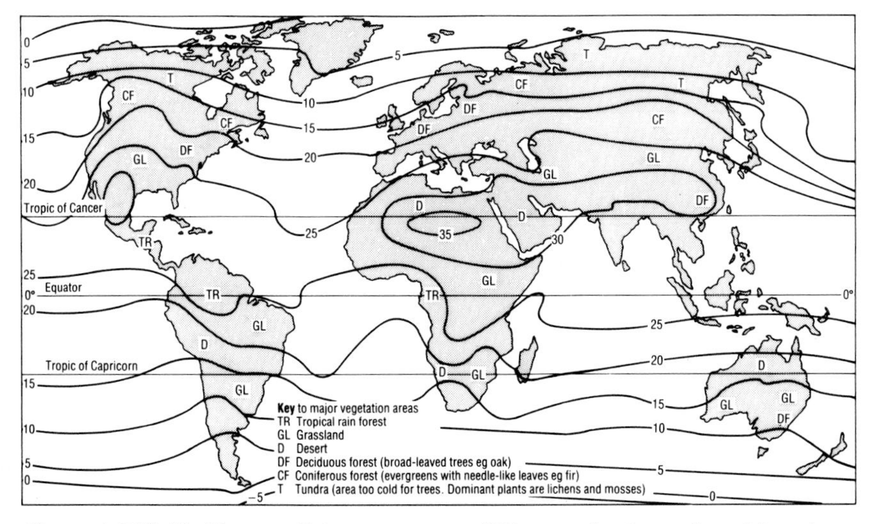

Figure A.7 Worldwide mean July temperatures (°C) at sea level, together with major vegetation areas

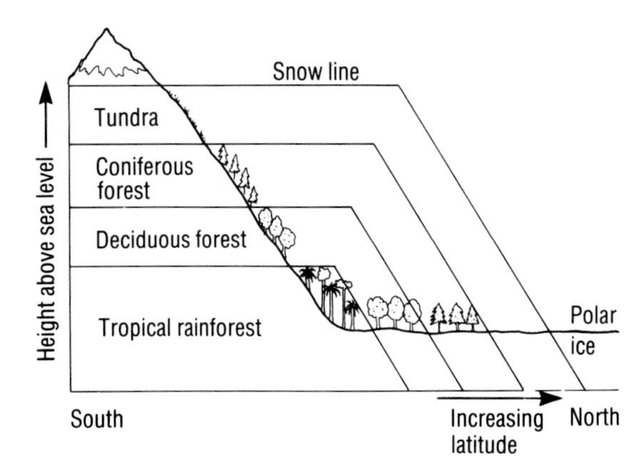

Figure A.8 Plant communities living up mountains near the Equator are similar to those found in more northerly and southerly latitudes

Availability of oxygen and carbon dioxide

Gas	%
Nitrogen	78
Oxygen	21
Carbon dioxide	0.03
Other gases, eg water vapour, argon and xenon	<1

Table A.1 Gases of the troposphere make up the air we breathe

The thin blanket of gases surrounding the Earth is called the **atmosphere**. Without it, life could not exist. The atmosphere can be divided into layers according to temperature.

The bottom layer of the atmosphere, called the troposphere, extends for about 16 km above sea level. The air temperature drops as it gets further away from the Earth's surface. That is why there is always snow on top of high mountains like Mount Everest. The troposphere is the most important for living things. It contains most of the atmosphere's gases, water vapour and dust particles (Table A.1). Oxygen is essential for the respiration of most organisms, carbon dioxide is needed for photosynthesis.

Above the troposphere is the stratosphere. Here oxygen molecules (O_2) break down and recombine to form a layer of ozone (O_3). This screens out harmful ultraviolet rays which would kill most forms of life if they reached the Earth. Figure A.9 shows the layers of the atmosphere.

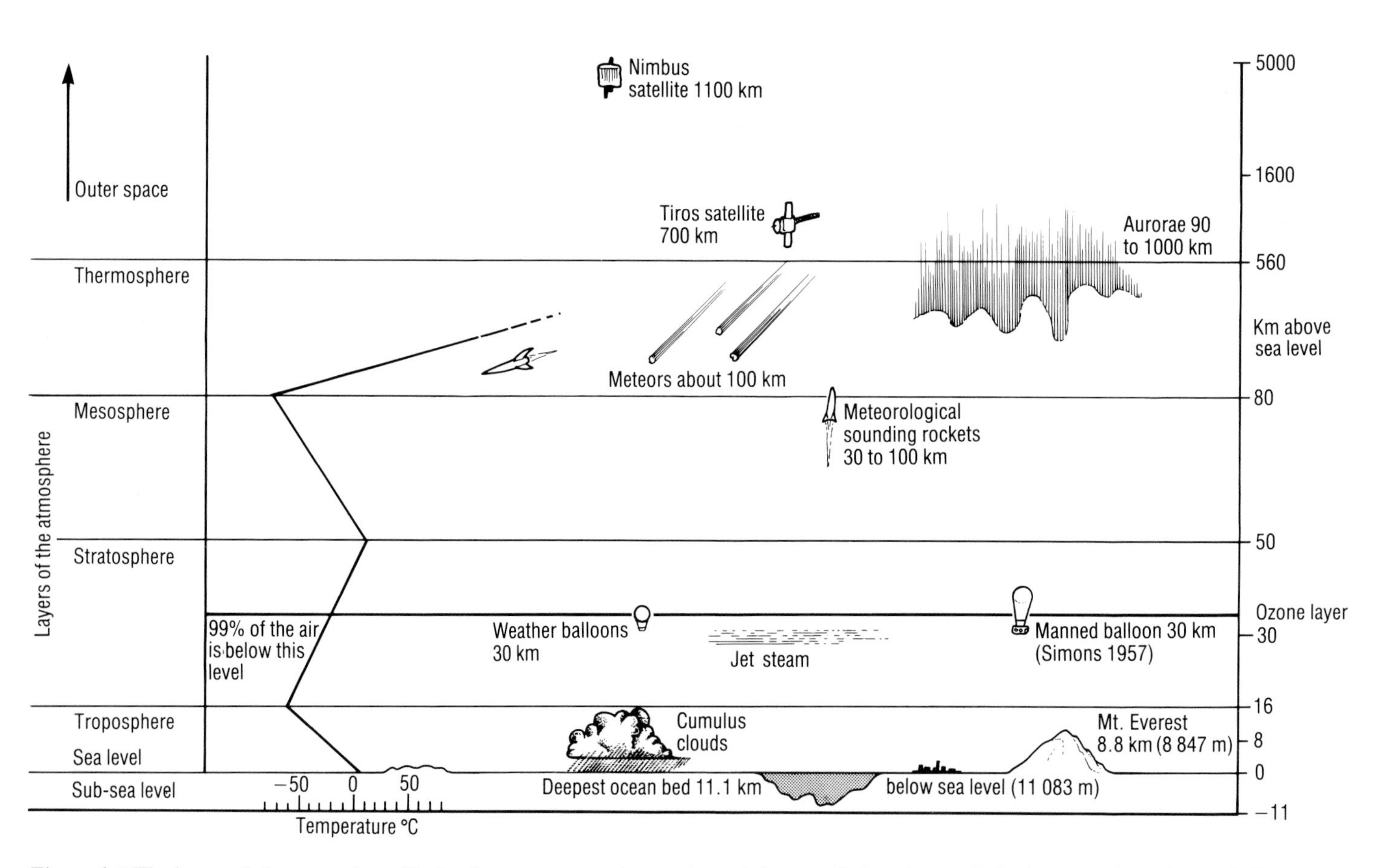

Figure A.9 The layers of the atmosphere. Notice the temperature changes in each layer and that almost all the features of weather and climate affecting living things are found in the troposphere

Figure A.10 shows sections through a pond. It shows some of the physical and chemical conditions of water that affect where pond organisms live. Notice that the processes of photosynthesis (see page 59) and aerobic respiration (page 36) help to maintain the balance of oxygen and carbon dioxide dissolved in the water.

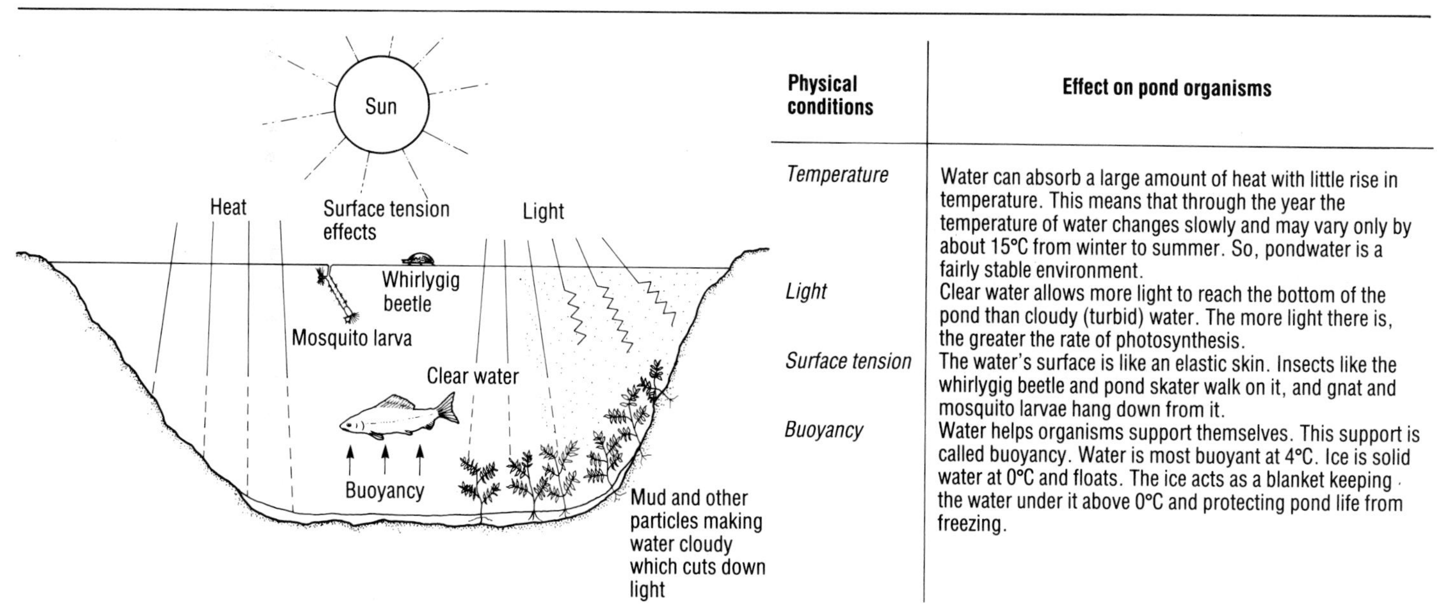

Physical conditions	Effect on pond organisms
Temperature	Water can absorb a large amount of heat with little rise in temperature. This means that through the year the temperature of water changes slowly and may vary only by about 15°C from winter to summer. So, pondwater is a fairly stable environment.
Light	Clear water allows more light to reach the bottom of the pond than cloudy (turbid) water. The more light there is, the greater the rate of photosynthesis.
Surface tension	The water's surface is like an elastic skin. Insects like the whirlygig beetle and pond skater walk on it, and gnat and mosquito larvae hang down from it.
Buoyancy	Water helps organisms support themselves. This support is called buoyancy. Water is most buoyant at 4°C. Ice is solid water at 0°C and floats. The ice acts as a blanket keeping the water under it above 0°C and protecting pond life from freezing.

Figure A.10(a) Physical conditions in a pond

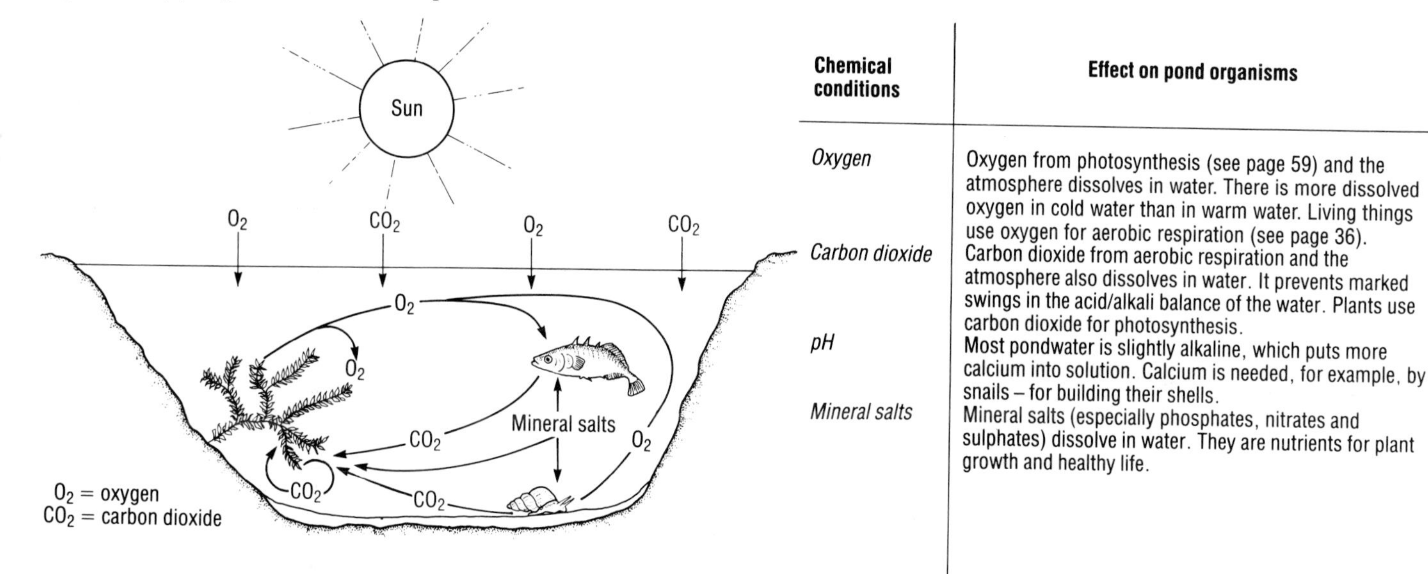

Chemical conditions	Effect on pond organisms
Oxygen	Oxygen from photosynthesis (see page 59) and the atmosphere dissolves in water. There is more dissolved oxygen in cold water than in warm water. Living things use oxygen for aerobic respiration (see page 36).
Carbon dioxide	Carbon dioxide from aerobic respiration and the atmosphere also dissolves in water. It prevents marked swings in the acid/alkali balance of the water. Plants use carbon dioxide for photosynthesis.
pH	Most pondwater is slightly alkaline, which puts more calcium into solution. Calcium is needed, for example, by snails – for building their shells.
Mineral salts	Mineral salts (especially phosphates, nitrates and sulphates) dissolve in water. They are nutrients for plant growth and healthy life.

Figure A.10(b) Chemical conditions in a pond

The physical and chemical conditions at the surface and bottom of a pond are different. Over 24 hours, conditions change at the surface but remain more-or-less the same (except for light) at the bottom. At night-time, for example, the surface temperature drops but the temperature at the bottom of the pond remains fairly steady. So at night, water at the bottom of the pond is warmer than water at the top. This switch in temperature is called a **temperature inversion**. At night-time, too, there is less oxygen in the water, particularly at the surface, because plants no longer release oxygen from photosynthesis but aerobic respiration, which uses oxygen, continues. Figure A.11 on page 111 shows these changes.

Figure A.12 shows annual changes in the amount of mineral salts in solution. In autumn and winter decomposers break down the dead remains of plants and animals (detritus) lining the bottom of the pond. This releases minerals and gases. In spring and early summer, longer daylight hours bring a burst of plant growth which uses the mineral salts built up over the winter months.

Activity A.1

The oxygen/carbon dioxide cycle in water

Water contains oxygen and carbon dioxide in solution. The gases come from the atmosphere and organisms living in water. Plants use carbon dioxide for photosynthesis. This produces the oxygen needed for aerobic respiration. Aerobic respiration produces the carbon dioxide for photosynthesis. The balance between oxygen and carbon dioxide in solution (in a pond for example) depends, therefore, on the balance between photosynthesis and aerobic respiration. This experiment uses a bicarbonate indicator to show the effect of plants and animals on the oxygen/carbon dioxide balance in water. The indicator is sensitive to changes in the level of carbon dioxide in solution and changes colour.

You will need:

- 5 boiling tubes
- boiling tube rack
- 2 pieces of pond weed
- 4 small fresh water snails
- bicarbonate indicator solution
- silver foil
- clock
- drinking straw
- small measuring cylinder

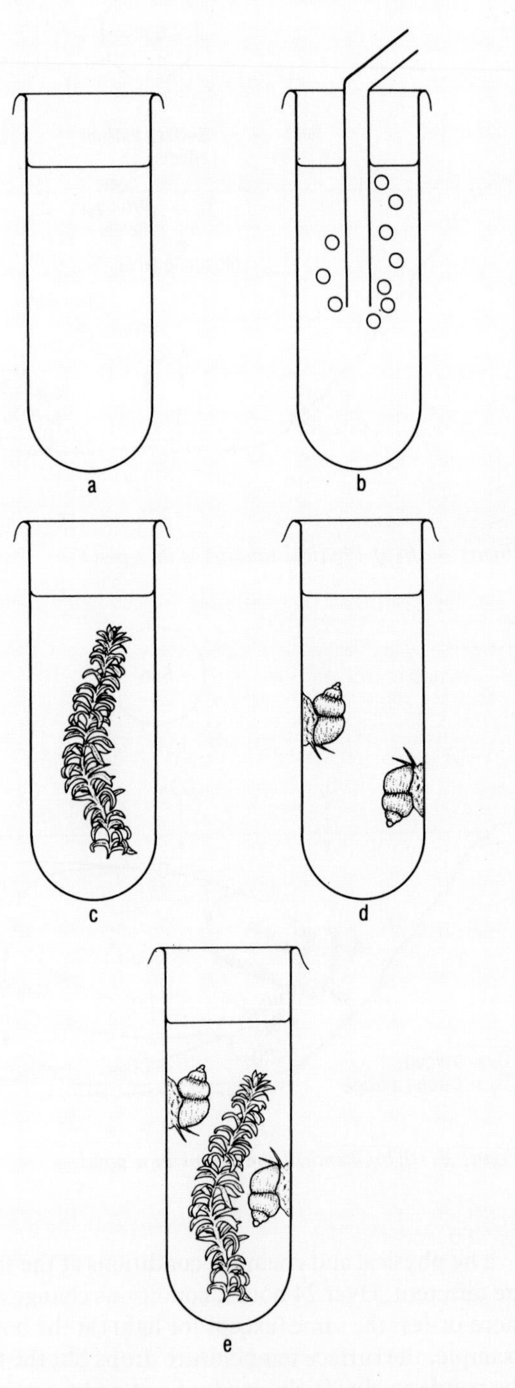

1 Copy out Table A.2 and record your results as you do the experiments.
2 Label the boiling tubes a – e.
3 Set up the boiling tubes in the rack as follows:
   a 25cm³ bicarbonate indicator solution
   b 25cm³ bicarbonate indicator solution, breathed into through a drinking straw
   c 25cm³ bicarbonate indicator solution with a piece of pond weed
   d 25cm³ bicarbonate indicator solution with two small fresh water snails
   e 25cm³ bicarbonate indicator solution with a piece of pond weed and two small fresh water snails
4 Cap the boiling tubes with aluminium foil.
5 Note the time (or start stop watch).
6 Watch the boiling tubes for changes in the colour of bicarbonate indicator solution.
7 Note how long it takes for any change in colour to begin and enter the time taken in column A of Table A.2.
8 Note how long it takes for the change in colour to fully develop and enter the time taken in column B of Table A.2.
9 If there is no colour change in the bicarbonate indicator solution then enter 'No colour change' in column C of Table A.2.

Tube	A Time taken for colour change to begin	B Time taken for colour change to fully develop	C Other comments
a			
b			
c			
d			
e			

Table A.2

Questions

a When you breathed through a straw, carbon dioxide bubbled into the bicarbonate indicator solution. What colour change took place?
b Why are the boiling tubes covered with aluminium foil?
c What is the reason for setting up tubes a and b?
d Arrange the results in columns A and B of Table A.2 as a bar chart.
e Briefly describe the effects of pond weed and snails on the oxygen/carbon dioxide balance in water.
f How well do you think the experiment imitates what happens in a pond?

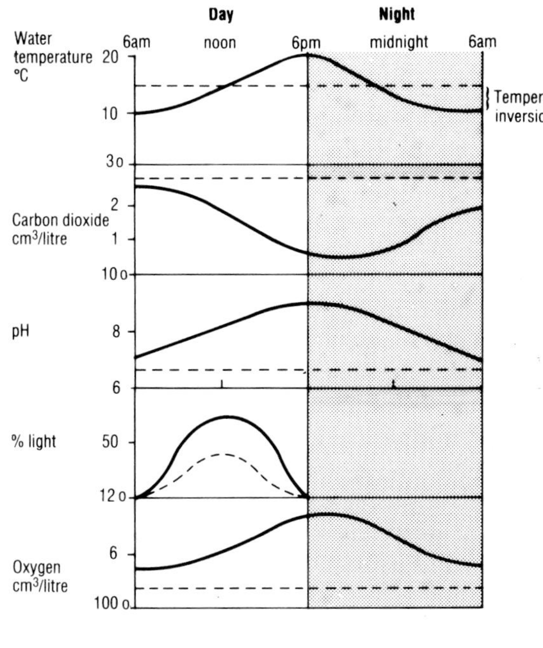

Figure A.11 Changes in some physical and chemical conditions at the surface and bottom of a pond

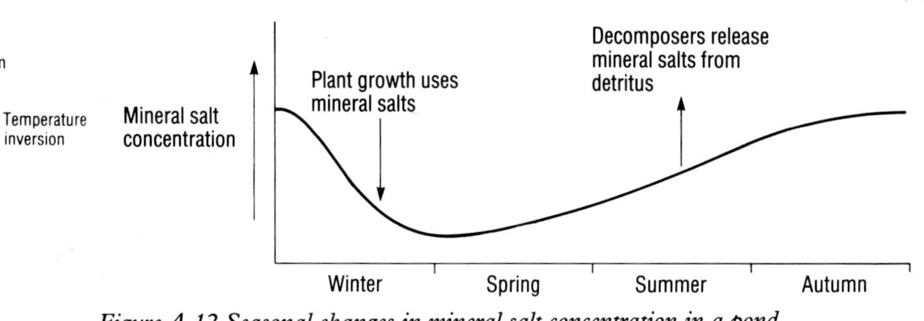

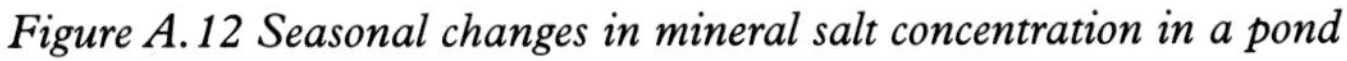

Figure A.12 Seasonal changes in mineral salt concentration in a pond

Summary

- Light, rain and temperature are important factors in the environment.
- The atmosphere helps to create an environment suitable for life.
- The physical and chemical conditions of water affect where pond organisms live.

Questions

1. Briefly discuss the effects of different physical conditions on the organisms living in a pond.
2. Why is the amount of light an important influence on life inside a wood?
3. Why can acacia plants survive the dry season of the savanna?
4. Study Figure A.5 (page 106). Describe and explain how the vegetation of West Africa changes along the transect A–B.
5. Name the layers of the atmosphere.
6. Why can whirlygig beetles walk on water?
7. Describe the seasonal changes in mineral sált concentration in a pond. Why do these occur?

Unit B
Competition and interaction

Figure B.1 National Water Sports Centre, Nottingham. Rowing eights battling it out in the finals of the National Championships

Figure B.2 Savanna in the dry season. Water is in short supply and the watering hole is a focus for animals living in the parched environment

Look at Figure B.1. All the competitors in the race want to gain first place, but only one of the crews will achieve it.

In nature, competition means much the same thing: competitors are rivals for something in short supply. In this case, the 'competitors' are organisms of the community and the prize is a resource like water, light, space, food or mates (Figure B.2).

Competition between individuals of the same species helps control **population size**. For example, cacti are widely spaced. They look as if they have been deliberately planted out. The pattern appears because of the shortage of water. Although many tiny cactus seedlings sprout in a particular area there is only enough water for one of them to grow into a mature plant. Growing cacti are the competitors and water is the resource in short supply.

Figure B.3 shows beech seedlings, saplings and trees in Epping Forest (on the border of north-east London and the county of Essex). Hundreds of seedlings sprout from thousands of seeds. Some grow into saplings but only a few grow to maturity. Figure B.4 shows why.

The branches of a full-grown beech tree spread about 12 metres in all directions. They touch the branches of neighbouring trees, forming a continuous layer. This canopy deprives the plants underneath of vital sunlight. When a tree is blown over or falls down, light floods through the gap in the canopy and stimulates vigorous plant growth on the forest floor. This sunlit clearing becomes an arena for intense competition between beech seedlings, and then saplings, in a slow but relentless race. Many competitors start out but there is limited space for the spreading branches; slower growing rivals may be overshadowed and die. The sapling that grows fastest will eventually fill the gap in the canopy, cutting off the sunlight that signalled the start of the race many years previously.

Competition between different species is usually greatest among individuals at the same trophic level. If both species are competing for the same resource, one may replace the other. For example, in an experiment two species of clover were grown. When they were planted separately, both grew well. However, when they were grown together, one species eventually replaced the other. The reason was that the successful species grew slightly taller than its competitor and overshadowed it.

In Scotland two species of barnacle are commonly found on rocky shores (Figure B.5). The species called *Chthalamus* is found higher up the shore than the other species called *Balanus*. The two species form zones which stand out clearly as Figure B.5 shows. Normally *Balanus* grows faster and crowds out *Chthalamus* from the lower zone. In an experiment, when *Balanus* was cleared from an area of rock, *Chthalamus* settled and flourished. However, when an area of *Chthalamus* was cleared, *Balanus* could not live in the upper zone and did not move up shore to take its place. So *Chthalamus* is driven from the lower zone through competition with *Balanus* but lives higher up the shore where *Balanus* cannot survive.

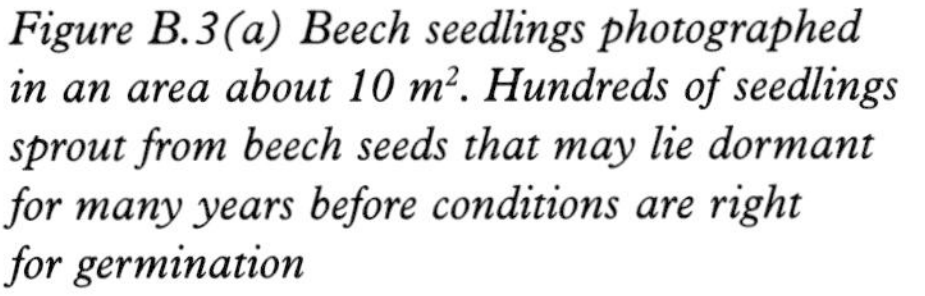

Figure B.3(a) Beech seedlings photographed in an area about 10 m². Hundreds of seedlings sprout from beech seeds that may lie dormant for many years before conditions are right for germination

Figure B.3(b) Beech saplings: about 20 years of growth produces a thicket of saplings

Figure B.3(c) Beech trees: a tree takes about 100 years to grow to full height. The photograph shows that very few saplings (see b) achieve this. Most perish, overshadowed by one or two that grow the fastest

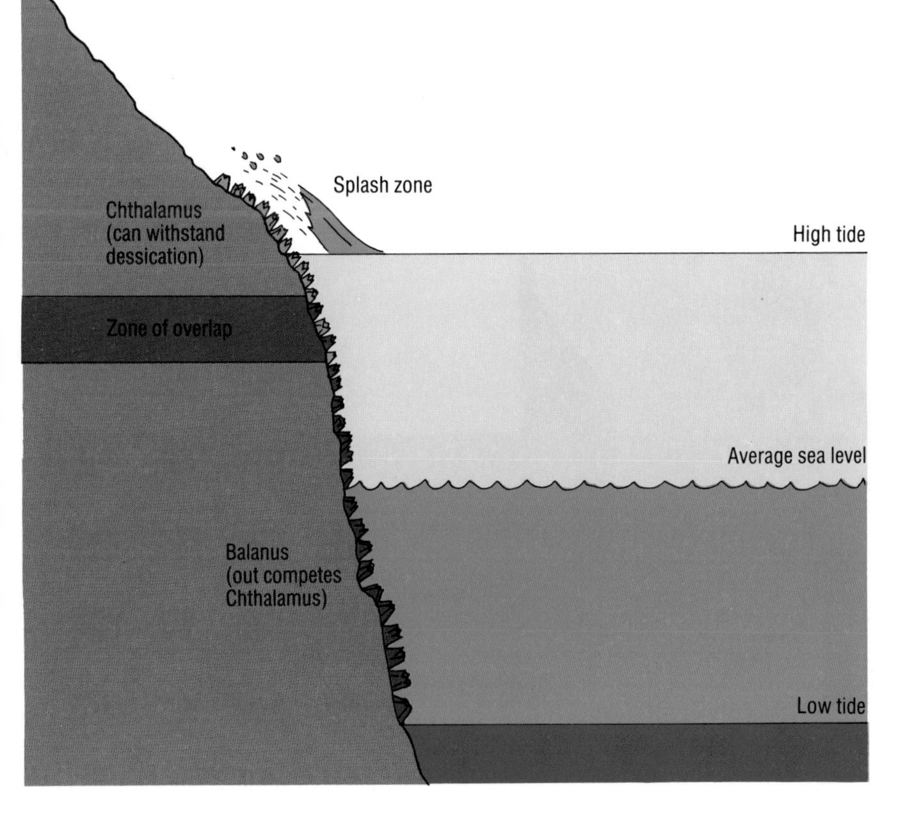

Figure B.5 Chthalamus *and* Balanus *occupy different zones on the seashore*

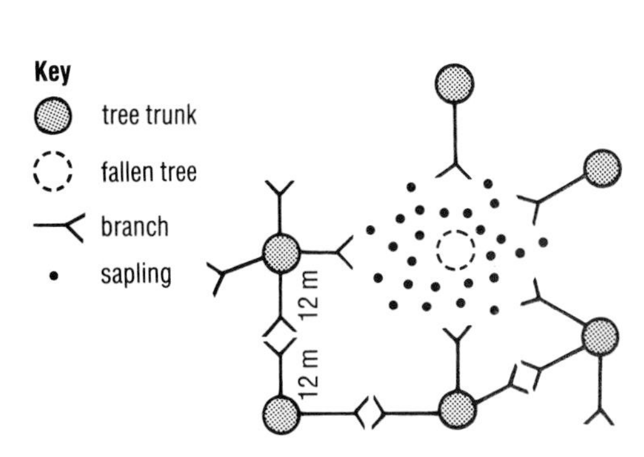

Figure B.4 Which saplings will survive to fill the gap? Length of branches decides distance between neighbouring trees. This produces a continuous canopy and a regular pattern of tree trunks. The sunlit clearing caused by the fallen tree breaks the pattern and provides opportunities for new plant growth. The saplings compete to complete the canopy once more

The removal of one species through competition means that only one species occupies a given niche at any particular time. In the case of the clover experiment one species was removed completely; with the barnacles, *Chthalamus* was forced into another niche higher up the shore.

Because niches are separate from each other the available space, food, water etc can be shared between different species. This means that limited resources will not be over-exploited (Figure B.6).

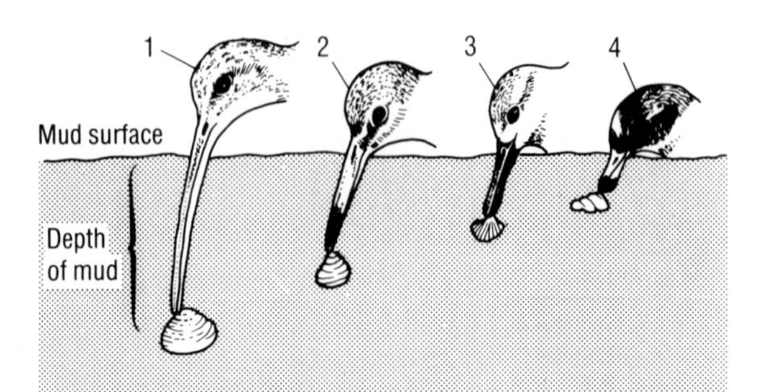

Figure B.6 Seashore birds probe for shellfish living at different depths in the mud. Length of beak allows each bird species to exploit food in a particular niche. Niches do not overlap, so resources are shared and the seashore community can support a rich variety of birdlife

So, competitive interactions maintain the community as a whole. 'Interaction' means the way in which a species affects another. Competition is an example, as we have seen. Other forms of interactions are shown in Figure B.7.

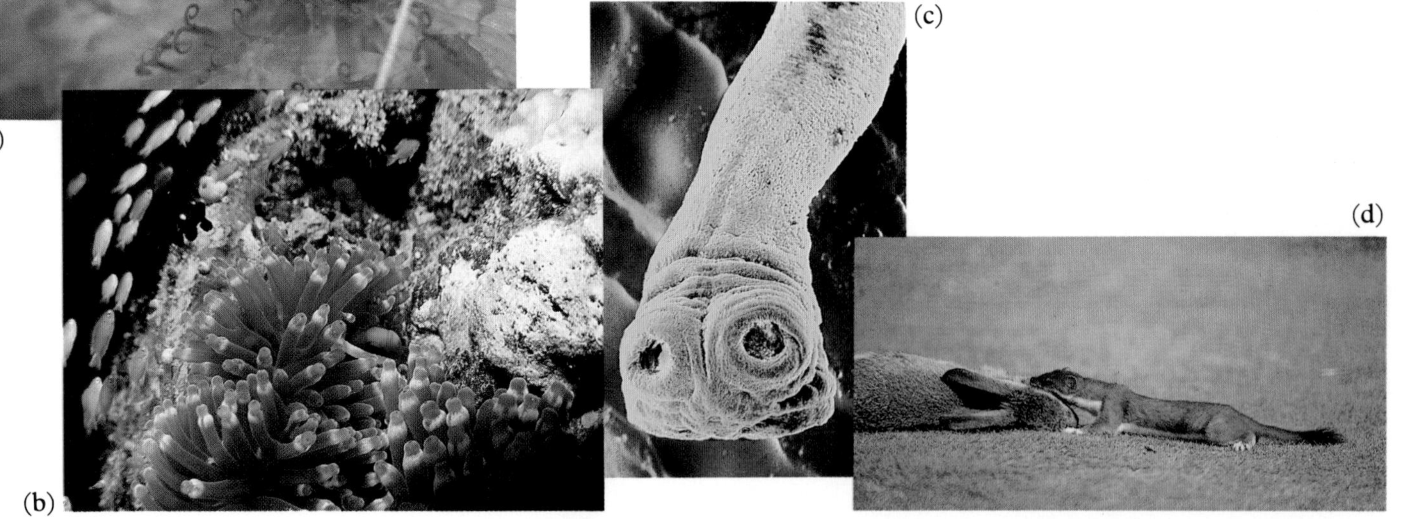

Figure B.7
(a) MUTUALISM – two or more species benefit from their close relationship. Pollination in flowering plants often depends on insects. In return, the flower's visitors benefit from the sweet-tasting nectar stored in the nectaries.
(b) COMMENSALISM – an association between one species that benefits (the commensal) and another which is unaffected either way. The fish is immune to the sting cells on the tentacles of the sea anemone and gains protection from predators. The sea anemone is unaffected by the fish.
(c) PARASITISM – an association between one species that benefits (parasite) and another which is harmed (host). The parasite is usually much smaller than the host. Beef tapeworms live in the human intestines surrounded by digested food. They absorb food through the body wall. The host is deprived of food and the intestine may become blocked. The tapeworm's wastes cause illness.
(d) PREDATION – one animal (the predator) kills other animals (the prey) for food. Rabbits, voles and other animals are the prey of the stoat, which is a predator.

Predators are adapted to catch prey and prey are adapted to escape predators. For example, the cheetah's speed, combined with its teeth and claws, help it bring down an antelope. However, if the antelope can avoid the first attack it may gallop off to a safer place because the cheetah soon tires (Figure B.8).

Look at Figures B.7 and B.8 again. Interaction between predator and prey is intense. Mutualism, commensalism and parasitism are long-term associations between species. Predation is instant and fatal!

Figure B.8 The cheetah is the fastest runner on Earth. Its thin, long body and flexible spine, which lets its front and hind legs overlap for full stretch, allows bursts of over 110 kph. The muscular effort involved means that it cannot sustain this speed for more than a few hundred metres. The antelope, though not so fast, dodges the cheetah's first rush. Its long, lever-like legs then let it run off to safety. Antelopes in poor physical condition are more likely to be killed

Summary

- Organisms compete for resources in the community.
- Competition ensures that resources are not over-exploited.
- Mutualism, commensalism, parasitism and predation are other ways in which organisms affect one another.

Questions

1. What is competition?
2. Briefly explain why a given niche is occupied by only one species.
3. How is a cheetah adapted to catch an antelope? How is the antelope adapted to escape the cheetah?

Unit C
Populations

A population is a group of organisms of the same species living in a particular place at the same time. Its size can be affected by factors discussed on pages 112–115: availability of food, competition, predation, parasitism and disease all affect the survival chances of individuals, and therefore the size of a population. The size of a population can change through births (which add individuals to a population) and deaths (which take individuals from a population). It can also change when individuals move from one population to another. Movement out of the population is called **emigration**, while movement into the population is called **immigration**. When the growth rate of a population, due to births and immigration, is equal to the rate at which it is declining, due to deaths and emigration, the size of the population is **stable**.

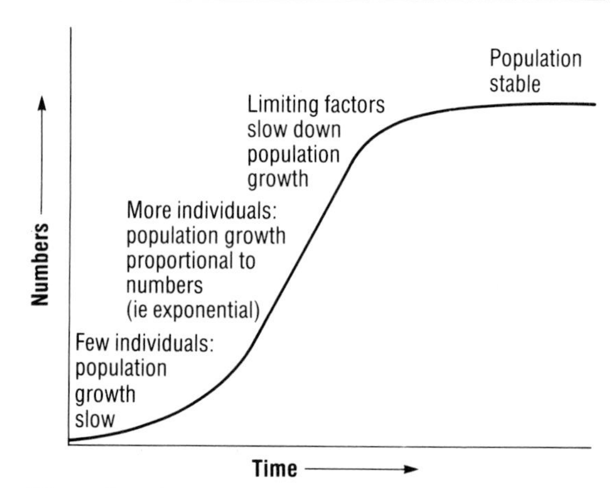

Figure C.1 Population growth curve

Population growth

To find out how populations increase in size scientists study the population growth of bacteria, fungi, plants and animals both in the laboratory and in their natural surroundings. The results of these studies show that populations grow in a particular way (Figure C.1). At first the population increase is small and the **rate of increase** (number of individuals added to the population in a given time) is slow. However, as the population grows so does the rate of increase in numbers and the curve of the graph becomes steeper. At this stage the rate of increase in the population is proportional to the number of individuals present and can be explained by a simple sequence of numbers:

2 4 8 16 32 64 128 256 512 1024 ... and so on.

Notice that the next number in sequence is double the previous number. If a population increases in this way it means that each generation is double the size of the previous generation. Scientists call this type of increase **exponential growth**.

Exponential growth does not go on for ever. Certain **limiting factors** stop it. These limiting factors include shortages of food, water, oxygen, light and shelter; build up of poisonous wastes; diseases; predators; and social factors. One or more of these factors will affect the rate of population growth so that eventually it levels off and numbers become stable, varying in a fairly regular way around an average. This is shown in Figure C.2 which shows the relationship between numbers of the snowshoe hare and numbers of its predator the Canadian lynx.

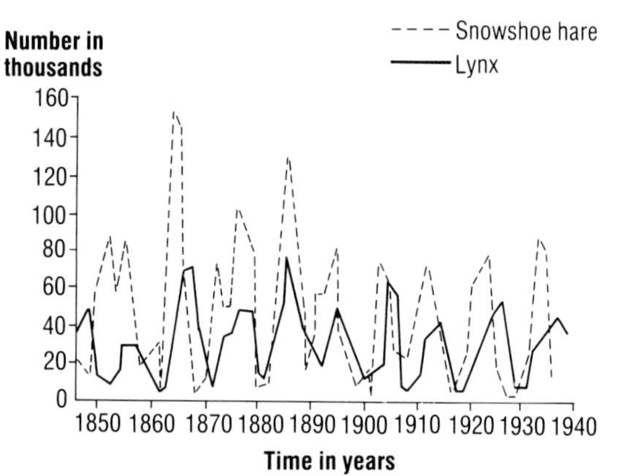

Figure C.2 Long-term changes in numbers of the snowshoe hare and Canadian lynx. Counts each year of skins brought to the Hudson Bay Fur Company of Canada by hunters give the figures for the graph. The highs and lows in the numbers of lynx lag behind the highs and lows of the snowshoe hare because when prey is scarce predator numbers fall, and when prey builds up predator numbers follow, with more food being available

Sometimes populations suddenly decrease. They do not level off and become stable. This crash may happen because the population has used up all its food or has been overcome by disease or predators (Figure C.3). Such populations are often only temporary or seasonal. For example, algae in a pond sometimes rapidly increase in numbers in spring (called a bloom) but decrease during the summer as the mineral salts needed for growth are used up.

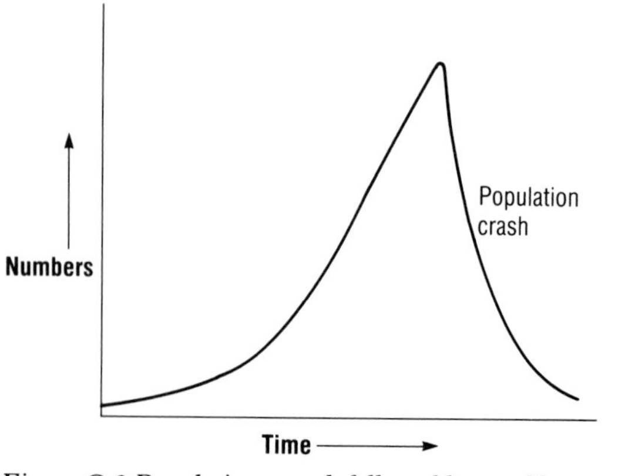

Figure C.3 Population growth followed by a sudden crash

Investigation C.1

Population changes in the grain weevil

This activity tests your ability to put ideas about changes in population size into a statement and then plan experiments which test the statement. Read through the whole investigation and then write down your statement of ideas before planning the experiments. What is the best way to test your ideas? Here are some hints:

- The grain weevil (*Sitophilous* spp.) is a type of beetle. It lives in stored grain. Grain seeds are its food and female weevils lay their eggs in the grain. We can say that stored grain is the weevil's ecosystem in which it lives and dies. The grain weevil is a serious pest of stored grain in many countries of the world.
- The grain weevil and its ecosystem make a useful laboratory model for testing ideas on the way populations change in size. List the advantages of working in the laboratory with a model ecosystem rather than in the natural environment: what are the disadvantages?

Describe how you conducted your experiments.

1 What conclusions can you arrive at from the results about changes in size of grain weevil populations?
2 Do you think the results imitate what happens in the natural environment? Briefly give reasons for your answer.
3 Do you think studies like this one help us find ways of controlling populations of insect pests?

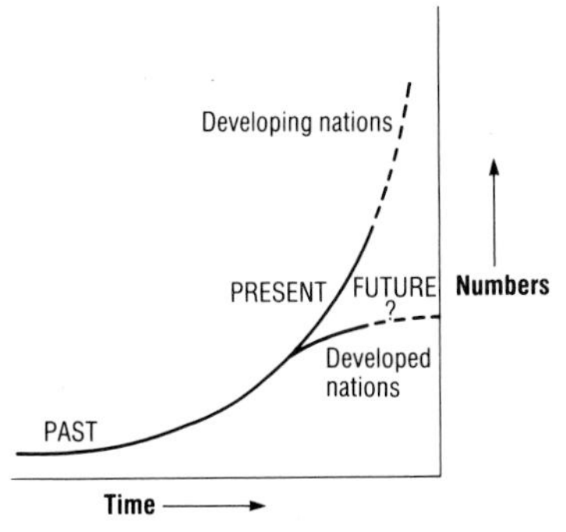

Figure C.4 Population growth in developed and developing nations. The dotted lines show possible future growth based on present trends

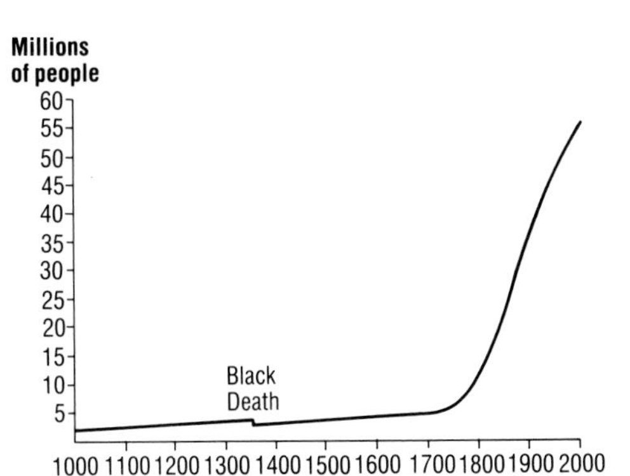

Figure C.5 Population growth in Britain from AD 1000 to present, showing a possible levelling-off in the near future. The dip in population in the mid-1300s was caused by the Black Death (bubonic plague)

The human population

For thousands of years the rate of growth of the world's human population was relatively slow. During the last 300 years it has increased dramatically. In the 19th century and the first part of the 20th century, the human populations of Europe and North America (the **developed nations**) grew exponentially. Now populations in these countries are levelling off. Population growth in Latin America, Asia and Africa (the **developing nations**) is still exponential and the worldwide trend is upwards (Figure C.4).

The reasons for world population growth are much the same as those that caused the population increase in Britain after the Agricultural and Industrial Revolutions of the 1700s.

1 Food production improved
2 There were more jobs (lack of work had limited the size of families)
3 Medicine made rapid advances (vaccination, new drugs to combat disease, aseptic surgery)
4 Public health improved (supplies of clean water and better disposal of sewage).

So, the balance between births and deaths was destroyed and population rapidly increased as Figure C.5 shows. Similar reasons are now causing the exponential growth of human populations in other countries, especially reduction in the death rate of children because of worldwide vaccination programmes against infectious diseases like measles and polio. At the present rate of growth the world's population is expected to reach around 6000 million by about the year 2000 (Figure C.6). This means that between now and then there will be a world increase in population of about 61 million people each year. Put another way, all those killed in the Second World War (estimated at 50 million) are replaced within 10 months at the present rate of increase.

Activity C.2

Estimating the size of a snail population

Snails are useful subjects for studying the size of a population. Insects and mini-beasts, as a naturalist once said, 'are here today and gone the next second,' but snails have the advantage of slow speed.

Snail populations alter fairly slowly and so rapid changes in population size over a short period of study are unlikely.

As well as finding the size of a snail population, this exercise tests your ability to follow instructions, present results and use simple mathematics to interpret them and make conclusions.

Snails are usually found in damp, overgrown places such as a hedgebank or rockery. They are more likely to be found when it is warm and damp. So, carefully choosing where and when you study a snail population will improve the chances of success.

Find a large group of a common species like the garden snail.

For this activity you will need:

- a small pot of bright (white or yellow) quick-drying emulsion paint (NOT oil-based which may damage the snail)
- a thin paint brush (or a twig will do)

1 Mark a number of snails (not less than 40) with a small dab of paint, returning each one to the spot where you found it.
2 Leave them for a day or so before returning to collect about the same number that you previously marked with paint. Make sure that you do *not* specially look out for marked snails.
3 Note the number of marked snails in your collection and present your results in a suitable way.
Calculate the size of the population using the following simple equation:

$$\text{POPULATION SIZE} = \frac{\text{number marked first time} \times \text{number found second time}}{\text{marked number found second time}}$$

Substitute your results in the equation and find the size of your snail population.
This method is called the Capture/Recapture method or the Lincoln Index (named after the American who worked it out).
4 An assumption is made in the equation. Can you think what it is? Here is a clue: what is the relationship between the number of snails marked as a proportion of the total population and the number of marked snails found later on as a proportion of the total collected? Do you think the assumption is a reasonable one?
5 Write up your results as a short report on methods of calculating population size.

Predicted numbers { 2000, 1990
Known numbers { 1975, 1950, 1900
World population
Number of people (× thousand million): 1, 2, 3, 4, 5, 6
Year AD: 0, 400, 800, 1200, 1600, 2000

Figure C.6 World population growth year 0 – AD 2000, based on present trends

Summary

- Populations change and grow in particular ways.
- Different factors limit growth and control eventual size.
- The effects of limiting factors such as disease and food availability on the human population have been reduced by improvements in medicine and agriculture.

Questions

1 What is a population?
2 Look at Figure C.2. Why do you think changes in the numbers of Canadian lynx are slightly behind changes in the numbers of snowshoe hare?
3 Give reasons for the rapid increase in human population over the last 400 years.

Unit D
Food chains and food webs

Figure D.1(a) The tiger's large pointed canine teeth grip and tear its struggling prey. Its carnassial teeth cut like scissor blades through flesh and bone

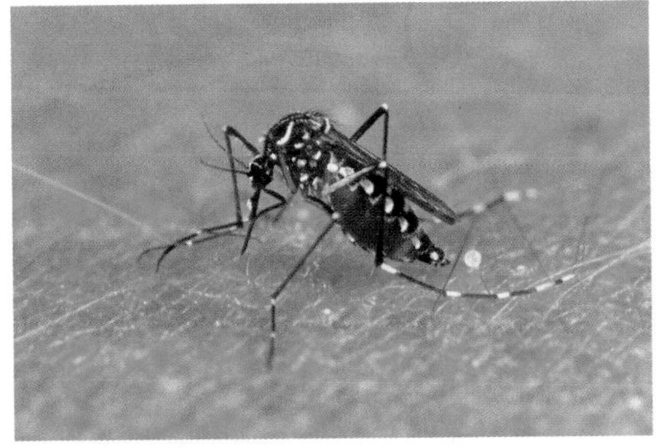

Figure D.1(b) Mosquito feeding; its long, hollow needle-like mouthparts push through the skin to pierce a blood vessel. Chemicals in the mosquito's saliva stop the blood from clotting as it is sucked up the mouthparts into the mosquito's mouth

One way of finding out how a community works is to look at the relationships between its different niches – in other words, find out 'who eats what or whom'. Looking at an animal's teeth or mouthparts helps tell us what it feeds on (see Figure D.1).

We can also
- watch animals feed;
- look at what is in the gut;
- carry out an experiment to find out what food an animal likes best; this is called a **food preference test**.

Food chains

Animals can be divided into three groups according to what they eat:

- *Carnivores* eat meat.
- *Herbivores* eat plants.
- *Omnivores* eat both meat and plants (human beings are omnivorous).

Carnivores are **predators**: they catch and eat other animals. The animals are known as their **prey** and are often herbivores.

Some animals are **scavengers**. They feed on the remains of prey left by predators, or on the bodies of animals that have died for other reasons. The links between predator, prey and scavenger can be shown as a **food chain**. Most food chains have no more than four links. Figure D.2 shows examples of food chains in a pond and an oakwood.

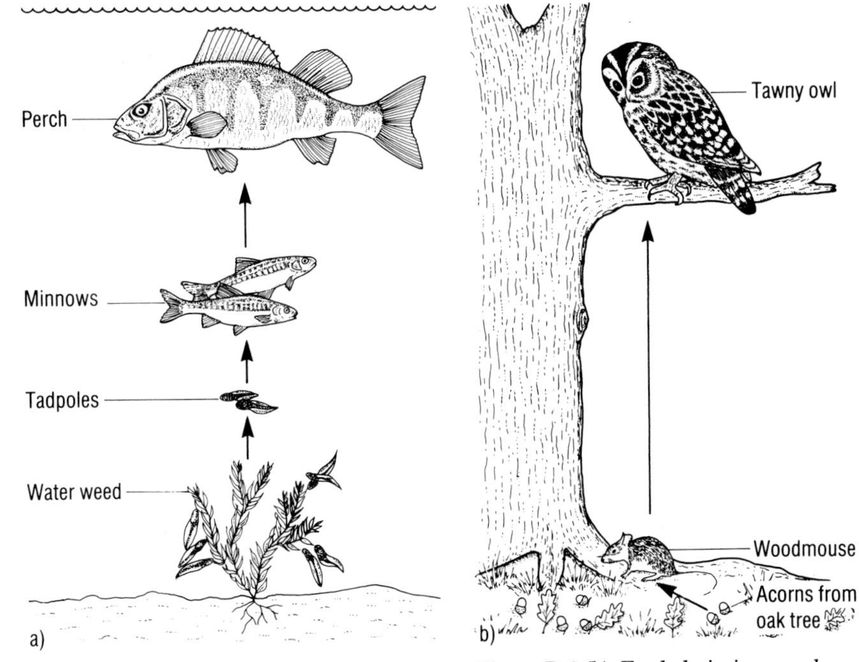

Figure D.2(a) Food chain in a pond *Figure D.2(b) Food chain in an oakwood*
Food chains are written with arrows pointing from the eaten to the eater

Another way of describing a food chain is as a pathway of 'food energy' through a community. Food provides the energy for the body's activity: growth, repair, movement. Foods rich in fats, oils and carbohydrates are a particularly good source of energy. You can find an experiment to measure the energy value of different sorts of food on page 10.

Food webs

Usually 'who eats what or whom' is not as simple as a food chain seems to suggest. Most animals eat more than one type of plant or other animal. **Food webs** are more accurate than food chains for describing food relationships in a community. Figure D.3 shows examples for a pond and an oakwood.

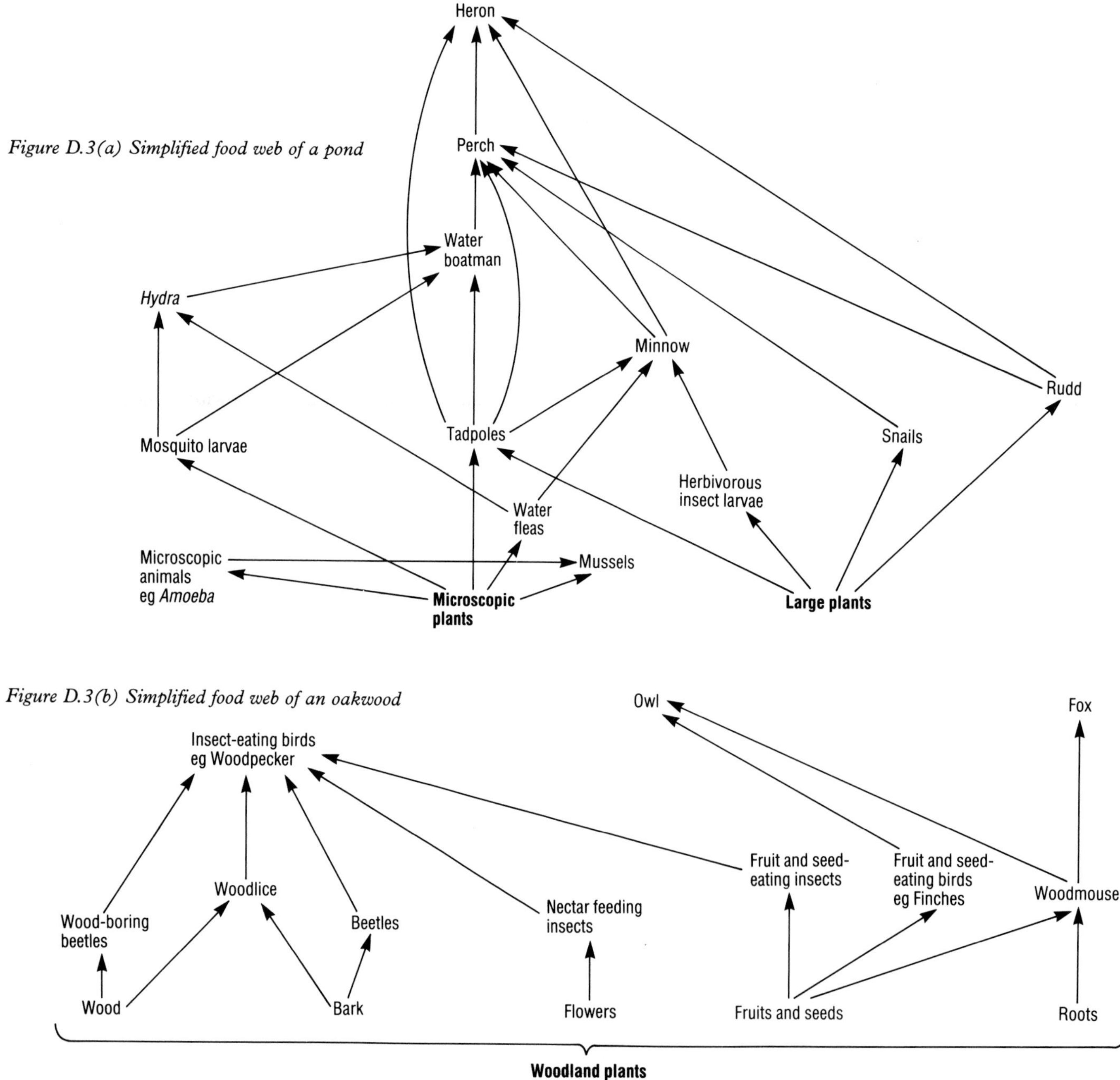

Figure D.3(a) Simplified food web of a pond

Figure D.3(b) Simplified food web of an oakwood

Photosynthesis

The Sun floods the Earth with heat and light energy. Some of the light energy is absorbed by chlorophyll (the green **pigment** in plant cells) and used in chemical reactions which convert carbon dioxide and water into sugars. These reactions are stages in the process of **photosynthesis**; they take place inside tiny structures called chloroplasts in plant cells.

Food chains and food webs always begin with plants (as Figures D.2 and D.3 show) because only plants can use sunlight to make food by photosynthesis. Animals use this food directly when they eat plants. Even when they eat other animals, predators depend on plant food indirectly, since somewhere along the line the prey has been a plant eater. Sunlight underpins life on Earth. Without sunlight – and photosynthesis which depends on it – living things would cease to exist.

Summary

- Feeding relationships between living things can be shown as food chains or food webs.
- Life on Earth depends ultimately on photosynthesis.

Questions

1 'Who eats whom' between predators and prey?
2 Why is a food web a more accurate description of feeding in a community than a food chain?
3 Why do food chains and food webs always begin with plants?

Unit E
Ecological pyramids

Another way of describing the feeding relationships in a community is by an ecological pyramid. Figure E.1 is a diagram of an ecological pyramid.

The pyramid has several feeding layers, called **trophic levels**. The first trophic level, at the base of the pyramid, consists of the plants of a community. They are called **producers** because they produce food by photosynthesis (see Unit D). The other trophic levels are made up of consumers, so-called because they feed on plants and/or on other consumers. Most consumers are animals. A few plants, such as the Venus fly trap (Figure E.2) which catches insects, are, in an ecological sense, consumers as well. Consumers in

- the second trophic level feed on plants and are called **herbivores**
- the third trophic level feed on herbivores and are called **first carnivores**
- the fourth trophic level feed on first carnivores and are called **second carnivores**

So each trophic level groups together organisms which have similar types of food. For example a snail and a cow are herbivores. Both belong to the second trophic level. Consumers which eat both plants and animals belong to more than one trophic level and are called omnivores.

The group of organisms in each trophic level is smaller than the one below it. This gives the pyramid its shape (see Figure E.1). So the number of trophic levels in a pyramid is limited and levels higher than second carnivores rarely exist because there is not enough food.

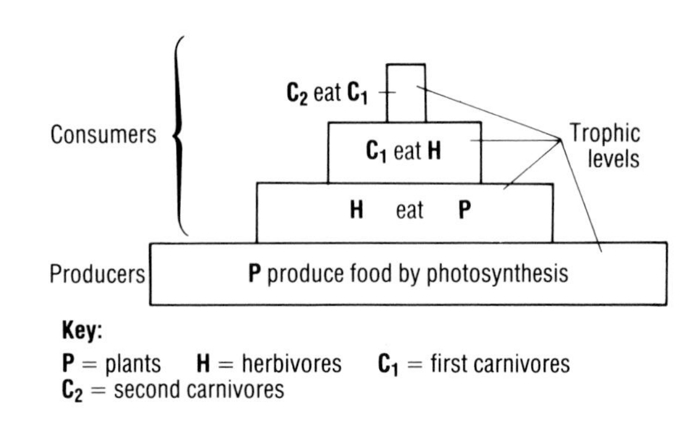

Figure E.1 Ecological pyramid

Figure E.2 The Venus fly trap lives in wet, boggy places where nutrients needed for growth are washed out of the soil. Insects provide the Venus fly trap with these scarce nutrients, especially nitrogen and phosphorus. The plant's leaves close around any insect that touches off the trigger 'hairs' in the middle of each leaf. Enzymes produced by the leaf digest the trapped insect

There are three types of ecological pyramid:

1. **Pyramid of numbers** which shows the number of organisms in each trophic level.
2. **Pyramid of biomass** which shows the amount of organic material (usually stated as dry weight) in each trophic level.
3. **Pyramid of energy** which shows the flow of energy through the community.

Figure E.3 shows how a pyramid of numbers is plotted. It is rather like a bar chart – but the bars are horizontal and stacked on top of each other.

Figure E.4 shows number pyramids for (a) grassland and (b) woodland communities. Notice the different shapes. In grassland (Figure E.4(a)) the producers (grasses) and consumers (mainly insects, spiders etc) are mostly small and numerous. A lot of plants support many herbivores, which in turn support fewer carnivores. Plotting the number of organisms in each trophic level gives us an upright pyramid, tapering to a point.

However, the pyramid for woodland (Figure E.4(b)) seems to point down as well as up. It shows that relatively few producers (trees) support a large community of herbivores and carnivores. You might think that woodland consumers are in danger of starvation! Of course this does not happen. Most woodland communities can exist for a long time.

The reason for the pyramid pointing down as well as up is that each tree is large and can meet the food needs of many different organisms. It

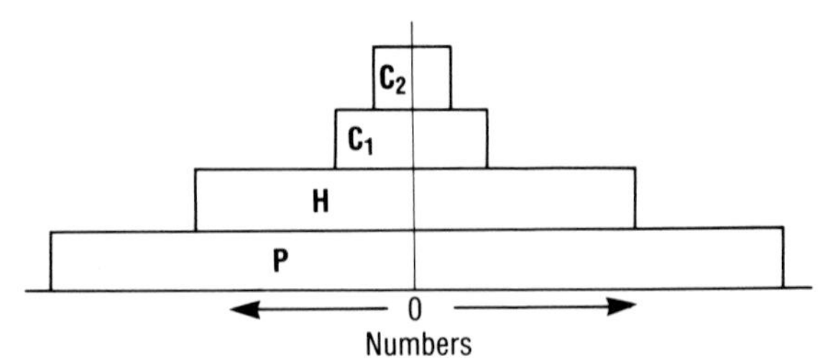

Figure E.3 How to plot a pyramid of numbers. Half the number of organisms in each trophic level is plotted on one side of the vertical line, the other half is plotted on the other side of the vertical line (letters are as for Figure E.1)

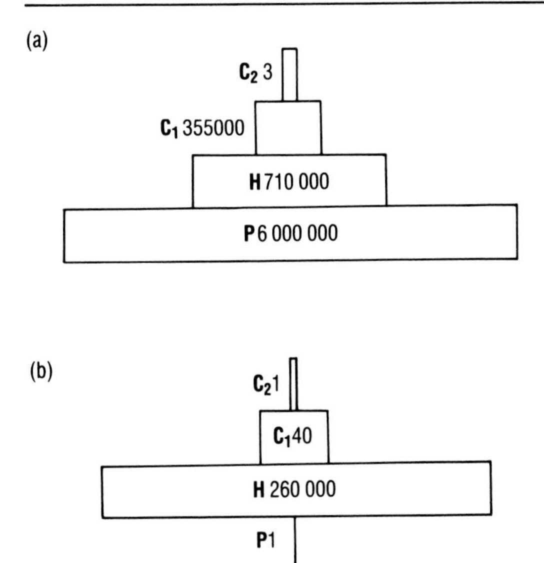

Figure E.4 Pyramids of numbers for (a) grassland and (b) woodland communities

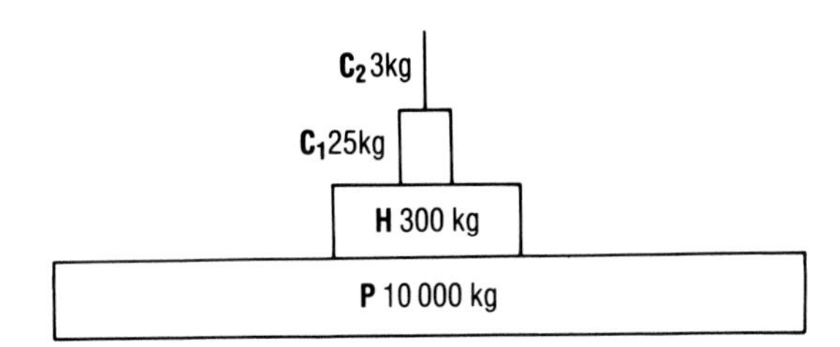

Figure E.5 Pyramid of biomass for woodland in kg/m²

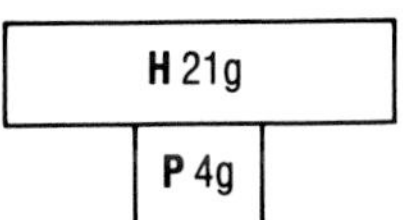

Figure E.6 Upside-down pyramid of biomass for the English Channel in g/m³. Microscopic algae (producers) only live for a few days but reproduce millions of offspring very quickly. Collecting them over a short period misses this rapid turnover of living material and results in a pyramid which suggests that the biomass of herbivores (H) is greater than that of the producers (P) they feed on

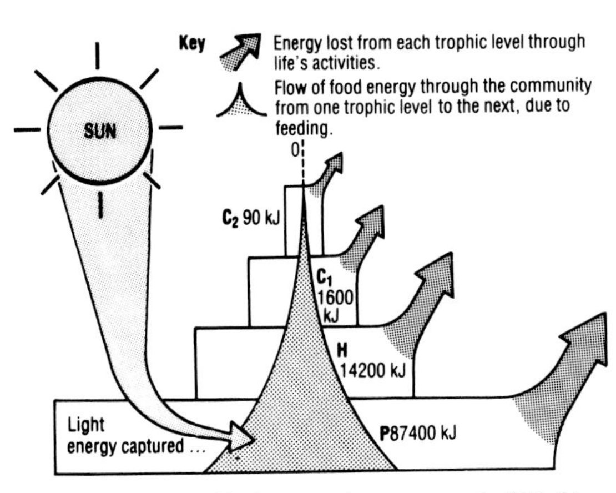

Figure E.7 Pyramid of energy for a stream in kJ/m²/year. Food energy in each trophic level is measured using a bomb calorimeter

takes tens of thousands of grass plants to equal one oak! The number pyramid therefore does not accurately describe the food chain relationships of the woodland community because it does not take into account differences in size of producers and consumers.

Investigation E.1

What factors affect the numbers of species in two different areas?

Explore a grassland community and a bush/shrub community. How you go about your exploration depends on your ideas. Here are some possibilities:

1. Most of the animals you find will be insects and spiders. Which are herbivores and which are carnivores? Will this help you place the animals into trophic levels? If not, plan another way of classifying animals into herbivores and carnivores. Give your reasons.
2. The grassland community may be a field, bank etc – far too large to count all the plants and animals present. Even a shrub or small tree will have too many animals for you to count. So you can only study a part of the community and assume that it is typical of the whole. The parts of the community studied are called samples; from them results for the whole community can be worked out. Plan how to do this.

Pyramid of biomass

Figure E.5 shows a biomass pyramid that measures the dry weight of organic material (**biomass**) in each trophic level of a woodland. Organisms collected from a community contain different amounts of water. They are dried out before weighing because their dry weight gives a more accurate reading of the amount of living material in each trophic level. When the total weight of individuals in each trophic level is plotted the pyramid is a more normal shape, giving a better description of the woodland community. Even so, biomass pyramids can also be upside down if producers are much smaller in size and change in numbers more quickly than the consumers they support as Figure E.6 explains.

Pyramid of energy

As we saw on page 121, the energy that flows through a community begins with sunlight, which is used by plants for photosynthesis. The food energy produced by photosynthesis at the first trophic level flows through the community because of the feeding activities of consumers. Food energy is therefore transferred from one level to the next, but the transfer is never 100% efficient. At each trophic level some energy is used to power life's activities – respiration, growth, movement, the production of heat and waste, the chemical activity of cells. This represents 'lost' energy which does not flow on to the next trophic level. So, the amount of energy decreases as it flows through the community.

Figure E.7 shows energy flow and energy loss. The pyramid tapers to a point. The number of links in the food chain is limited by the food energy available. As the amount of food energy becomes smaller, so does the amount of living material that can be supported in each trophic level. When food energy dwindles to nothing, trophic levels and links in the food chains can no longer exist.

The energy pyramid, therefore, gives the best picture of the relationships between producers and consumers. Whereas numbers and biomass record the organisms supported in each trophic level at any one time, the energy pyramid shows the amount of food being produced and consumed in a given time. Its shape therefore is not affected by differences in size or changes in numbers of individuals.

Keeping the body warm

The thousands of chemical reactions taking place inside cells (**metabolism**) release heat energy, which warms the animal's body. In particular, the high rate of metabolism of birds and mammals generates so much heat energy that their body temperatures remain at around 37–41 °C, even when the temperature of the environment changes. This is why birds and mammals are described as **warm-blooded**.

A high body temperature is especially important for birds, in which the energy demands of flying make a high metabolic rate an advantage. Most birds have a body temperature around 41 °C. Mammals do not depend on such a high metabolic rate and therefore have a body temperature around 37 °C.

Because the body temperature of birds and mammals is higher than the temperature of their surroundings, the loss of heat energy to the environment is particularly high. However, different mechanisms keep the temperature at the centre of the body (the **core temperature**) steady. Birds and mammals can therefore live in places where other animals would soon die because they cannot maintain a steady core temperature.

Summary

- Another way of describing feeding relationships between organisms is the ecological pyramid.
- Each trophic level of a pyramid groups together organisms that eat similar food.
- There are three types of ecological pyramid: pyramids of numbers, of biomass and of energy. The energy pyramid gives the best description of feeding relationships.

Questions

1 Why are plants called producers?
2 What are consumers?
3 What is meant by the word 'biomass'?
4 Why is the pyramid of biomass usually a better description of a community than the pyramid of numbers?

Extension questions

5 Why is the pyramid of energy the best description of all?
6 How does the description of the flow of energy through the community help us understand why there are a limited number of links in a food chain?

Unit F
On the farm

Food production

FOOD! We all need it; but how do farmers produce enough food for the world's growing population?

'Intensive farming' (see Figures F.1 and F.2) means that farmers try to produce as much food as possible from the land that is available for raising crops (arable farming) and animals (livestock farming).

Scientists and engineers have developed new technologies to help farmers in their work so that the land produces food in great quantities (Figure F.3).

Figure F.1 Intensive cereal production. The field of wheat has been grown with the help of chemicals that can kill pests, supply the soil with nutrients and strengthen the plants against wind and rain damage. The farmer balances the extra cost of buying the chemicals against the profits from an increased yield of grain. Most bread and pastry is made from cereals produced intensively

Figure F.2 Intensive egg production. Battery hens are kept in cages in artificial light. Their food is probably not grown on the farm, but bought in. The farmer balances the extra cost of feeding against the higher yield from battery hens compared with free-range poultry. Close confinement restricts movement and therefore reduces the heat energy lost through muscle contraction. Heat energy losses are further reduced by keeping the hens' environment at the temperature best suited for growth. Supplying heat is cheaper than supplying food!

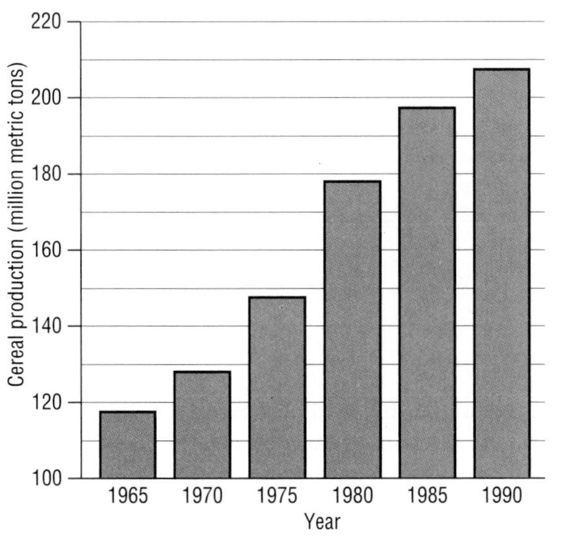

Figure F.3 Cereal production in Western Europe – cereals provide about half of people's food energy needs. Production has increased dramatically with yields of wheat increasing by three times over the period for which there is data (1965–1990) as land is farmed more and more intensively

Energy on the farm

Farms are environments with people as consumers in a food chain of crops and livestock. Farms are managed to produce as much food as possible but the amount of food produced depends on:

- the amount of energy entering the farm ecosystem
- the efficiency with which energy is converted into plant and animal tissue.

The energy entering the ecosystem is the 'input'. The food produced is the 'output'. Inputs and outputs have energy values. Figure F.4 shows the energy values of different inputs and outputs for a typical intensive farm of 460 hectrares in southern England. The farm is mostly arable but some livestock is also raised.

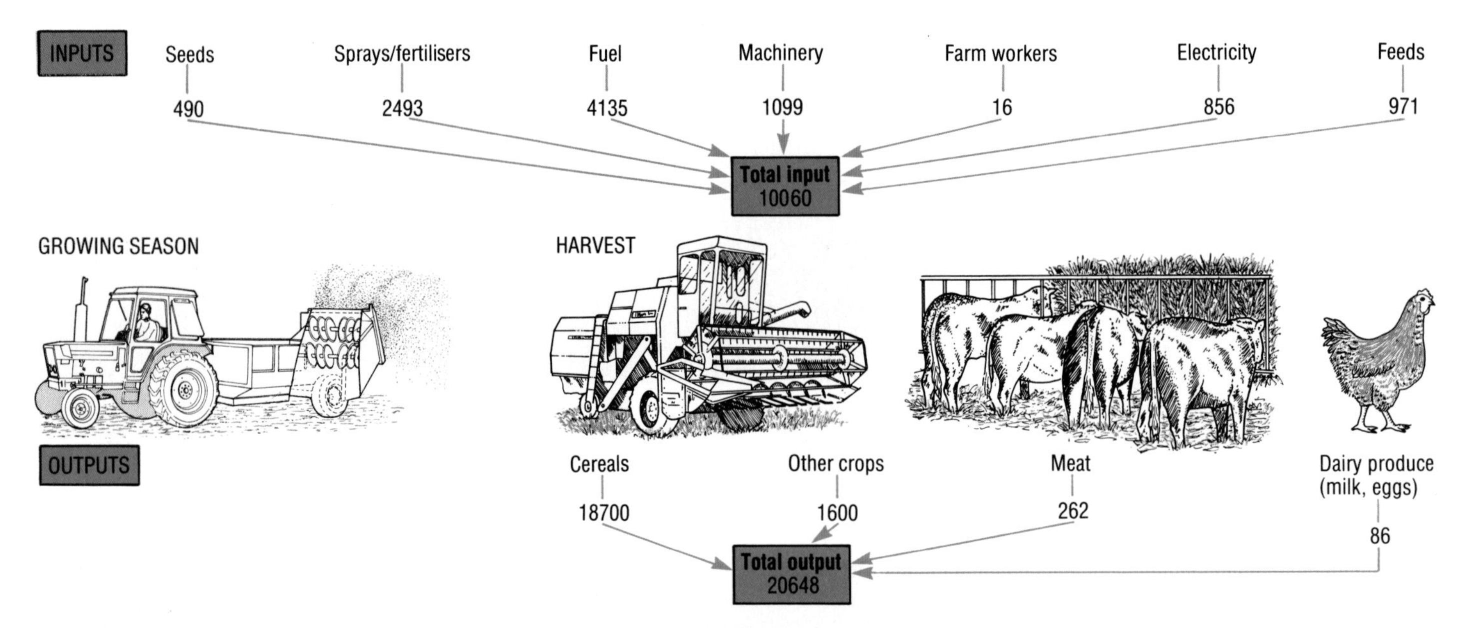

Figure F.4 Inputs and outputs on an intensive farm. Fuel accounts for nearly 99% of energy inputs; more than 40% directly as petrol and diesel fuel to power machinery, the rest indirectly because the manufacture of sprays, fertilisers, feeds etc and energy generation depend on fuel. Less than 0.2% energy input comes from manual work. The figures are in (MJ × 1000) where 1 MJ = 10^6 Joules (J)

How does today's energy picture on the farm compare with that of pre-intensive days? Figure F.5 shows the energy values of inputs and outputs in the 1820s on a farm of 460 hectares in southern England. The farmers grew mainly cereal crops but raised some livestock as well (the figures are worked out from historical records describing working hours and conditions on farms in the early 19th century).

The energy inputs illustrated in Figures F.4 and F.5 do not include the input from sunlight. Photosynthesis converts the energy of sunlight into the chemical energy of plant tissue. Without sunlight plants cannot grow. So, the amount of sunlight influences productivity through its effect on plant growth.

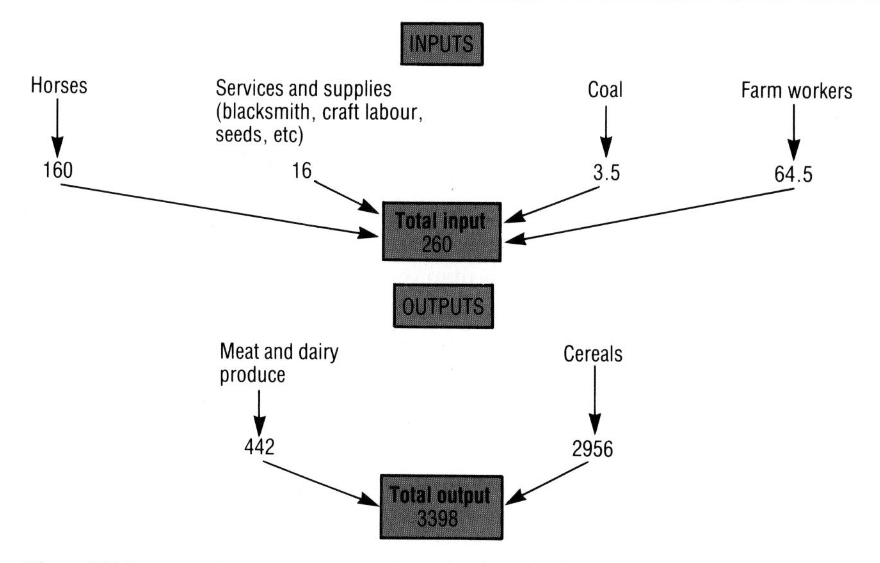

Figure F.5 Inputs and outputs on a pre-intensive farm in the 1820s. Energy values are worked out on the basis that a horse uses 8 MJ per hour at work; a farm worker uses 0.8 MJ per hour. More than 98% is done by horse and human muscle power. Fuel accounts for less than 2% of total energy inputs. Compare these figures with those for the intensive farm in Figure F.4. The figures are in (MJ × 1000)

We can probably assume that the amount of crop growth due to input from the sun is roughly the same today as it was in the 1820s. If so, then the difference between the non-sun energy inputs from the two farms accounts for the difference in output between them. We can therefore calculate the effect that intensive farming has had on food production (Table F.1).

Energy (MJ/year) ×1000	Intensive farm	Pre-intensive farm	Energy difference
Total non-sunlight input	10 044	260	9 800
Total output	20 648	3 398	17 250

Table F.1 Comparison of energy inputs and outputs between the intensive and pre-intensive farms illustrated in Figures F.4 and F.5

Today each farm worker produces 60 times more food than his equivalent of the 1820s. These levels of productivity supply the whole population of Britain with most of the food it needs. Indeed, if we ate less meat, butter and milk (as in World War II) then the country could EASILY become self-sufficient in food. How?

Put simply, massive inputs of oil underpin the success of intensive farming. Table F.1 shows that the six-fold increase in output has been won at the expense of a more than 40-fold increase in input – nearly all of it as oil. In other words, each year a farm worker uses an energy input equivalent to more than 11 tonnes of oil to achieve the level of personal productivity just mentioned. This puts the energy demands of agriculture in the same league as heavy industries such as steel production and ship building!

Fossil fuel is a non-renewable resource. As demand increases so reserves decline, creating an energy gap. This is as much a crisis for intensive farming as it is for other industries.

Activity F.1

Input from the sun

1 Use the following data to calculate the energy value of the sun's input for a farm of 460 hectares.

Energy received at ground level	= 0.35 MJ/cm²/year
100 million cm²	= 1 hectare
Size of farm	= 460 hectares
Total amount of energy received each year at ground level on the farm	= ?

2 Use this answer with the following data to calculate the amount of light energy converted into plant tissue. Check your calculations.

Efficiency with which photosynthesis converts light energy into chemical energy	= 0.8%
Length of time that plants cover soil each year	= 6 months
Average amount of soil covered by plants	= 50%
Total amount of sunlight energy converted into plant tissue each year on the farm	= ?

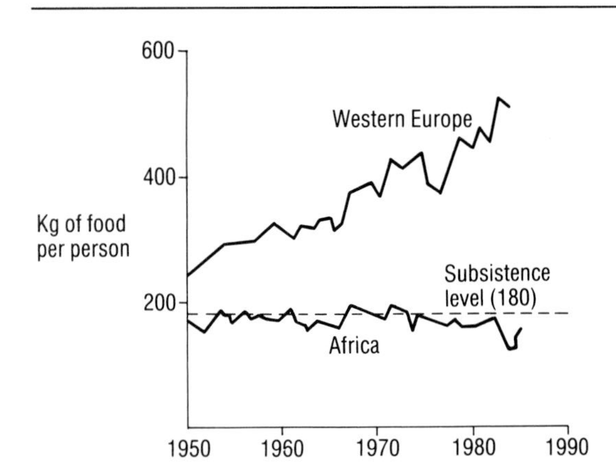

Figure F.6 Food production in Western Europe compared with Africa. Europe's population is almost stable and food production per person has risen sharply. However, Africa's population is increasing faster than the rate of food production, and millions of people eat less food than the minimum needed to stay healthy (subsistence level). Notice that the gap in food production between Western Europe and Africa has increased since 1950. Can you think of reasons why this has happened?

The food problem

Farmers in Britain produce more food than we need, yet they continue to grow more and more, encouraged by the high prices governments guarantee for their produce. Crops, meat and dairy products end up in store creating very expensive grain, beef and butter 'mountains'. Food production in Western Europe as a whole has steadily increased since 1950. Also, with so much food available we eat more than we need yet in poorer countries millions of people do not have enough food (Figure F.6).

Why don't we send our surplus food supplies to countries that need it? This might seem a sensible solution to world hunger, but it does not provide a real answer – except in emergencies like the famine of 1984–85 in North Africa:

- transporting massive amounts of food is difficult;
- the food may not suit the customs and traditional diet of the people who receive it;
- massive food imports flood the market and destroy the local economy, making it difficult for local farmers to sell their produce.

Nor is intensive farming itself, with its dependence on fossil fuels, a way for poorer countries to feed themselves. The technology is too expensive and often not appropriate for the societies it is intended to help.

Instead, encouraging results are coming from the development of non-intensive traditional systems. Remember, fossil fuel is a non-renewable resource: as demand increases, so reserves go down. At present fuel is relatively cheap, but as it becomes scarce the price will go up until the cost of inputs makes food too expensive. New discoveries of fossil fuel and energy conservation postpone the problem for now, but it lurks in the background to be faced by the intensive farming industry some time in the future.

Shortening the food chain

Scientists estimate that for every 100 g of plant material eaten only 10 g ends up as herbivore biomass. In other words, approximately 90% of the energy transferring between trophic levels is lost. Energy transfer between trophic levels is inefficient because:

- some of the plant material is not digested and passes out of the herbivore's body as faeces
- the herbivore uses energy to stay alive
- when the herbivore dies its body represents 'locked up' energy, some of which transfers to decomposers.

Similar losses of energy occur between higher trophic levels:

herbivores → first carnivores → second carnivores

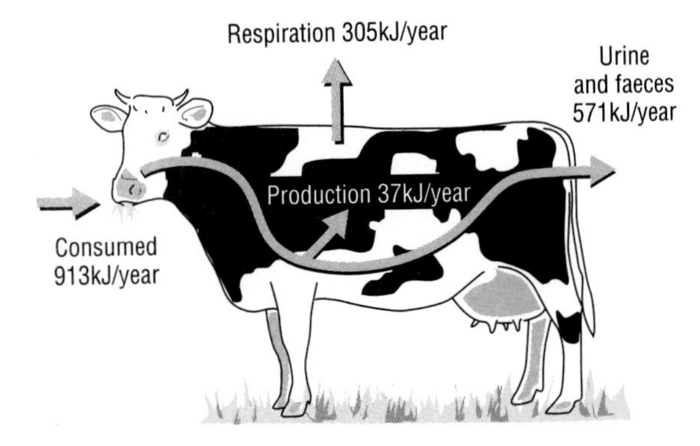

Figure F.7 The energy intake and output of a cow

Figure F.7 shows the energy intake and output of a cow. Notice that more than 50% of the energy content of the grass that the cow eats is lost in its urine and faeces. Energy is also lost from the body as heat produced in respiration (see page 36).

The energy budget of the cow can be summarised in the following equation:

Energy intake = Energy transfer in respiration + Energy transfer in the production of cow's biomass + Energy in urine + Energy in faeces

What is the message for feeding people? The nearer a consumer is to the original plant source in a food chain, the greater the amount of energy available to the consumer population. In other words, the fewer the links in a food chain, the more of the original energy there is available for the end consumer. World hunger is more to do with war, political unrest, difficulties of transport, damage to the environment from poor farming methods and inequalities of income, rather than too many people to feed. Shortening the food chain would help make sure that there is enough food for everyone.

Summary

- Intensive farming aims to produce as much food as possible from the land available.
- Energy inputs and outputs on intensive farms are much higher than on non-intensive farms.
- There is a wide gap in food production between rich and poor countries.
- The efficiency of food production is improved by reducing the number of links in farming food chains and by controlling energy losses from livestock.

Extension questions

1 Look at Figures F.4 and F.5. Comment on the energy budgets for the farms illustrated.

2 Look at Figure F.6. Why do you think the gap between food produced per person in Western Europe and Africa has increased since 1950?

3 Look at Figure F.7.

a Do you think the cow converts grass efficiently into biomass (meat, milk, etc)?

b What percentage of the energy intake is excreted in faeces and urine?

c What percentage of the energy intake is used up in respiration?

d In the light of your answers so far, why do you think cows spend a great deal of their time eating grass?

Unit G
Investigating crop growth

How do you carry out investigations? Trying out some of the investigations on the next few pages will help you deal with a few of the questions and problems facing scientists engaged in research.

The job of scientists is to try to find out more about the world and the universe. Scientific research is very varied, so any one scientist will concentrate on one part of science.

In science the best interpretation possible is made, using all the evidence available. At a later date, further evidence may show that the original interpretation was wrong. Scientific ideas are not fixed, but change as more is discovered. The changing explanations of how plants respond to light (page 78) show how scientists must be prepared to alter their views when new evidence is provided.

Activity G.1

Investigating lettuce growth

Read through the steps opposite which suggest how you could work through an imaginary investigation of lettuce growth. Find each step in the diagram of the investigation cycle (Figure G.1). In a group discuss how you would tackle each step.

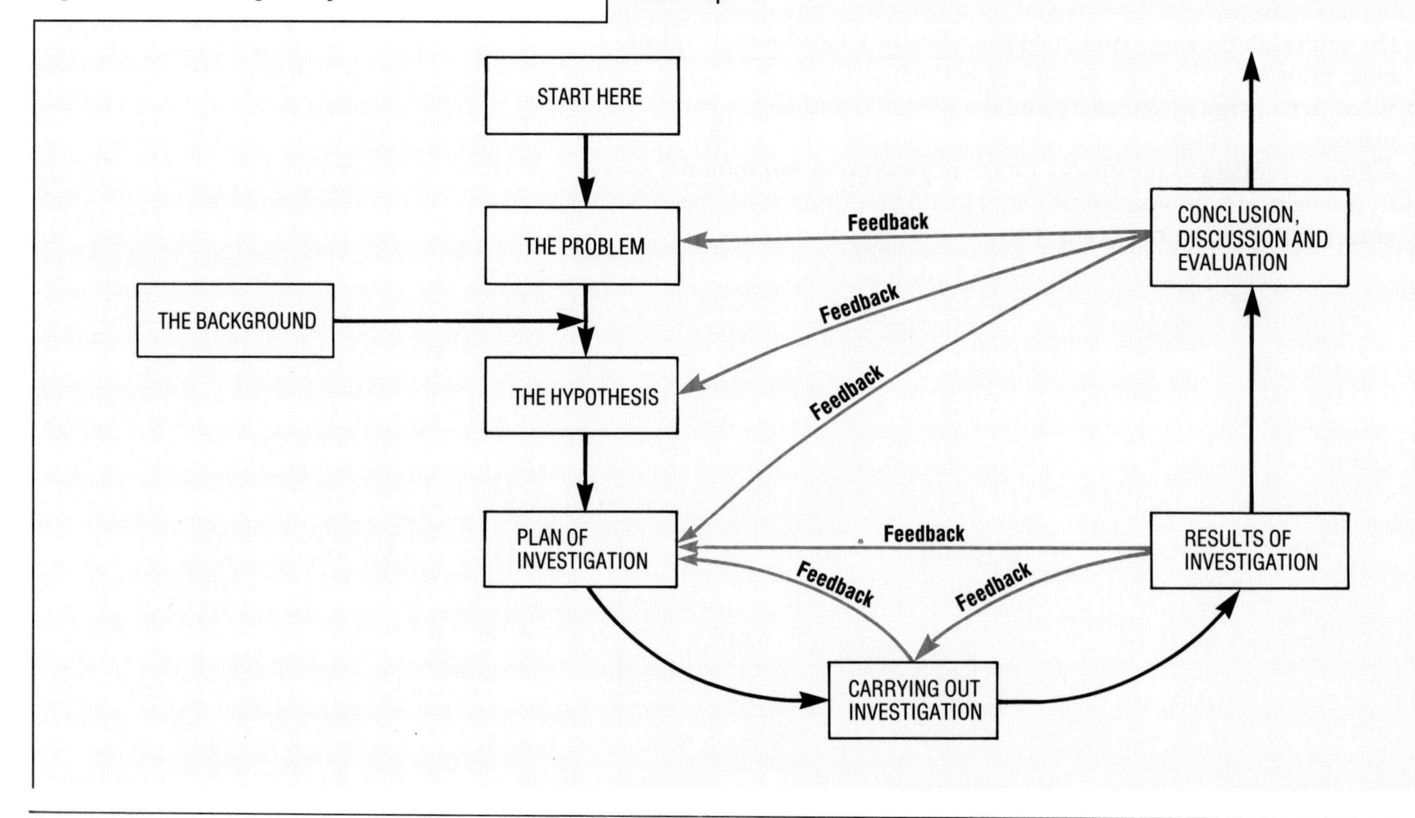

Figure G.1 The investigation cycle

Step 1: Decide exactly what you want to discover
Imagine that you want to discover how far apart lettuces should be grown. Is this a good enough description of the problem?

Step 2: Background knowledge and reading
What do you know about lettuce growth already? Probably, your answer is nothing! What can you find out by reading?

There is plenty written about lettuce growing, but is there anything which helps you with your problem? Select what is relevant for your study. What spacing do the seed packets and gardening books recommend? Has anyone done experiments on the problem? Has anyone investigated the problem with a different crop plant? Can you discover any ideas about the best method for your investigation?

Step 3: Deciding on a hypothesis
What are you going to test? The hypothesis or idea you decide to test might be: 'The growth of lettuce plants is affected by their distance from each other.' The investigation will test whether the hypothesis seems to be true.

Step 4: Planning the investigation
This involves thinking about many aspects.

a The **independent variable** is the condition you will vary, in this case the spacing distance. How many different spacings will you try? How many plants will you need for each trial? How many trial plots will you need to test each spacing?

b The **controlled variables** are the conditions you will keep constant (the same) for all the plants in all the plots. In this investigation you need to think about many of these controlled variables, including:

- the lettuce plants – are they all the same type or variety and age? Have they all had the same previous treatment?
- the growing environment – soil, light, moisture.

c The **dependent variable** is what you will measure. In this investigation it is the growth of the lettuces. This will be dependent upon the independent variable – the original spacing you choose. How will you measure the lettuce growth? You could measure wet mass or dry mass or diameter. Will you measure every plant or a sample taken at random? If the latter, how big will your sample need to be? How will you ensure that your sample is random?

Step 5: Carry out the investigation

Step 6: Record and display the results
Describe clearly what you observed and measured. Use tables, histograms, pie charts and graphs, as appropriate, to display your results.

Step 7: Conclusions, discussion and evaluation
Do the results of your experiment really show what you set out to investigate? Are the results reliable? Could they be reproduced by someone else? Could you improve the methods and experimental design you used? What further experiments would be helpful? Would the same spacing be best if you used different controlled conditions? Do your results agree with the recommended spacing for lettuces? Do they agree with the results of other people's investigations?

Many investigations have the steps shown in this investigation. The investigation cycle may help you to remember them. Sometimes, as you carry out your investigation, you will find that you need to modify your approach as you go along. The red arrows in Figure G.1 show you some points where this **feedback** may occur.

Fleming's 'chance' discovery

Sir Alexander Fleming's discovery in 1928 of the effect of penicillin on the growth of bacteria is well-known. The bacteria he was growing became contaminated with the mould *Penicillium* which prevented the growth of bacteria around it. Fleming realised that the mould was producing something which inhibited bacterial growth. This was a very important observation, but it took nearly 15 years for penicillin to be developed for use in treating diseases caused by bacteria.

Fleming was not the first to notice that some micro-organisms had anti-bacterial activity. But, he was the first to produce a clear idea of how the effect was produced. In his description of his discovery in the *British Journal of Experimental Physiology* in 1929 he described a method for testing the strength of anti-bacterial activity of different substances, including penicillin (Figure G.2). We now know that many organisms produce substances which inhibit the growth of others. For example, some weed and crop plants release chemicals into the soil which affect the growth of other plants.

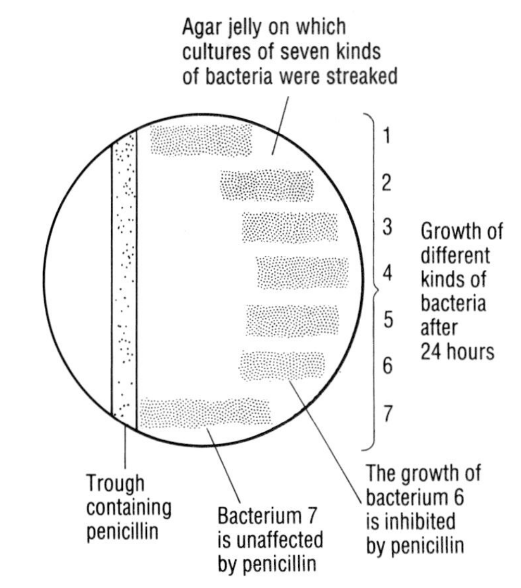

Figure G.2 Sir Alexander Fleming's method for showing the effect of penicillin on different bacteria

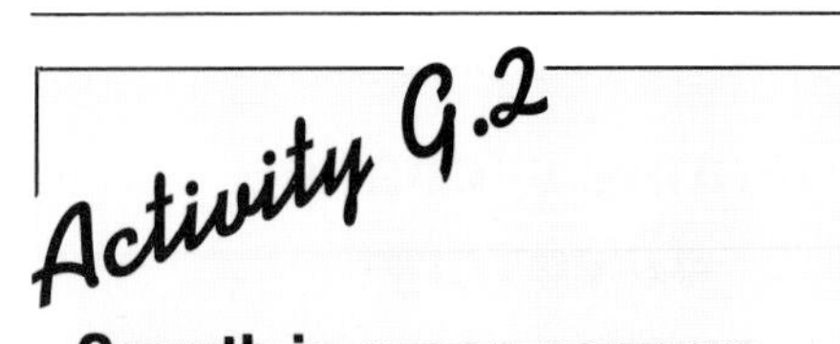

Growth in green peppers

When a plant grows, different parts grow at different rates — they grow *differentially.* It is often useful to discover what percentage of the plant becomes useful crop, and what becomes leaves, stem, root, and so on. The method used is called **partitioning**.

A student grew a green pepper plant in a greenhouse. When it was fully grown and had produced four fruits, she found the total mass of the plant, T. She partitioned, or split up, the plant into root, stem, leaves, fruit flesh and seeds. The soil was washed off the roots which were then blotted dry. She then found the mass of each part M.

Figure G.3 A green pepper plant: (a) in flower; (b) the fruit is beginning to form; (c) the growing pepper fruit; (d) a fully grown green pepper

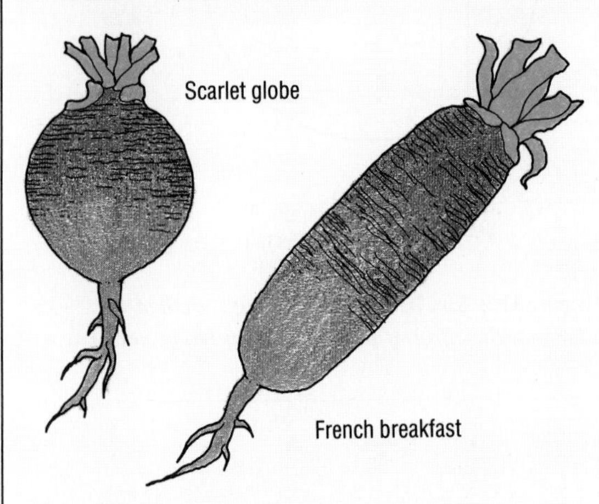

Figure G.4 Two varieties of radish: Scarlet Globe and French Breakfast

Team investigations and chance discoveries

Many investigations are best done by a team of people who co-operate together. All the scientists in the team discuss the problem and the planning. The investigation itself can be split between them. If they are using the same techniques they must make sure that they all carry out the technique in exactly the same way. Their methods must be **standardised**.

If you are working with other people on an investigation, you will probably find that you produce more ideas if you discuss the problem together. Some of the ideas may seem unreasonable but others may be really worthwhile. Co-operation between the team members will be necessary in tackling the practical part of the investigation. Afterwards, the whole group will be better than a single person at evaluating the methods and results.

At research stations such as Rothamsted in Hertfordshire (part of the Institute of Arable Crops Research) and Wellesbourne in Warwickshire (part of the Institute of Horticultural Research), teams of people work together on large and long-term studies of plant growth. Sometimes, several teams may be involved in the investigation of a problem. For instance, discovering a new pesticide might involve:

- the chemists who produce and work out the properties of possible pesticide chemicals
- the entomologists who test the effect of the pesticide on different insects
- the ecologists who investigate what happens to the pesticide in the environment and in the food chain.

Sometimes, a chance observation leads to a great discovery. However, progress usually arises because scientists are well-prepared and are making well-planned observations, measurements or investigations.
observations, measurements or investigations.

Scientific work must be **communicated** to others – it has no value if it is not shared. There are many scientific meetings and journals which allow scientists to describe their research to others.

	Fresh mass, M (g)	Percentage of plant (%)
Leaves	23	
Stem	60	
Flesh of fruit	92	
Seeds	138	
Roots	190	
Total, T	503	

From these figures, she worked out the percentage of each plant component in relation to the whole plant, using the formula

$$\text{percentage} = \frac{\text{mass of part} \times 100\%}{\text{total mass}} = \frac{M \times 100\%}{T}$$

1. Complete the table of results.
2. Write the conclusions, discussion and evaluation for this investigation. (Consider the following points. Does one plant give a good idea of how green peppers grow? Were the conditions in which the plant was grown important?)
3. Plan an investigation, using partitioning, to compare percentage of the plant which forms the radish in two varieties of radish, Scarlet Globe and French Breakfast (Figure G.4).

Crowded carrots?

Carrots will grow very densely, up to 30 plants in a 10 cm square if they are given fertilisers. At Wellesbourne, scientists are examining the effect of each plant on its neighbours. This will provide information on:

- whether the plants compete for mineral, light and water requirements
- whether it is beneficial to thin out the seedlings.

The scientists want to discover which arrangement gives the best and most saleable yield of carrots. In their research, small plots of ground are used to investigate the effect of different spacings on the growth of carrot plants. In some plots larger spaces have been left (Figure G.5). Several plots for each experimental set-up are used. **Replications** (copies) of each treatment provide an important way of ensuring that results are reliable.

What variables must be controlled to ensure that the treatments in the different plots are **comparable**? How could the yield and quality of the carrots be measured?

Figure G.5 A trial plot of carrots

Summary

The steps in an investigation include:

- reading the background information
- deciding upon the problem
- producing a hypothesis
- planning
- carrying out the investigation
- presenting the results, writing a conclusion
- discussion and evaluation
- feedback.

In many investigations the independent variable is changed and the dependent variable is measured. Other conditions or variables are controlled.

Activity G.3

Do onions protect carrots?

Your carrots have been badly damaged by carrot flies. Their maggots leave small, dark tunnels in the carrots. A friend claims that interplanting carrots and onions is helpful. She says that you should grow four rows of onions between rows of carrots. How would you investigate the truth of this claim? What other factors would you have to consider?

Activity G.4

Design your own growing compost

At present, many growing composts include peat. However, obtaining peat from the environment has damaged some very interesting and valuable sites for wildlife throughout the British Isles.

Can you produce a good compost for growing geranium cuttings without using any peat? Plan your work, and, if possible, produce your trial composts and test them for growing geranium cuttings. Prepare a leaflet about your new compost and its properties.

Extension questions

1 In a study of the effect of nitrogen fertiliser on onions, seeds were sown in three plots, A, B and C.

Plot	Nitrogen fertiliser added	Appearance of plants
A	none	lower leaves yellow
B	20 g/hectare	normal
C	40 g/hectare	normal

a What was the independent variable in this experiment?
b What was the dependent variable?
c i What factors would you expect to be controlled in this experiment?
ii Why would it be important to find out how much nitrogen there was in the soil before beginning the experiment?

2 Explain the meaning of these terms: **a** horticulture, **b** arable crop, **c** hypothesis, **d** replication, **e** reliability, **f** reproducibility, **g** sample, **h** random, **i** variability of living material.

3 Suggest two reasons why it is important to include the date in the report of an investigation of plant growth.

Unit H
Managing food production

Plants worth millions

Figure H.1 Flowers are high-value crops and are often grown under glass

Gardeners grow plants for their own pleasure or to provide their own food. Farmers and market gardeners produce fresh, disease-free food or decorative plants (Figure H.1) to provide themselves with an income. 80 per cent of the fresh, outdoor-grown vegetables sold in the United Kingdom are grown here. Their market value is hundreds of millions of pounds each year. On top of this is the value of all the farm crops, flowers and decorative plants which are grown for sale. Growing plants commercially is big business. The grower tries to obtain the best crop yield possible for the least cost of time, effort, equipment and materials.

Cultivating plants commercially requires suitable conditions. New and better varieties of crops, which give better yields or better quality products, are being developed all the time. Each variety of plant is a unique biological system with its own peculiarities and requirements. For each crop, the grower considers:

- soil and environmental conditions
- requirements for germination and growth
- weed and pest control
- time taken to produce a crop
- harvesting
- storage and marketing requirements.

The types of crops which are grown change as demand and conditions change. During the past 30 years many people have changed from eating butter to eating margarine made from vegetable oils. Oil can be extracted from the seeds of oilseed rape. Once, fields of the crop were a rarity, now they are commonplace (Figure H.2). Oilseed rape has become the third most commonly grown crop in the United Kingdom, covering around 1

Figure H.2 A field of oilseed rape, and a close-up of the plants

Figure H.3 Sunflowers – will they become a common crop in the United Kingdom?

per cent of the land. If global warming occurs this will affect the crops grown. Sunflowers (Figure H.3) may become a regular sight in our countryside. Oil can also be extracted from the seeds of sunflowers. After oil is extracted from the seeds, the protein-rich residue can be sold for animal food.

Growing tomatoes

Tomato plants need care, but are interesting to grow and produce a useful crop. Most people grow them indoors, beginning with seed sowing in February or with small plants. A heated greenhouse gives the best yield, but they can be grown on window-sills.

For outdoor growth, seeds should be sown indoors in March or April and the young plants transplanted out when they are at least 15 cm tall and frosts have ceased. The seeds are sown thinly in a tray of moist seed compost and covered with 0.5 cm of compost. The compost can be moistened by standing the seed tray in shallow water, and letting the water soak up. Then the water is removed and the seed tray drained. All the seed trays should be clearly labelled. Tomatoes need to be kept at about 18 °C for about a week. After germination, the temperature can be lowered.

When the seedlings have their seed leaves and two more leaves they are pricked out (Figure H.4). This means transferring them to potting compost in 75 mm pots. A dibber (or plant label) is used to lift the roots in a lump of soil, while holding a leaf gently. The delicate roots and stem should never be handled. The stem should be covered with compost up to the seed leaves. Pricking out gives the young plant more space, light and nutrients.

Perhaps the easiest way to begin to grow tomatoes is to use specially prepared plastic growbags containing growing compost (Figure H.5). Small tomato plants can be bought in April and planted in the bags. The compost will need a good watering the day before you transplant the tomato plants.

As they grow, the plants need to be supported. They are usually tied to canes or to wires strung across the glasshouse. The plants need warmth to grow well and good ventilation helps to avoid fungus diseases. The lowest leaves, below the first tomatoes, are usually cut away to allow the air to circulate. Regular light watering is necessary until the first group (truss) of tomatoes is well-formed. From then onwards, the soil must always be moist, which means watering twice a day.

How tomatoes develop

Giving tomato plants a fine spray of water helps pollination. The pollen can also be transferred from flower to flower with a paintbrush. The petals will drop off after fertilisation but the sepals remain and the ovary swells to form the red fruit containing its seeds.

All the new sideshoots which grow should be removed with a sharp knife. What differences do you think it would make if these shoots were allowed to grow?

Soon after the fruits begin to form, it is useful to start to add a suitable fertiliser to the water. How could you test whether this had any effect on the final tomato crop?

How good is the tomato crop?

A record of the number and mass of tomatoes produced by one plant can be kept. The total mass produced by one plant is called its yield. Typically, one plant in a glasshouse with some heating may produce 4 kg of tomatoes, and outdoors the yield may be 2 kg per plant. It may also be valuable to know for how long tomatoes are produced. In a greenhouse, the first ripe tomatoes may appear in July, the last one in October. From outdoor plants, you will be lucky to have tomatoes ripening for many weeks of August and September.

Another important factor in choosing your variety of tomato is the quality of the fruit, their size, shape, taste, skin and flesh texture, and regularity. The rest of the tomato plant is poisonous, although it can be composted.

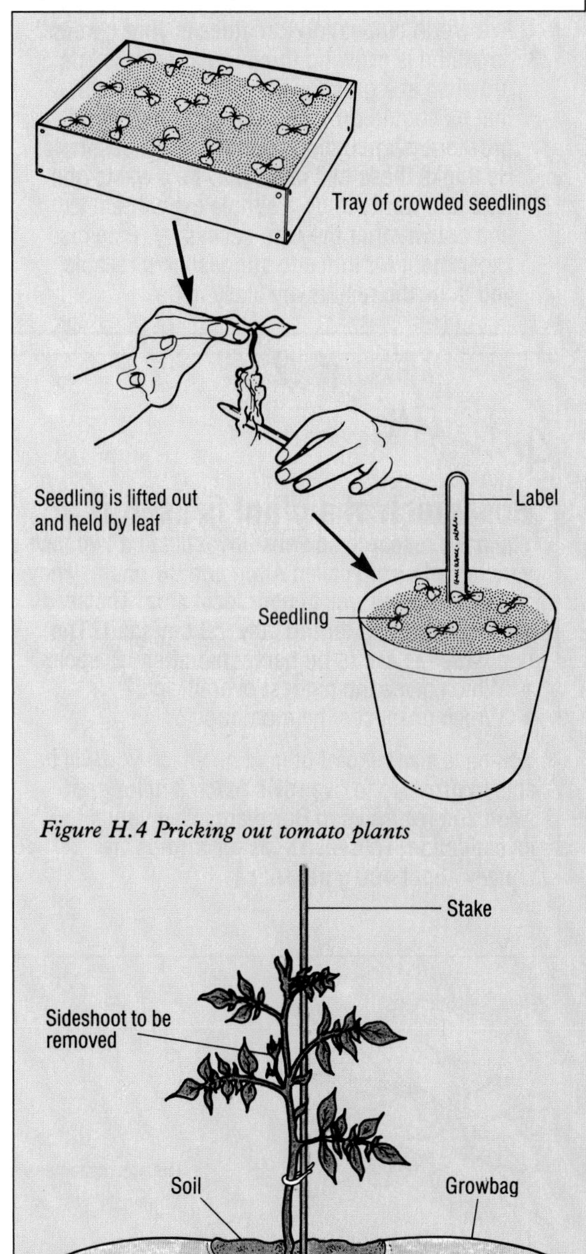

Figure H.4 Pricking out tomato plants

Figure H.5 A tomato plant in a growbag

How much of the plant is useful?

If we gathered all of a crop plant we would find all the yield of new plant material. This is the total **productivity** of the plant, the result of its photosynthesis. It is possible to compare the productivity of two crop plants by evaluating a metre square of each plant. But, not all parts of the plants are valuable as a crop. A useful method of comparing two crops is to use a modification of the partitioning method (page 132). You can compare the ratio of the mass of the edible portion with the mass of the whole plant. You can use this ratio to compare the yields of cabbage with Brussels sprout or kohlrabi plants, for example.

$$\text{ratio of useful crop} = \frac{\text{mass of edible portion}}{\text{mass of whole plant}}$$

Often, some of the rest of the plant has a use, perhaps for animal bedding, compost or mulching.

Fruit trees

Unlike the previous crops we have mentioned, apple trees are woody perennials, which do not die down in winter and which produce food year after year. Fruit production (Figure H.6) depends on successful pollination. Many apple varieties, for instance the famous Cox's Orange Pippin, are self-sterile and cannot pollinate their own flowers. Other varieties give a better yield if cross-pollinated. They depend on honey bees to transfer pollen from the flowers of another variety of apple to their stigmas.

It is helpful to plant together two or more apple varieties that flower at about the same time. The variety James Grieve makes a good pollen source for Cox's Orange Pippin. A crab apple, such as the variety John Downie, will also provide suitable pollen for many eating apples. Catalogues often indicate which varieties flower together.

Meeting market requirements

Crops are expected to be **homogeneous**, similar in size and quality. Supermarkets know their customers want perfect apples and carrots. Factories which freeze vegetables and manufacturers who process crops want products that are reliable in quality and which fit their equipment. It is also much easier to harvest a crop if the plants are very similar and are all ready together.

Plants which are genetically alike and are grown in a carefully managed environment will meet these requirements. As far as possible the soil and climatic conditions across a whole field must be similar. The plants should be sown at the recommended depth and spacing. This should ensure good germination and later growth. The recommended spacing will reduce the competition for space, water and minerals to a minimum. The crop hygiene (control of weeds, pests and disease) is also important. However, the farmer will try to keep the cost of herbicides and pesticides as low as possible.

Monoculture, the growing of a single crop in large fields, makes it easier to manage the crop well. Very little land is wasted and the conditions can be controlled. It is hence easier to produce high quality products. Monoculture may also bring some problems, however. Large fields can be affected by the

Activity H.1

Tomato growing

1. Obtain some seed catalogues or packets. Find out information about three varieties of tomato plant and make a table to compare their characteristics. Which variety would you choose and why?
2. What costs are there in growing four tomato plants indoors and outdoors? What is the approximate value of the crop you will obtain? Is it worth cultivating tomatoes in your garden?
3. Jonathon is growing three tomato plants in a growbag in a greenhouse. He has found out that he should provide support as the plants grow and should pinch out the new sideshoots. He thinks these two tasks may be a waste of time and wants to do a simple experiment to find out whether they are necessary. Plan his experiment for him and suggest how reliable you think the results are likely to be.

Activity H.2

How much of a plant is useful?

You are a research scientist investigating two new varieties of turnip called Alton and Burnham. They are said to grow well in your local area. The seeds are to be planted during July, 30 cm apart. The two varieties are to be harvested after 12 weeks.

- Which gives the highest overall yield?
- Which produces the most food?

You have a small plot of land on which you will be able to grow up to five rows (each 3 m long) of Alton and five rows of Burnham. Plan your investigation. (**Note:** Yields for turnips are usually about 400 g per root.)

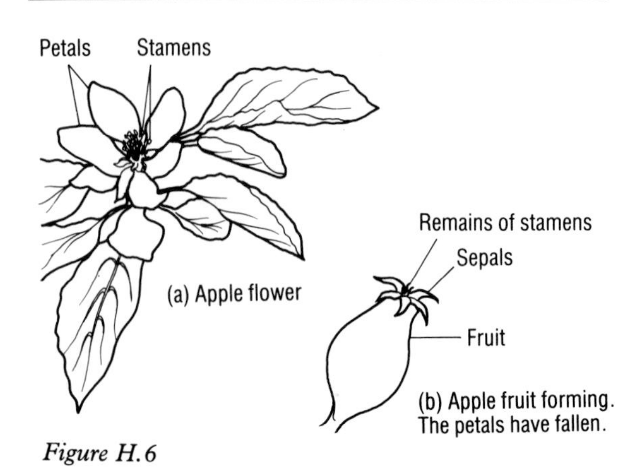

Figure H.6

Activity H.3

The ideal seed drill

Use the information below and Figure H.7 to produce a leaflet advertising and describing the seed drill.

Figure H.7 The seed drill

1. Seed drills are designed to be attached to tractors and to sow the seed at the correct spacing.
2. The spacing of the seeds is selected so that good germination and a high final yield is achieved.
3. The seed hopper is the funnel where the farmer places the seeds.
4. The principle of this drill is simple. The seeds drop onto a belt which has holes in it. Both the spacing of the holes and the speed of rotation of the belt can be adjusted to provide the required spacing of the seeds.
5. The drill is strong and easy to maintain.
6. As a safety factor, there is an electronic beeper which lets you know when the belt is rotating.

Activity H.4

Growing crops: Inputs and outputs

Choose a crop. List all the inputs and outputs required to produce the crop. Underline the items which involve energy inputs and outputs. If possible obtain figures and produce balance sheets for costs and profits and for energy inputs and outputs.

wind which may damage the crop or cause the soil to blow away while the land is bare. If a pest or disease does affect the crop, the losses may be huge. This is in contrast to crop rotation which helps to prevent a particular crop being affected by a pest remaining in the soil from the previous year.

Glasshouses, coldframes and cloches

Glass and clear plastic allow the wavelengths of sunlight energy to pass through. Underneath, the light energy changes to heat, which has a different wavelength and which cannot escape. Thus, inside glasshouses, coldframes and cloches the temperature becomes higher than outside.

The climate often limits which plants will germinate or grow successfully. The raised temperature under glass allows plants to survive which would otherwise die outside. In summer, the temperatures may become very high and ventilation allowing air movement may help to keep conditions cool.

The simplest kind of protection is a movable cloche (Figure H.8). Cloches help the soil to warm up. Some plants, such as runner beans, will not germinate in the cold soils of March and April. At the same time they need a long growing season to produce a good crop. With a cloche it is possible to sow the seeds successfully earlier in the year. A plastic tunnel acts in the same way. Cloches and tunnels also protect against wind and heavy rain.

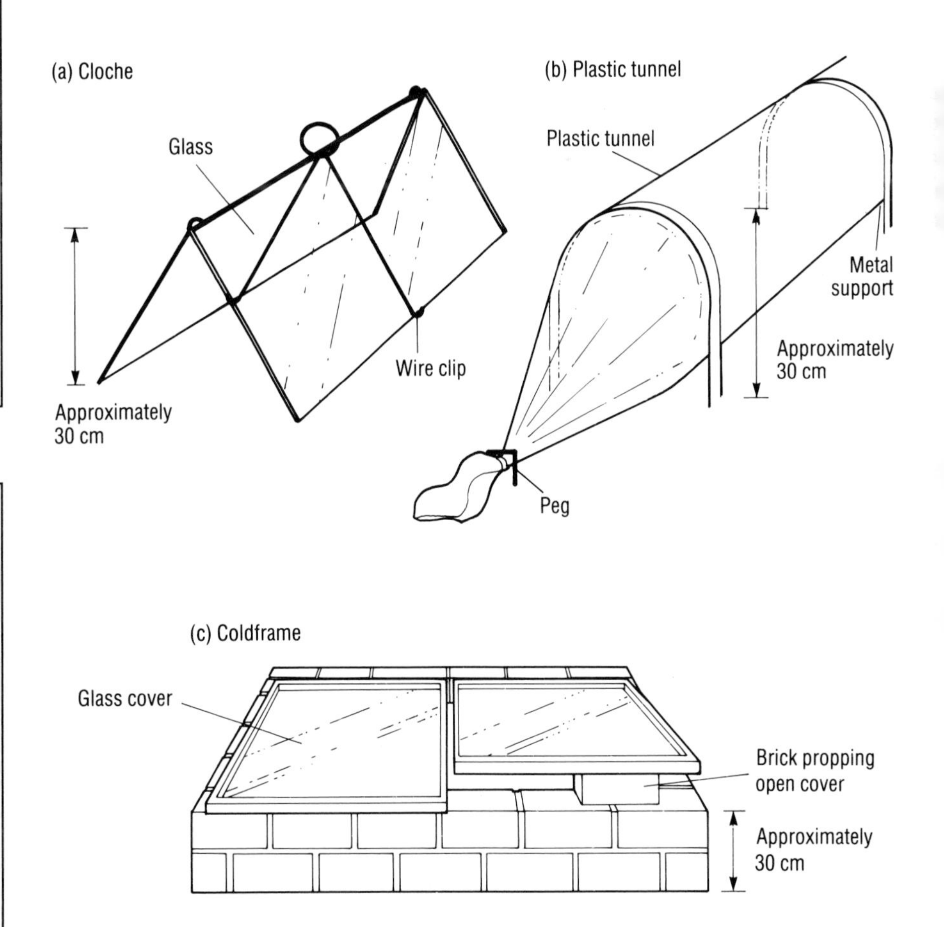

Figure H.8 Coldframes, plastic tunnels and cloches protect delicate plants

Growing sugar for profit

Sugar crops are grown for profit. The grower has to balance expenses against income. This involves considering:

- capital outlay (eg expenses for equipment)
- recurrent expenses (eg cost of seeds, fuel and energy)
- fertilisers and pesticides, irrigation costs, cost of workers
- marketing costs (eg costs of containers, transport)
- what is paid for the products.

If a grower provides the right conditions he can exploit one thing which is free – sunlight energy. Plants build up their own food using sunlight energy in photosynthesis. The rate of production of new material will differ in different situations.

60 per cent of the world's sugar comes from sugar cane (Figure H.9). Sugar cane is a plant of the tropics and subtropics. The rate of production of new material varies. One variety of sugar cane was measured for two years. It produced:

Figure H.9 Sugar cane

- 100 tonnes/hectare/year of cane
- 12 tonnes/ hectare/year of sucrose sugar.

In fact the average production of cane for all sugar producers is under 60 tonnes per hectare. This is because much sugar cane is grown by relatively poor farmers who provide less favourable conditions.

Some of the rest of the plant may be used locally for fuel. The sucrose has to be extracted from the cane and refined. Sugar is sold for use as it is or used as a raw material for chemical processes such as production of ethanol. The sweet byproduct – molasses – is most often used for producing animal food.

Since the product has to be processed before use, the grower receives only a small proportion of the money you pay for a kilogram of sugar. Fertilisers, pesticides, and cultivation and irrigation costs mean that this is not all profit. Some growers use the equivalent of 1500 MJ (megajoules) of energy per tonne of cane produced!

Figure H.10 Sugar beet

The other 40 per cent of the world's sugar comes from sugar beet (Figures H.10 and H.11) which is more suited to temperate climates. In the growing season (May to October) a crop of sugar beet growing in the United Kingdom may receive about 2.415 MJ per hectare. The yield at harvest may be:

- 76 tonnes/hectare of roots
- 12.2 tonnes/hectare of sucrose.

The yield of sugar is surprisingly similar to a good crop of sugar cane. (1 g sugar contains 1.68 kJ of energy.)

Therefore, 1 tonne contains $1.68 \times 1\ 000\ 000$ kJ

$$= 1.68 \times \frac{1\ 000\ 000}{1000} \text{ MJ}$$

The sugar produced on 1 hectare contains $1.68 \times 12.2 \times 1000$ MJ

$= 20\ 496$ MJ energy.

Figure H.11 Sugar is extracted from the roots of sugar beet

Activity H.5
Growing your own peppers

Sweet peppers (see page 132) are best grown in a protected environment in the United Kingdom. On a large scale, they are grown in glasshouses but you can grow them on a windowsill in pots. You can buy seed or use seed from a ripe pepper. Sow the seeds in pots of seed compost in the spring. Prick the seedlings out into separate pots of growing compost. Provide supporting stakes. You should have some peppers ready to eat by the summer.

Again there are expenses for the sugar beet grower. He needs a tractor, plough, drills for sowing seed, spray equipment, irrigation equipment, harvesting equipment, carts and sheds. He has to buy seeds, fuel, fertiliser, herbicides, pesticides and water. Farm workers have to be paid. Each United Kingdom grower will have a quota which fixes the amount of sugar that can be grown. The sugar beet will be sold directly to one of the large sugar-processing factories.

The farmer's profit will depend on whether he has the skill to obtain a good quality crop which exactly meets the quota. The farmer cannot control sunlight, rainfall and the number of pests around. But the other factors are under the farmer's control. The challenge is to keep the expenses as low as possible.

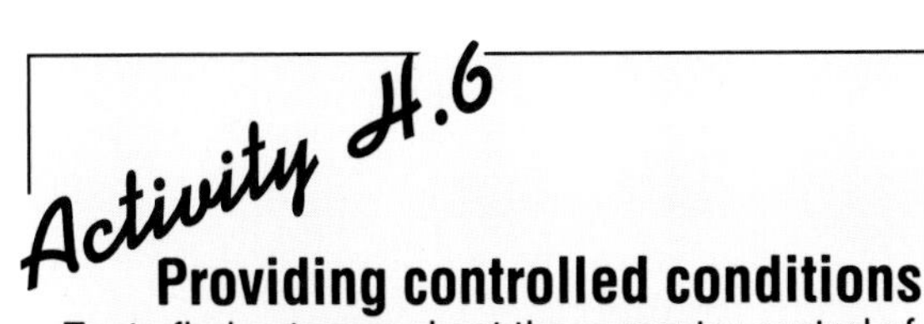

Providing controlled conditions

Try to find out more about the computer control of glasshouse conditions. If possible use datalogging and control equipment to monitor and control the conditions in a seed incubator or plant growth chamber.

Figure H.12 Heating ducts of a glasshouse

Figure H.13 Lighting in a glasshouse can be very elaborate

Coldframes are more substantial. The lid can be raised when the weather is suitable to provide some ventilation. Coldframes may be used to germinate seedlings. They are also used to 'force' or speed up the growth of plants such as lettuce so that the crop season is longer. Sometimes, plants are moved into an unheated coldframe from a heated glasshouse to 'harden off' before being planted outside.

A glasshouse is a more elaborate form of protection. In a glasshouse it is possible to control conditions very precisely. The protected, controlled growing of plants in commercial glasshouses sometimes closely resembles a factory production line.

Modern commercial glasshouses are heated (Figure H.12) and have thermostatic control. They also have automatically controlled lighting (Figure H.13), ventilation, water and mineral supply. The water and dissolved minerals may circulate around the roots of the plants (see hydroponics, page 66). The conditions in the glasshouse may be monitored by sensors leading to a computer. The computer processes the information and alters the controls to provide the correct growing conditions.

Crops and carbon dioxide

The law of limiting factors (page 116) states that the speed of a biological or chemical process will be affected by the factor in lowest supply.

Photosynthesis requires carbon dioxide and light. In the bright daylight conditions of a glasshouse, the concentration of carbon dioxide in the atmosphere limits the rate of photosynthesis. There is normally about 0.03 per cent or 330 volumes per million of carbon dioxide in the air. An increase to 1000 volumes per million will allow a doubling of photosynthesis in bright light. This will lead to a large increase in a crop such as tomatoes. There will be a greater yield of bigger tomatoes.

It is possible to increase the carbon dioxide artificially. Burning high-quality fuels will increase the carbon dioxide without producing unwanted fumes. This also increases the temperature which is useful in winter, but may be a disadvantage in summer. It is also possible to release carbon dioxide from cylinders or bulk storage tanks.

Summary

- Commercial growing requires controlled growing conditions. Plants and cultivation methods are developed to provide high yield and reliable quality.
- In glasshouses it is possible to control the growing environment very precisely.

Extension questions

1 What conditions must be provided when sowing seeds, pricking out seedlings and transplanting plants to obtain healthy plants?

2 **a** Why do orchards usually contain more than one variety of apple?
b Your uncle wants to plant two apple trees. He asks your advice about which varieties to choose. Write a letter to him explaining about choosing apple varieties. Mention pollination and size of tree.

3 Outline the advantages and disadvantages of
a crop rotation;
b monoculture.

Unit I
Decomposition

Fungi and bacteria, together with worms, woodlice, insects and other mini-beasts, feed on dead organic material and break it down. Figure I.1 shows some of them at work in the dead wood of a fallen tree. Their activities release minerals into the environment. These minerals are absorbed, in solution, by new generations of plants and passed to animals through their feeding activities (Figure I.2).

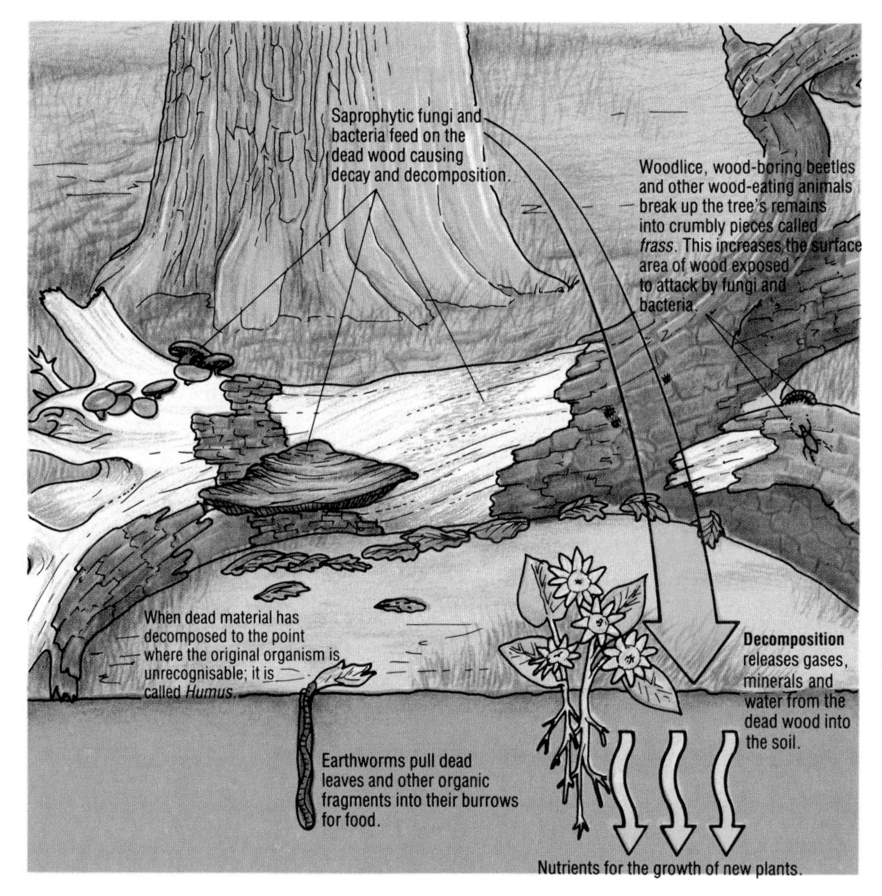

Figure I.1 Humus is a dark fibrous material. It contains the decomposed remains of plants and animals and is formed through the feeding activities of decomposers, so called because they break down dead organic matter. Soil fertility depends on their activities

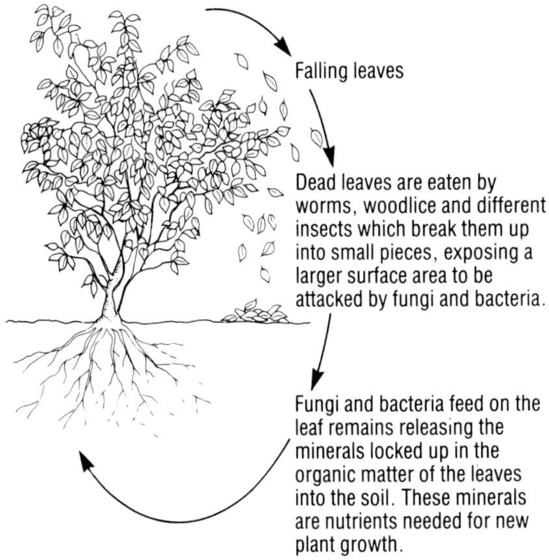

Falling leaves

Dead leaves are eaten by worms, woodlice and different insects which break them up into small pieces, exposing a larger surface area to be attacked by fungi and bacteria.

Fungi and bacteria feed on the leaf remains releasing the minerals locked up in the organic matter of the leaves into the soil. These minerals are nutrients needed for new plant growth.

Figure I.2 Minerals in organic matter are released into the environment by the feeding activities of fungi and bacteria. They become part of new organic matter when absorbed in solution by growing plants

Fungi and bacteria are everywhere; in soil, air and water. Figure I.3 shows what happens when bread is left uncovered for a few days.

Fungal **spores** in the air settle on suitable material and grow by sending out tiny tube-like **hyphae**. These eventually form a 'mat' called the **mycelium**. The tips of the hyphae produce enzymes that digest the bread and convert it into food substances which are absorbed by the growing mycelium.

Look at Figure I.3 again. Differences in colour show where the cells of different species of bacteria have settled and multiplied. The bacteria also produce enzymes that convert the bread into food which they can absorb. The result is decay and decomposition which turns the bread into a rotten, unrecognisable mass. This is why fungi and bacteria are also called **decomposers**. They are nature's 'refuse officers', breaking down dead remains and clearing the way for new life.

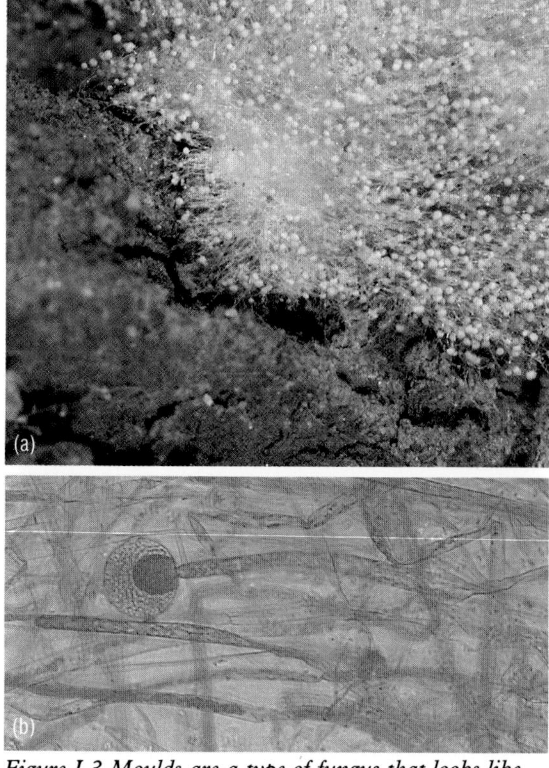

Figure I.3 Moulds are a type of fungus that looks like fluffy cotton wool. Here Mucor, *the black pin mould, has smothered a piece of uncovered bread (a). The black shiny capsules containing spores are carried upwards on stalk-like aerial hyphae (b). When ripe, the capsules burst and thousands of spores are wafted away on air currents. If the spores settle on food or somewhere else suitable, they grow into a new mould*

Elements for life

All matter is made of **chemical elements**. Almost 100 different elements occur naturally in the physical environment. They make up soil, water and the gases of the atmosphere. The most common are also those that make up living matter. This is not surprising since the elements that organisms need come from the physical environment.

About 20 elements are essential for life. Of these, six make up more than 95% of living matter. They are,

- carbon (C)
- hydrogen (H)
- nitrogen (N)
- oxygen (O)
- phosphorus (P)
- sulphur (S)

(If their letter symbols are arranged in order they can be easily remembered as CHNOPS.)

The elements CHNOPS combine in different ways to form **compounds**. Important compounds in living things are:

- *glucose* – a major source of energy
- *fats* – a store of energy
- *proteins* – for building bodies

Obviously these are not all the compounds essential for healthy living but they are the basic ones needed for life's structures and activities.

The carbon cycle

Respiration and decomposition are nature's way of cycling elements between the environment and living things. Figure I.4 shows how these processes operate in the carbon cycle.

The atmosphere is a reservoir of carbon in the form of carbon dioxide. Each year, land plants remove about 25 billion (thousand million) tonnes of carbon (as carbon dioxide) from the atmosphere. They convert it into food (mostly starch) through photosynthesis. In the oceans single-celled algae use a further 40 billion (thousand million) tonnes of carbon from carbon dioxide in solution.

Once it is part of living tissue, carbon is transferred from one organism to another as animals feed on plants and other animals. Carbon is returned to the atmosphere as carbon dioxide through respiration and the decomposition of dead organic matter. Burning wood and fossil fuels (coal, oil and gas) releases large quantities of carbon dioxide into the atmosphere. Each year the amount increases, which adds to the greenhouse effect.

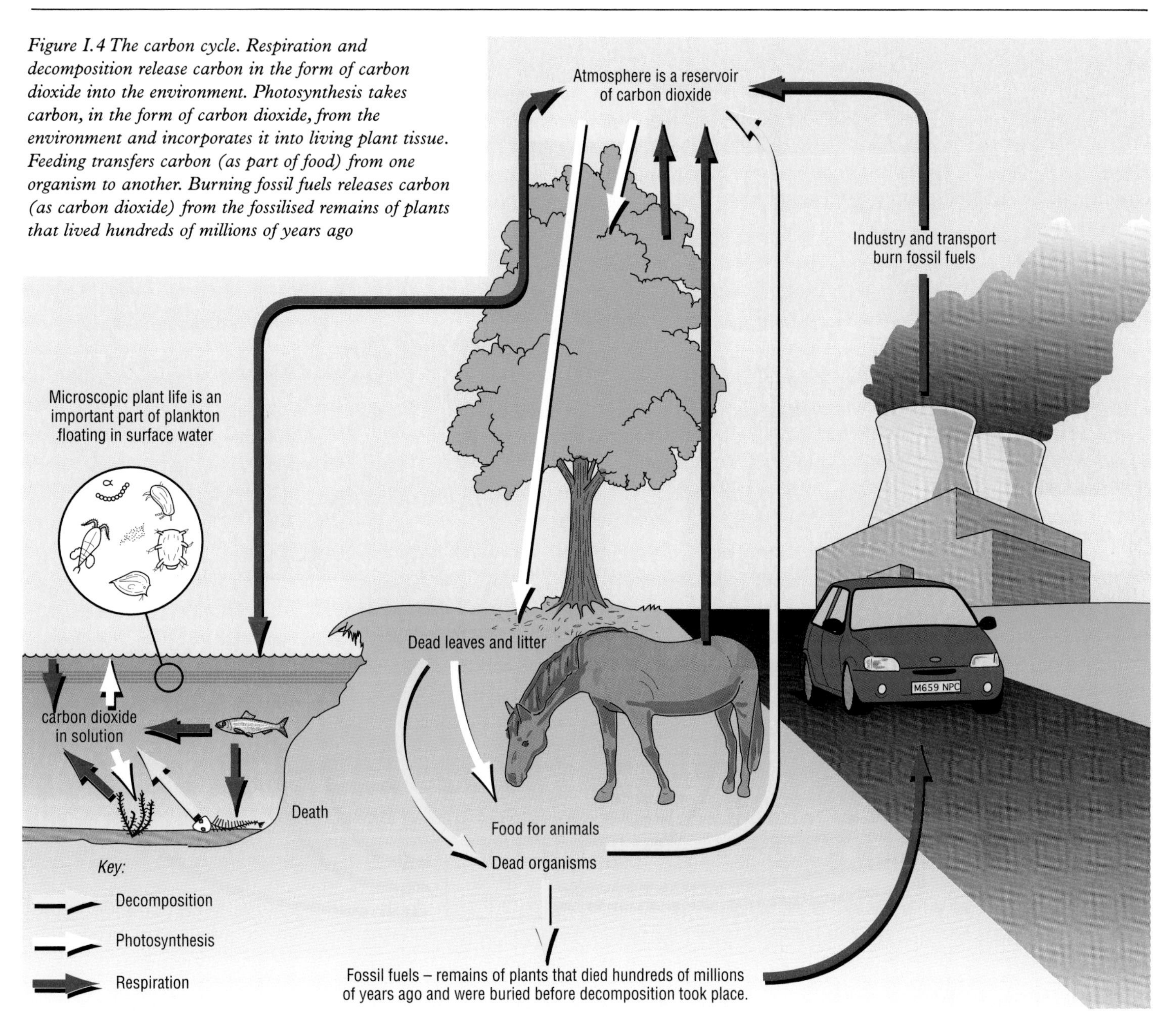

Figure I.4 The carbon cycle. Respiration and decomposition release carbon in the form of carbon dioxide into the environment. Photosynthesis takes carbon, in the form of carbon dioxide, from the environment and incorporates it into living plant tissue. Feeding transfers carbon (as part of food) from one organism to another. Burning fossil fuels releases carbon (as carbon dioxide) from the fossilised remains of plants that lived hundreds of millions of years ago

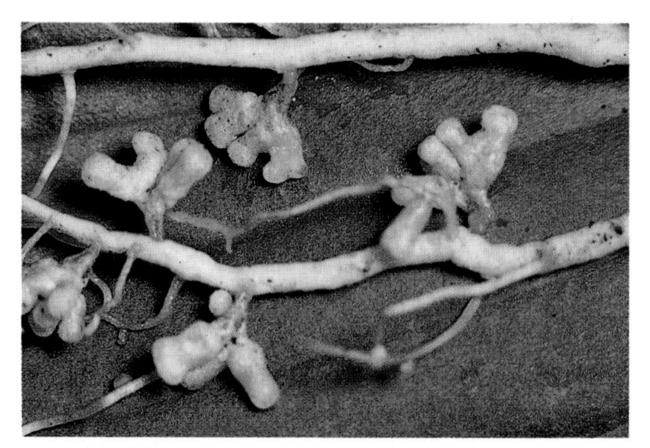

Figure I.5 Nodules on the roots of a pea plant contain bacteria which fix nitrogen from the atmosphere and convert it into nitrates that the plant can use

The nitrogen cycle

In the soil there are many organisms, which feed upon or decompose the remains of living things in the soil. These organisms decompose humus. In compost heaps, they convert plant and other waste to material which can be used as a natural fertiliser (see page 147).

Among the soil bacteria are some which **fix** nitrogen gas in the soil air, changing it to nitrogen compounds, and eventually producing nitrates. Plants can absorb these nitrates and use them to produce their own amino acids, proteins, nucleic acids and other compounds.

Other nitrogen-fixing bacteria live in swellings or nodules on the roots of some plants (Figure I.5). Nodules are found in **legumes**, plants with their seeds in pods (eg peas, beans, lupins and clover). The bacteria living in the nodules use nitrogen gas to form proteins which the plant can use.

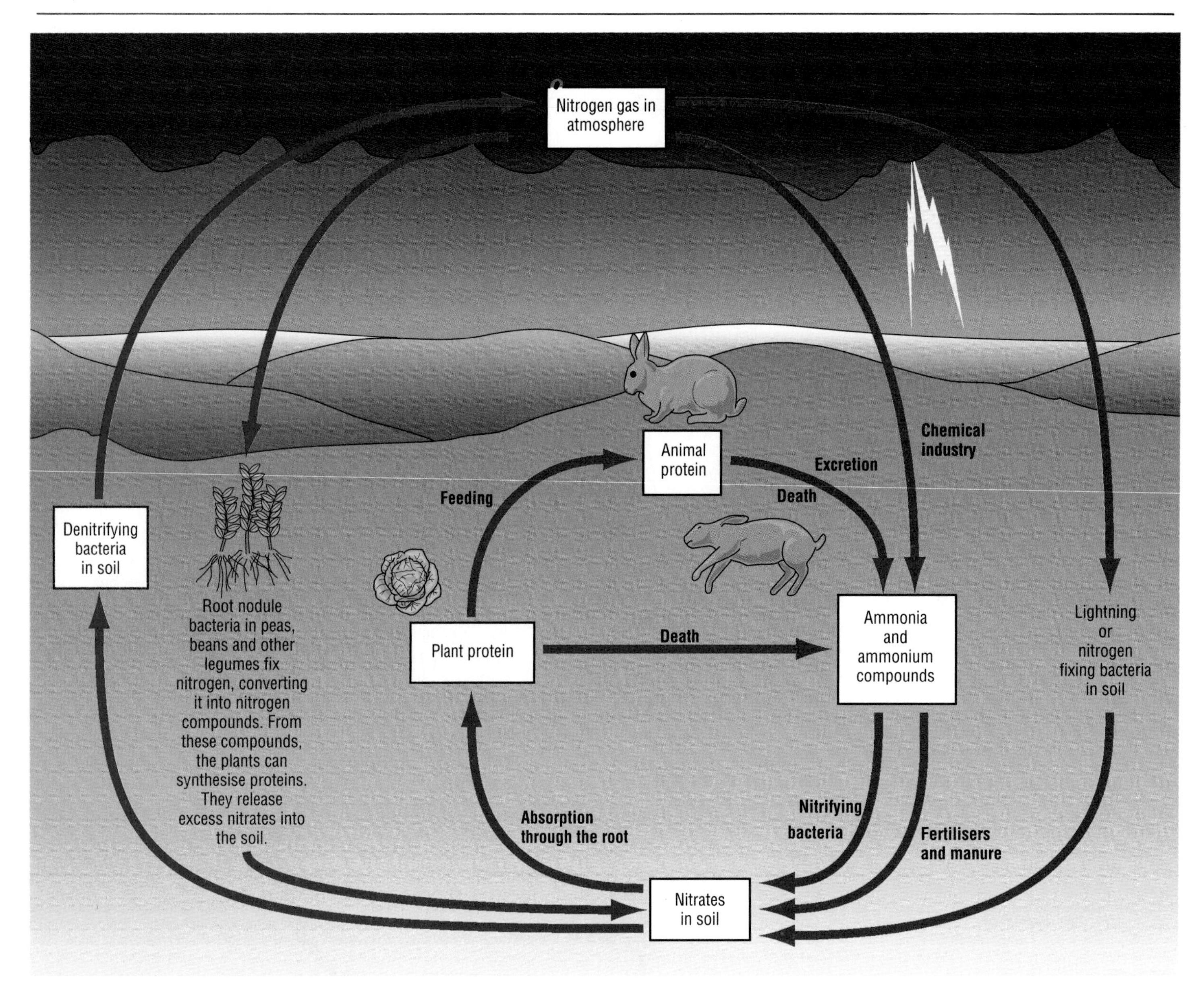

Figure I.6 The nitrogen cycle – the chemical industry makes ammonia from nitrogen. From ammonia, ammonium salts are made. They are used as fertilisers

A plant's nitrogen compounds can pass to animals when they feed on the plant material. Otherwise, the compounds return to the soil when the plant dies and decomposes. In the decay products of plants and animals, and in the excreta of animals, **ammonium** salts are present. **Nitrifying** bacteria in the soil convert ammonium salts into nitrates. Other bacteria, called **denitrifying** bacteria, can then change nitrogen compounds back to nitrogen gas. There is a continuous circulation of nitrogen in the soil (see Figure I.6).

In the past, farmers planted different crops on each field in rotation. Often, a legume crop was part of the rotation and its stem and roots were ploughed into the soil at the end of the season. This increased the nitrogen content of the soil. The use of nitrogen fertilisers, which nitrifying bacteria convert into nitrates, led farmers to abandon crop rotation. However, crop rotation also helps to prevent pests being a nuisance year after year. Many of the pests that affect a particular crop will have died out by the time that crop is planted again.

Activity I.1

Looking at root nodules

Examine the roots of a legume, looking for nodules.

In acid peat bogs where nitrates are not available, some plants have evolved methods of supplementing their nitrogen intake by digesting insects. Sundews and butterworts grow in this country. Elsewhere, there are Venus flytraps and pitcher plants (Figure I.7).

Figure I.7 An American pitcher plant – the cup-shaped leaves trap, digest and absorb insects

Summary

- Fungi and bacteria break down dead remains and release minerals and other nutrients into the environment. These materials are absorbed by plants and passed along food chains.
- Carbon and nitrogen are examples of substances being continuously cycled through the environment.

Questions

1 What is decomposition?
2 Outline the stages in the carbon cycle.

Extension questions

3 Plants capture light energy and use it to make food. Feeding transfers food energy along food chains. Explain how decomposition completes the transfer of all of the energy originally captured by plants.
4 Why do farmers sow leguminous crops like peas and beans in 'rotation' with other crops?

Activity I.2

Studying the activity of soil bacteria

Figure I.8 Inserting a cotton strip in soil

Scientists often need to compare the activity of bacteria in different soils. The **cotton strip assay technique** is used to compare the activity of bacteria which digest cellulose in plant cell walls. The technique involves burying pieces of a specially made cotton material in the soil (Figure I.8). At intervals, the fabric is dug up. Its appearance is noted. The tensile strength of the material is tested (Figure I.9) as a means of assessing the amount of digestion by bacteria. As a safety precaution, this testing has to be carried out in a special container to protect the scientists against any fungal spores on the cotton. These might cause allergies in some people.

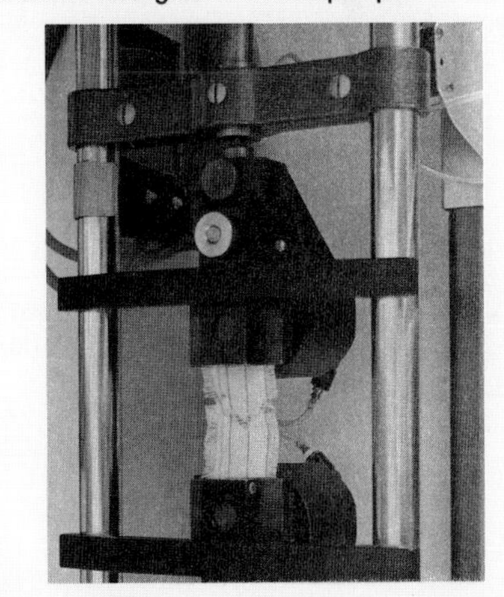

Figure I.9 Testing the tensile strength of cotton that has been buried

The cotton strip assay technique has been standardised so that it will give results which can be repeated. What features of the method would have to be controlled to ensure that results of different studies could be compared? Do you think the method is a good way of showing how soil bacteria digest cellulose?

Unit J
Sewage

Humans excrete approximately one and a half litres of urine and faeces a day. When the population was much smaller the waste would go into open sewers or directly into rivers or the sea. By the early 1800s in the UK the population was too large for this system of sewage disposal. Over 20 000 people died from the disease cholera in epidemics in 1831 and 1848. Eventually it was discovered that cholera and typhoid were due to untreated human sewage polluting the water supply. Parliament decided that something would have to be done.

Sewage contains organic wastes which, if left untreated, attract microbes and insects which help to spread disease. If the organic waste gets into rivers, lakes and seas it breaks down using much of the dissolved oxygen. Water without oxygen cannot support any life.

Modern sewage treatment relies on microbes to feed off the waste substances under controlled conditions. Carbon dioxide and methane gases are given off as the treatment proceeds. The solids left are used as fertilisers or soil conditioners. Some solids are dumped at sea.

The modern sewage works

Beckton, in the London Borough of Newham, is Britain's largest and most modern sewage works. It treats the sewage from some 2 500 000 people living within an area of 114 square miles. The works deal with an average **daily flow** of 1101 million litres (242 million gallons) of liquid. The untreated sewage goes through certain stages shown in Figure J.1.

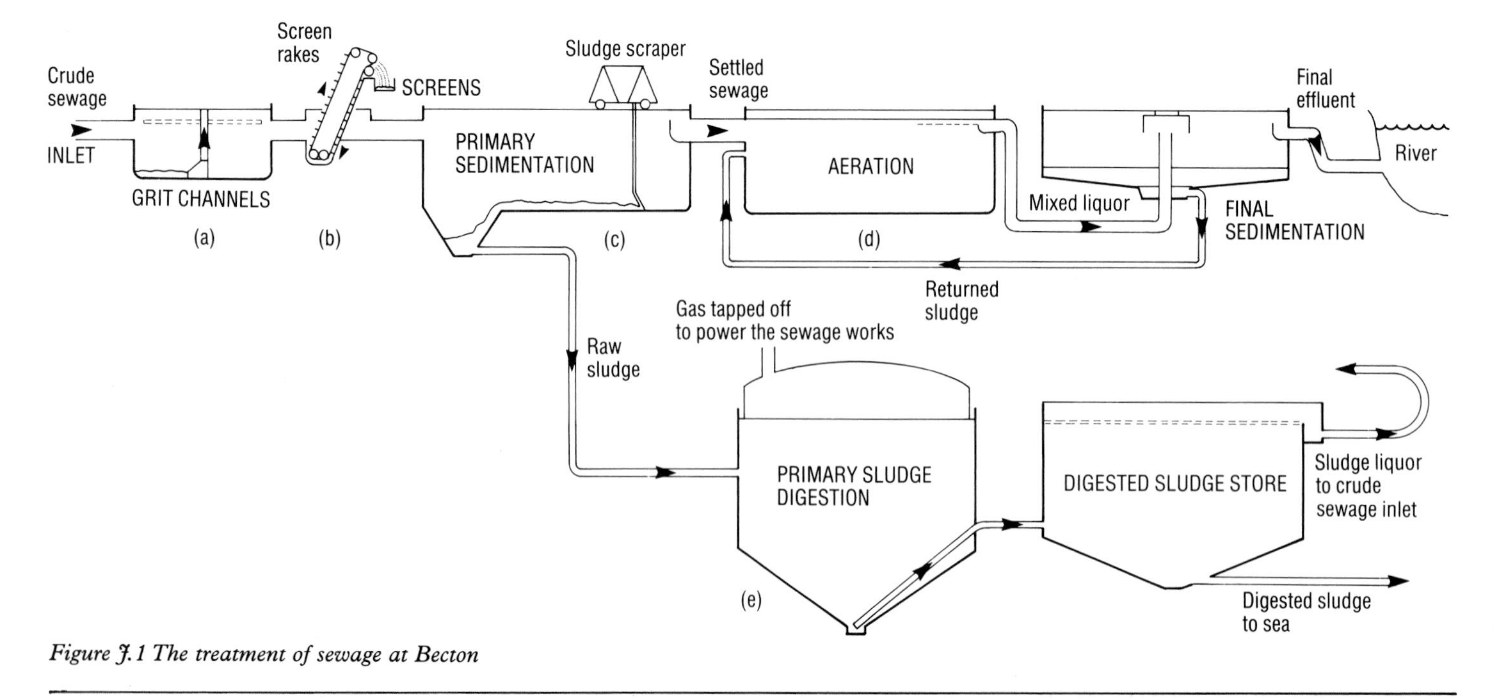

Figure J.1 The treatment of sewage at Becton

(a) **Grit removal channels**: The grit gets into sewage from roads and gardens. For this stage the sewage is made to flow very slowly so that the grit falls to the bottom. The grit is then removed.
(b) **Screens** remove any large solids, such as wood and rags.
(c) In the **primary sedimentation tanks** the heavy organic solids in the sewage settle out as primary sludge. This sludge is pumped to the primary sludge digester.
(d) The liquid is aerated by having compressed air forced through it in the presence of bacteria (see Figure J.2). It becomes a jelly-like mass with **aerobic bacteria** feeding on the organic matter with the help of oxygen gas. In eight hours most of the impurities have been destroyed and turned into carbon dioxide and water. The product from the aeration process passes into the **final sedimentation vessel**. Any sediment formed there goes back to the aeration tank for a second treatment. The liquid is then safe to pour into the river or sea.

In less modern and smaller sewage works this aeration stage involves bacteria in a different situation. The liquid is sprayed from rotating arms and falls onto a circular base of stones or coke. As the liquid percolates through the 2 metre thick base it is attacked by bacteria living among the stones and rendered harmless. Figure J.3 shows this familiar sight.
(e) **Primary sludge digestion**: The primary sludge is warmed and mixed with bacteria that can live without oxygen (called **anaerobic bacteria**). The bacteria feed on part of the sludge and make it harmless whilst generating methane and carbon dioxide. The methane is used as a fuel to heat the primary sludge digester. Warmth increases the activity of the bacteria. The digested sludge is stored and then dumped at sea or on land, where the rest of the sludge is degraded. In England and Wales one million tonnes of sewage sludge (dry weight) are disposed of each year.

If we consider only sewage that is returned to inland rivers, 80% has been fully treated in a similar manner to that described above. Raw sewage and sewage where the solids have been separated from the liquid, make up the remaining 20%. This 20% will be highly polluting for rivers.

Figure J.2 Aeration of sewage

Figure J.3 Circular biological filters

The compost heap

Uneaten food and outer leaves from vegetables and garden weeds can all be made into manure by composting them. Many keen gardeners have a compost heap to recycle vegetable scraps and garden waste. Bacteria, fungi and worms break down the waste into a compost, which can be used as fertiliser on the garden after a few months. The compost improves soil structure, retains moisture and smothers annual weeds around plants and between rows of vegetables.

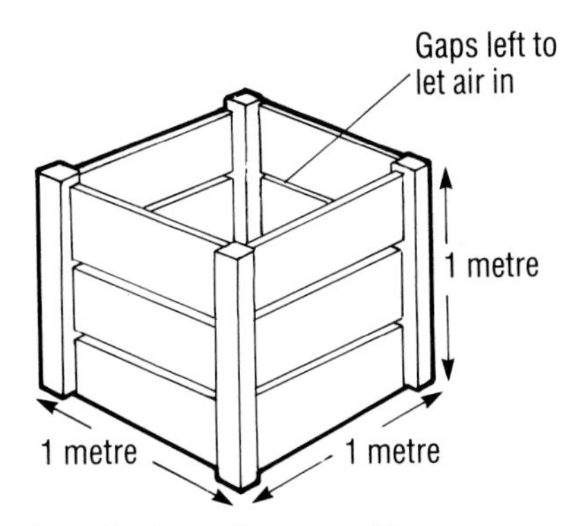

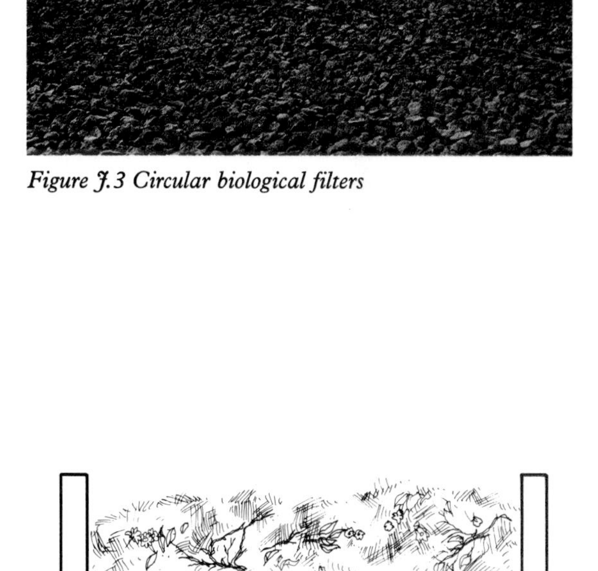

Figure J.4 Recycling vegetable waste

Activity J.1

To make a compost heap at school or in your garden

1. Construct a wooden framework one metre square to contain the waste (Figure J.4).
2. Site the heap near the bottom of the garden where any smells of rotting material can be ignored. Do not choose particularly wet or shady areas.
3. Make a layer of garden waste such as grass-cuttings, weeds, vegetable waste and left-over fruit and vegetables. Add a commercial preparation made to accelerate decomposition if you wish.
4. Cover the layer with manure or straw plus fertiliser if desired. Continue the alternate layers of waste and straw.
5. Water the heap in dry weather and turn layers over to get air to them.

Summary

- Raw sewage is a health hazard.
- Bacteria break down sewage waste into harmless and even useful substances.
- Sewage works are designed to deal with large volumes of waste.
- Compost heaps are homes for bacteria, fungi and worms which convert waste plant material into valuable manure.

Questions

1. What damage can untreated sewage do to the wildlife in rivers?
2. How do bacteria help in the modern treatment of sewage?
3. Summarise the stages in the treatment of sewage in a modern sewage works.
4. What are the benefits of spreading compost on the garden soil?

Unit K
The human impact and conservation

Pressure on the environment

Ten thousand years ago about 12 million people lived in the world. Today, the world population is approaching 6 000 million. It seems set to reach 10 000 million before the end of the 21st century. All of these people will need food, homes, hospitals, schools ... and many of them will want a wide range of manufactured goods.

These human needs and wants make great demands on the world's resources, and have an impact on the environments from which these resources come. Figure K.1 (overleaf) shows how the human impact on the environment has changed over time.

Resources are the raw materials needed to satisfy human demands. They fall into two main categories: renewable and non-renewable. Renewable resources come from living things and can be replaced, for example, food (crops and livestock); timber; and raw materials for clothes (cotton, flax and wool). Non-renewable resources cannot be replaced when used up. For example, there are only limited amounts of fossil fuels (coal, oil, natural gas) and metals in the form of their naturally-occurring ores.

The most extreme example of humankind's impact on the environment is nuclear power. Never before has human action been able to change the environment on such a large scale. Nuclear weapons give us the power to wipe out life on Earth.

Activity K.2

A matter for concern

Every day, newspaper reports highlight ways in which human activities threaten wildlife, habitats and our own health. Are these fears justified?

Carefully examine the newspaper reports in this unit. Then look at the graphs and other data about the same topics. Answer the questions under *Assessing the evidence* for each story.
Do the accounts agree? Answer the following:

1. Taking into account the extra information given, do you think the newspaper stories are true? Give reasons for your answer.
2. Does the data give the whole picture? Give reasons for your answer.
3. Do you think the headlines are misleading? In what way?
4. What are your main sources of information about the environment (for example, television, radio, newspapers, books, teachers, parents, friends)? Put them in order of importance to you. How could you check the information you obtain from these sources?
5. Do you think the answers to the survey (Figure K.2) are justified in the light of the newspaper reports and other information? Give your reasons.

Activity K.1

Environmental survey

In a recent survey a random sample of people in England and Wales were shown a list of environmental problems. They were asked: 'How worried do you feel about each of these problems?' Figure K.2 shows their answers.

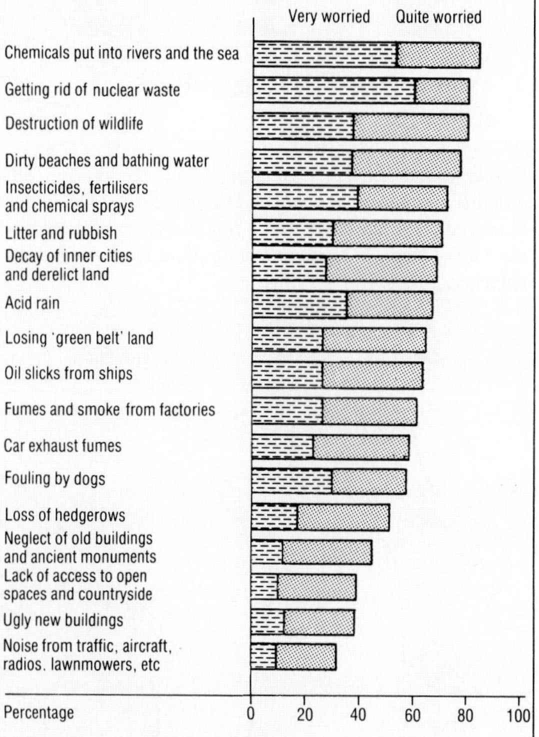

Figure K.2 Percentage of people concerned about different environmental problems

1. What were the three most worrying problems, according to the survey?
2. Which three problems were people least worried about?
3. List the problems in rank order of how worried *you* feel about them, with the problem that worries you most as 1, and so on.

Figure K.1 The human impact on the environment over 1.5 million years

(a)

About 1.5 million years ago bands of Homo erectus *(an early type of human) moved from place to place in search of food. They camped in one site to hunt animals and plants, moving on in a few days. They had no more impact on the environment than other medium-sized animals. They left few traces other than seeds, animal bones and a few sharp-edged stone tools used to cut up meat. Even these few remains could mislead us into thinking that early humans ate a lot of meat, because collecting plant foods needs almost no tools – except perhaps a sharpened stick for digging up roots, which would leave no evidence.*

(b)

Around 10,000 years ago people like these in the Middle East harvested wild wheat and other grains. When the grain was ripe a family could gather over a year's supply in just a few weeks. This store of grain meant that they did not have to move about in search of food. Harvesting and storing wild cereal grasses gave people a more settled way of life; they began to live in small villages.

Archaeologists investigating a site in modern Syria found stone blades and sickles, and stones for grinding corn into flour, as well as seeds and animal bones left over from people's food. There is no evidence that plants or animals were deliberately reared, so people had little impact on the environment beyond their village.

(c)
By about AD 100 people had started to farm. Perhaps they began by watering wild crops – then sowing seed. The crops improved and provided more than enough for people to eat. This meant that not everybody needed to work on the land. People developed skills in crafts such as weaving, pottery and tool-making. They provided things the community needed, possibly in exchange for food.
Villages became larger and some grew into towns. People had a much greater impact on the environment: farming the land, using raw materials.

(d)
People in Europe, North America, Japan and Australia (the developed countries) mostly live in big cities. Food for these people is provided by relatively few farmers. Industry and technology offer improved living standards and increasing leisure time; they use raw materials from the less-developed countries of Africa, South America and Asia. Environments in these countries may be stripped of the resources they need for development.
Environments in the developed countries are destroyed through pressure for living space, roads, railways, airports and other services. Compare the picture today with that of a few thousand years ago. How has the human impact on the environment changed?

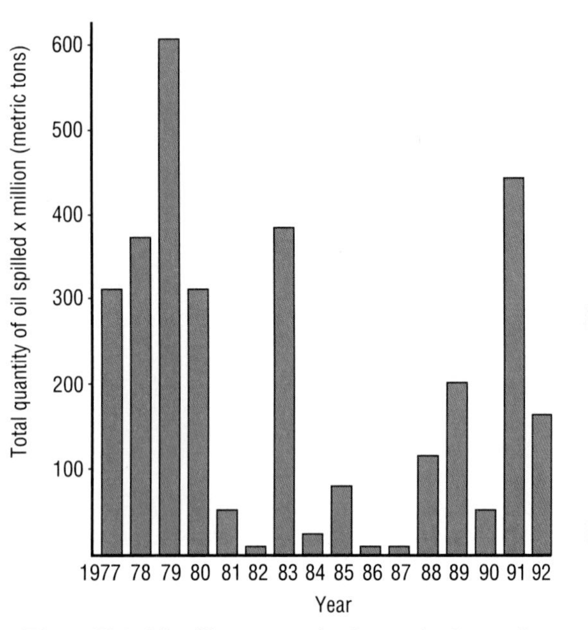

Figure K.3 Oil spills at sea – the data only shows oil lost through accidents at sea. It does not include oil lost from cleaning out ships' tanks, during off-shore drilling or from operations at oil refineries. These losses are thought to be as much as the amount lost during accidents

BLACK DEATH

Our correspondent in Alaska reports
Mystery still surrounds the events which caused the 987ft oil tanker *Exxon Valdez* to plough into Bligh Reef on Good Friday, causing the worst spillage in American history.

Conditions were good as the tanker left Port Valdez and entered Prince William Sound. 20 minutes later the *Valdez* was seriously off course and heading for the rocks. The bridge was manned at the time.

The *Exxon Valdez* was badly holed by the rocks. Two of the 13 oil tanks were punctured and oil began to pour into the sea. Attempts by the crew to get the ship off the Reef made matters worse. A massive hole was ripped in the bottom of the hull, puncturing six more oil tanks.

240,000 barrels (10m gallons) of oil gushed out of the *Valdez*. The slick has now spread out to cover 1000 square miles of Prince William Sound – twice the size of Greater London. Exxon admit that they are not equipped to deal with a disaster on this scale – they will be lucky to clear up even half the oil.

Thousands of animals are dead or dying, including some of the world's rarest species. Environmentalists fear that millions more could perish in the next few weeks, unless the oil is cleared quickly. Millions of migrating geese, duck and other birds arrive in the sound at the end of April, on their way to the north and east of Alaska. Experts fear that these, too, may be doomed by the spreading slick.

Disaster threatens the livelihoods of the 2000 people who live along the coast. Damage to fish stocks means that the £57m-a-year fishing industry could be decimated. Millions of wild salmon and herring – the main catch – will be contaminated. 675m juvenile salmon are about to be released from hatcheries into the main feeding grounds – even if these can be saved, public confidence in the Alaskan fisheries may have been lost.

Assessing the evidence

- How did oil escape from the tanker *Exxon Valdez*?
- How many barrels of oil escaped? (One barrel = 0.14 tonnes) Compare the loss with other oil spills shown in Figure K.3.
- How does the oil spill threaten the livelihoods of people living in the villages along the shore of Prince William Sound?
- Briefly describe three ways in which wildlife is at risk from the oil.

The wastes produced inside nuclear reactors are a 'cocktail' of different radioactive elements. These decay (give off radiation) at constant but different rates. The time taken for the radioactivity in a particular element to decay to half its original level is called its half-life. Table K.1 shows the half-life of typical radioactive elements, some of which are produced inside nuclear reactors.

Before assessing the evidence, read pages 239–241. The information will help you answer the questions.

Chemical	Half-life
Carbon-14	5760 years
Phosphorus-32	14.3 days
Sulphur-35	87.2 days
Calcium-45	165 days
Strontium-90	28 years
Iodine-131	8.04 days
Caesium-137	30 years
Plutonium-239	24400 years

Table K.1

Assessing the evidence

- What is meant by an element's half-life?
- Find out about alternative methods for storing nuclear waste. Why are different methods more suitable for some types of nuclear waste but not other types.
- Rate each method of storing nuclear waste for safety. Explain how you used the evidence to establish safety ratings.
- Look at Table K.1. Is 100 500 years long enough to store waste? Why/why not?

Save our beauty spots!

from our Environmental Correspondent

The Council for the Protection of Rural England warned today that more than 50% of Britain's 5000 sites of special scientific interest are at risk from agriculture and building development. The Council called for EC action to protect wildlife, habitats and landscape.

Sites are designated for protection by the Nature Conservancy Council. Since 1983, an average of nearly 70 sites a year have been destroyed or seriously damaged. The damage is often due to illegal agricultural or forestry practices, a spokesman said today.

A further cause for concern is the destruction of sites for building development. Government figures reveal that since 1981 about 20 000 acres of moorland have been lost in Northumberland, Cumbria and Durham, about 5000 acres in Lancashire and Cheshire, and 12 000 in the South-west.

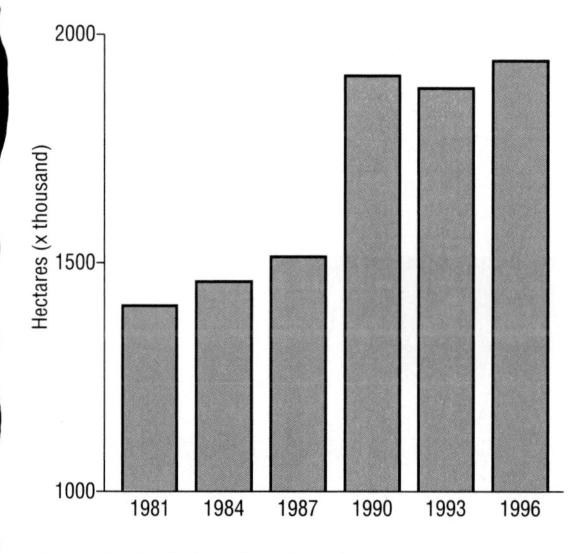

Figure K.4 UK sites of special scientific interest

Sites of special scientific interest (SSSIs) are parts of the countryside listed by the Nature Conservancy Council as areas with well-developed natural environments, geological features or habitats containing rare wildlife. Landowners who have SSSIs on their property are paid compensation for not developing them, and can be fined for damaging them. However, profits from development are often far greater than the fines.

Assessing the evidence

- What is a site of special scientific interest?
- Do you think SSSIs are important? If so, why?

  Figure K.4 shows that the land area covered by SSSIs is increasing, yet the newspaper reports that they are under threat. How would you account for the differences?

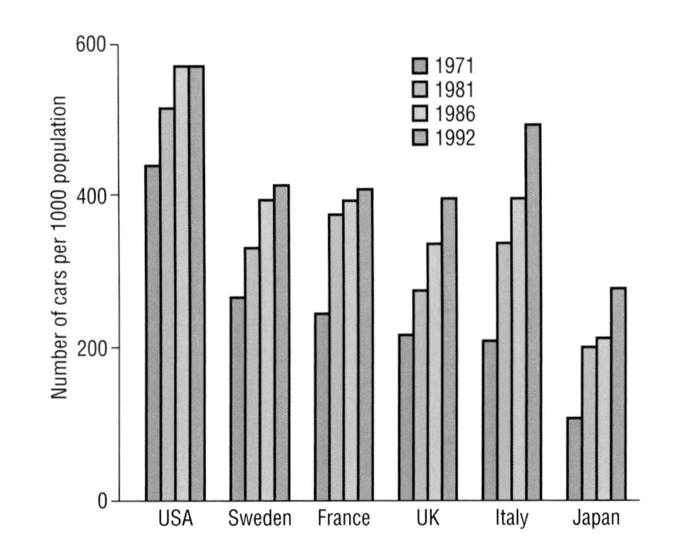

Figure K.5 International comparison of the growth in car ownership. The rate of growth in different countries varies. In the UK, the number of cars per thousand people rose by almost 50% between 1971 and 1986. US ownership rose by only 25% over the same period. However, there are many more car owners in the USA than there are in other countries.

Car ownership is an indicator of individual wealth and buying power – and thus of demand for goods and services which use up environmental resources. However, the graph should be interpreted with caution. Japan has one of the world's most successful economies, but it has the least car owners of the countries illustrated. Can you explain this?

Conservation

Our well-being depends on keeping a balance between using and protecting environments and resources. Conservation involves working to achieve this balance and care for the environment. Conservation means different things to different people, as Figure K.6 shows.

Figure K.6 What is conservation?
(a) Preserving wildlife and places of natural beauty.
(b) Maintaining the balance between living things (including humans) and resources.
(c) Providing attractive places to live and opportunities for leisure.
(d) Returning to the simple life.

Many people feel very strongly about conservation. Organisations like Friends of the Earth and Greenpeace inform people about environmental issues and try to persuade governments to adopt conservationist policies. But environmental issues are not just the concern of governments – after all, they are rooted in the long-term processes of ecology and outlive successive governments. Good management of the environment involves many day-to-day decisions that enable human beings to live in the world without damaging it.

Figure K.7(a) Grasses are different from other plants in that the point on the stem from which growth occurs is near the ground. Mowing the lawn kills those plants that cannot tolerate regular topping, but grasses flourish because their growing points are out of the way of whirling blades

Managing the garden environment

Have you ever thought about what you are doing when mowing the lawn? No, not that it's hard work or what's for dinner, but the effect of cutting down all those plants. One way of finding out is to leave part of the lawn uncut and watch what happens. Soon daisies and dandelions appear, followed by a range of other flowers (Figure K.7).

Leave a patch of lawn in your garden uncut for a season, then carefully record the changes that take place. Identify the different plants as they appear and note for how long they flower before dying back. Estimate the abundance of each species. You will notice that insects and other visitors are attracted to the flowers as your patch of uncut lawn develops into a haven for wildlife.

Since 1949 almost all flower-rich meadows have been lost to intensive farming. So-called 'improved' grassland used for grazing livestock is a desert for meadow flowers (Figures K.8 and K.9).

Our gardens can become a refuge for wildlife and plants with a few simple changes in the style of management. You can still do this in a small garden by choosing just one way to manage it for wildlife, and at the same time give yourself endless fun and interest.

Figure K.7(b) Leaving the lawn uncut gives other plants a chance to grow and flower. If left for a few years, woody plants grow up and overshadow the grasses which die out

Figure K.8 Unimproved meadows are home to a variety of wild flowers

Summary

- An ever-increasing human population demands resources. These demands bring about the human impact on the environment.
- Conservation aims to balance use with protecting the environment.

Figure K.9 Improved grassland is sown with one or two species of grass that grow abundantly with a dressing of artificial fertiliser

Questions

1 Describe differences between the human impact on the environment today and 10 000 years ago.
2 Briefly explain what conservation of the environment means.
3 What is 'improved' grassland?

Unit L
Land use

Impact of intensive farming

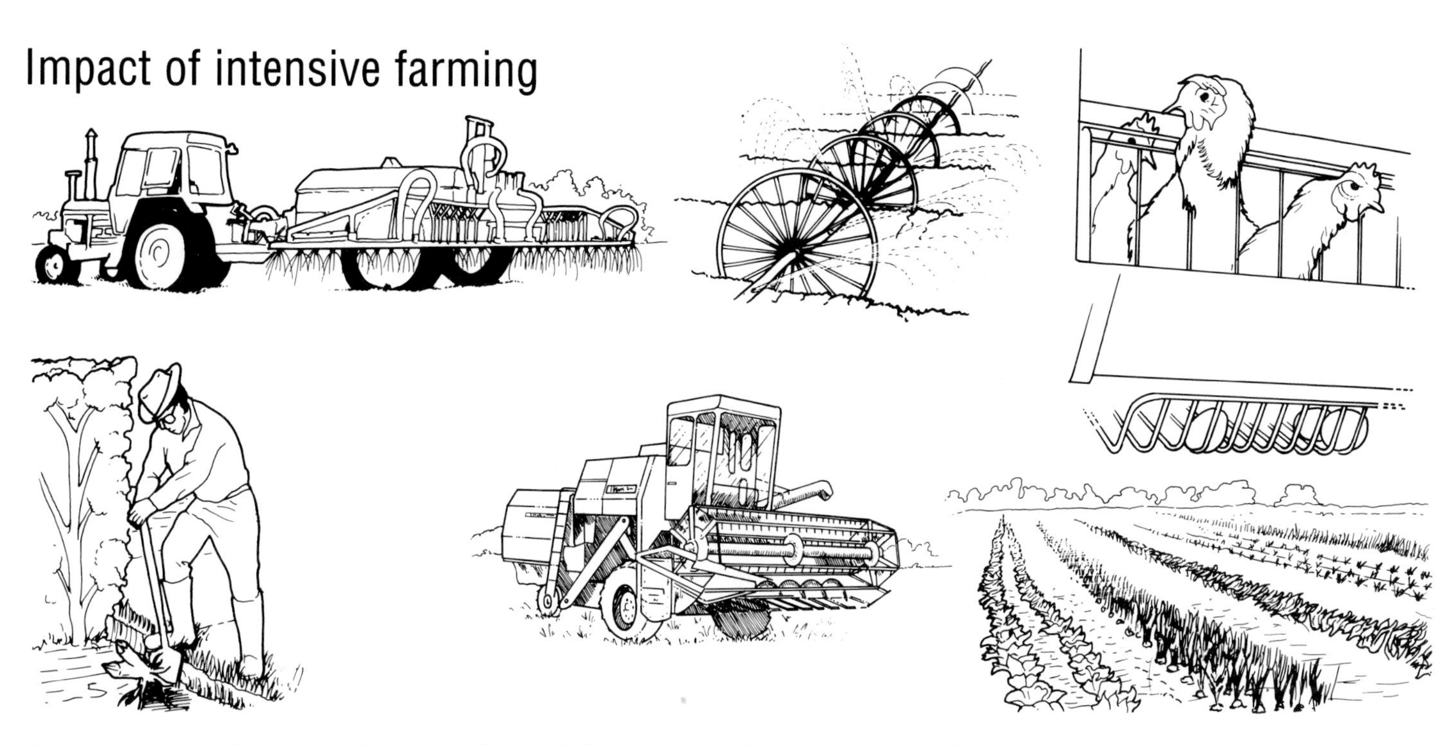

Figure L.1 Intensive farming – different technologies help farmers produce large quantities of food

Pest control

Pests are plants and animals that destroy crops and livestock, or prevent land from being used for farming. **Pesticides** are chemicals used to kill pests. There are three types of pesticides:

1 Insecticides – chemicals that kill insects
2 Herbicides – chemicals that kill weeds (plants that compete with crops for space, light and nutrients)
3 Fungicides – chemicals that kill fungi.

Farmers apply pesticides as sprays, dusts, dips, fogs or granules. In 1985 nearly 10 000 tonnes of pesticides were used on farmland in Britain. More than 800 000 tonnes were used worldwide, at a cost of about £28 billion (see Figure L.2). However, data are not available for all countries, so it is likely that these figures are an underestimate.

Developed countries use about 80% of all pesticides worldwide, with the USA using the most. If food production continues to increase at its present rate, then developed countries will use 2–4% more pesticides each year and developing countries 7–8% each year.

Table L.1 looks at the costs and benefits of using pesticides.

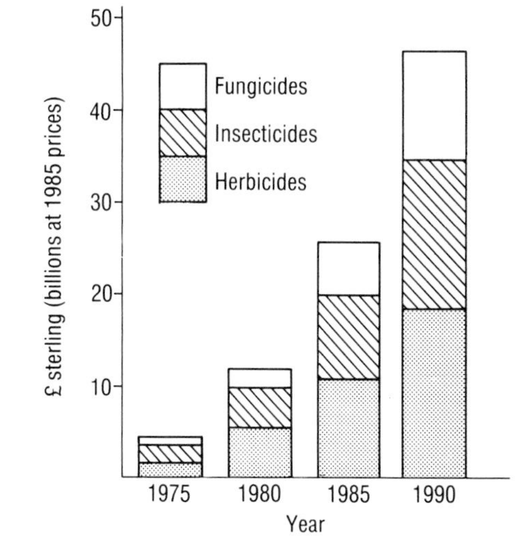

Figure L.2 World sales of pesticides. By how much have worldwide sales of pesticides increased between 1985 and 1990? Is this increase matched, for example, by the increase in cereal production in Western Europe over the same period? Comment on the comparison of figures

Costs	Benefits
• *salinisation:* sun's heat evaporates water, increasing the concentration of mineral salts in soil. Eventually land becomes too 'salty' for crops to grow; this reduces crop yields • *overwatering:* land becomes waterlogged, reducing crop yields. In warm countries this can threaten public health because it creates the habitat for the spread of water-borne diseases such as malaria and schistosomiasis	• brings water to land • food production increases

Table L.2 Irrigation: costs and benefits

Costs	Benefits
• spray is carried in the wind (drift) which can harm wildlife • pesticide run-off seeps into ground water, eventually draining into ponds and river systems, possibly contaminating drinking water and harming wildlife	• kills pests which damage crops • food production increases

Table L.1 Pesticides: costs and benefits

Irrigation

Irrigation brings water to land that would otherwise be too dry to grow crops. It also helps improve crop yields in areas where rainfall is low. Today, over 2 million sq km of irrigated land produces about 30% of the world's food. Experts predict that over 3 million sq km of land will be irrigated by the first decade of the 21st century. Table L.2 looks at the costs and benefits of irrigation.

Mechanisation

In the developed world modern farm machinery has replaced traditional methods of farming, although these still exist in many developing countries. Machinery is powered by fuel oil, replacing the muscle power of humans and animals. Modern farm machinery works best in large fields without hedges, fences or other obstructions. Table L.3 looks at the costs and benefits of mechanisation.

Costs	Benefits
• crop provides unlimited niche for consumer populations which expand to pest proportions • expensive because crops must be sprayed with pesticides to control pests • soil loses its nutrients so costs rise because artificial fertilisers are needed to replace lost nutrients • soil left bare between crops risking erosion from wind and rain	• efficient use of expensive machinery • reduces labour costs • high yields mean farmer can take advantage of selling in bulk to obtain best prices for produce

Table L.4 Monoculture: costs and benefits

Costs	Benefits
• land cleared of woods and hedges to make fields larger – large machinery is more efficient in big fields • habitats destroyed with loss of wildlife • wheels and tracks pack soil causing waterlogging	• powerful machinery needs few men to work it – cuts down on labour – wage costs reduced • more land can be used for farming • crops harvested more quickly

Table L.3 Mechanisation: costs and benefits

Monoculture

Growing a large area of a single crop is known as **monoculture**. The same crop is often grown year after year. Intensive arable farms specialise in monoculture of a limited number of crops (Table L.4).

Manure

Animal dung contains the undigested remains of food. It forms manure, which is a natural fertiliser. Large numbers of animals reared intensively indoors produce colossal amounts of manure. Most intensive farms specialise (ie they either grow crops or rear animals) – so a farm which rears animals intensively will have no arable land to put the manure on. It is washed into pits, called lagoons, where it forms a liquid slurry (Table L.5).

Costs	Benefits
• water needed to wash manure to lagoon • slurry leaks from lagoon seeping into streams and rivers: this uses up oxygen, killing fish and other wildlife and can contaminate drinking water	• None

Table L.5 Manure: costs and benefits

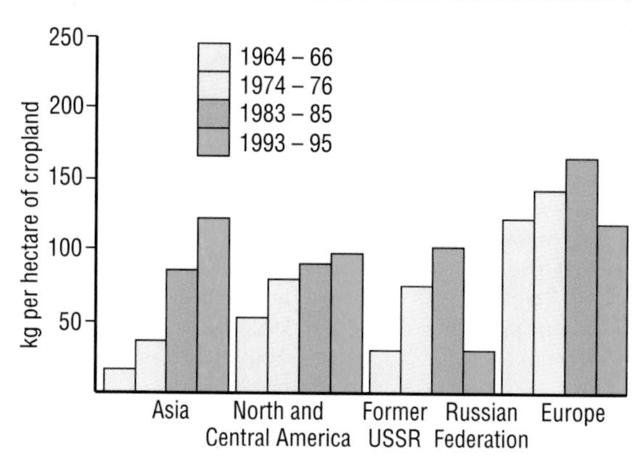

Figure L.3 Use of artificial fertiliser in different parts of the world. Notice the difference in the use of artificial fertiliser in Europe 1993–95. Can you explain this?

Artificial fertilisers

As we have seen, plants need certain nutrients for growth. In the wild, plants obtain these nutrients from dead organic matter, as it decomposes. Artificial fertilisers enable farmers to supply nutrients directly to their crops. Worldwide, the use of artificial fertilisers has increased by 30% in the last ten years (Figure L.3).

For example, plants need nitrogen, but they cannot obtain it directly from the air. Artificial fertilisers are produced by converting atmospheric nitrogen into nitrates which plants can use:

nitrogen + hydrogen → ammonia
ammonia + oxygen → nitrates

Farmers apply nitrate fertilisers as granules or sprays. Table L.6 looks at the costs and benefits of artificial fertilisers.

Costs	Benefits
• manufacture of fertilisers uses a lot of fuel • does not add humus to the soil so soil structure deteriorates and soil erosion increases • surplus fertiliser runs off into streams and rivers. This encourages population explosion of algae (called blooms) which use up oxygen in the water, killing wildlife. It can also contaminate drinking water with excess nitrates which are a health hazard	• land can be used continuously for growing crops • farmers do not need to keep animals for manure • efficiency increased by specialising in growing one or two crops each year • food production worldwide has increased by 19% over the past 10 years, helping feed the world's growing population

Table L.6 Artificial fertiliser: costs and benefits

Habitats and wildlife losses

Table L.7 shows estimated losses of wildlife habitats since 1945. These losses are due mainly to changes from traditional farming practices to intensive methods. Table L.8 shows the reduction in wildlife species found on modernised farms. Since 1945, 200 000 km of hedges in England and Wales have been removed. Table L.9 looks at the costs and benefits of hedge removal.

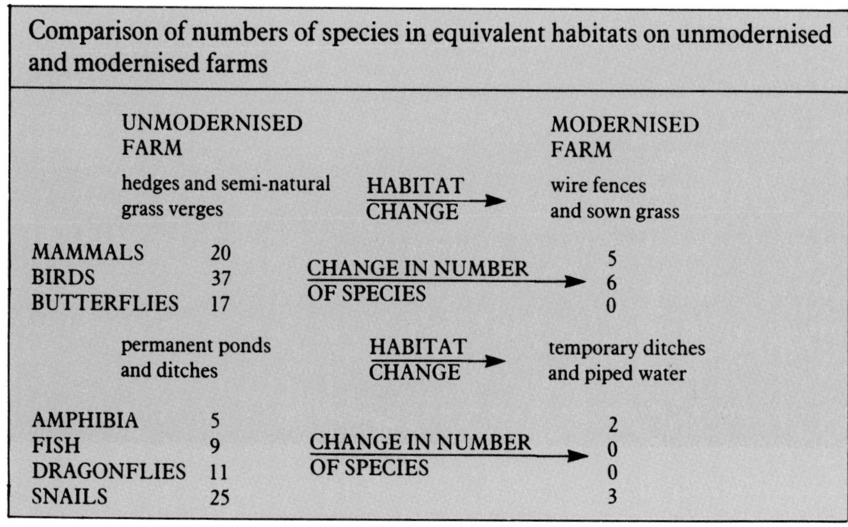

Comparison of numbers of species in equivalent habitats on unmodernised and modernised farms

	UNMODERNISED FARM		MODERNISED FARM
	hedges and semi-natural grass verges	HABITAT CHANGE →	wire fences and sown grass
MAMMALS	20	CHANGE IN NUMBER OF SPECIES →	5
BIRDS	37		6
BUTTERFLIES	17		0
	permanent ponds and ditches	HABITAT CHANGE →	temporary ditches and piped water
AMPHIBIA	5	CHANGE IN NUMBER OF SPECIES →	2
FISH	9		0
DRAGONFLIES	11		0
SNAILS	25		3

Table L.8 Reduction in wildlife species on modernised farms

Habitat	% Loss
Lowland unimproved grassland including flower-rich hay meadows	95
Lowland sheep walks on chalk and limestone	80
Lowland acid heaths	40
Ancient lowland woods	40
Lowland wet lands	50
Upland unimproved grasslands, heaths and blanket-bogs	30

Table L.7 Losses of wildlife habitats since 1945

Costs	Benefits
• habitats lost • loss of windbreaks leaving soil unprotected (i) evaporation of soil water increases (ii) topsoil blown away (erosion) • variety of wildlife decreases	• fields larger, so that machinery can be used more efficiently • land drainage can be improved more easily • original boundary hedges not needed when fewer • large farms are formed from merger of smaller ones

Table L.9 Hedge removal: costs and benefits

Impact of building, quarrying and dumping waste

There are 24 million hectares of land in the UK. Around 19 million hectares are used for agriculture. **Building, quarrying** and **waste disposal** are some of the other human activities that occur on the rest of the land (Figure L.4), and have a huge environmental impact.

Figure L.4(a) Nearly 90% of household waste is dumped into large holes or pits in the ground. The operation is called landfill

Figure L.4(b) Housing covers 1.7 million hectares of land in the UK

Figure L.4(c) Each year, quarrying in the UK produces around 300 million tonnes of gravel, limestone, sand and sandstone for concrete and other building materials

Dumping waste from long-term mining activities has an obvious visible impact on the environment. The waste from coal mines has, for centuries, been piled up into pit heaps, which completely remodel the landscape. Power stations produce vast quantities of pulverised ash, which also has to be dumped (Figure L.5).

Figure L.5(a) Estimates put the amount of waste lying in pit heaps around UK coal fields at around 200 million tonnes

Figure L.5(b) Coke and coal-fired power stations in the UK produce 10 million tonnes of pulverised ash each year

Questions

1 **a** Weigh up the benefits in food production of intensive farming against the costs to the environment.
b Do you think the benefits outweigh the costs? Give reasons for your answer, taking into account economic as well as environmental costs.

2 Crushed limestone is used for road building. Find out and assess the impact on the environment of quarrying the limestone, crushing it and then laying it as part of a new road.

3 Describe how the rubbish collected from your home each week is disposed of.

Summary

- Farming accounts for around 79% of land use in the UK.
- Intensive methods of farming can damage the environment.
- Building, quarrying and waste disposal also use land and affect the landscape.

Unit M
Water pollution

Polluting water: the North Sea

Some of Europe's major rivers flow into the North Sea. These rivers carry chemical wastes, which pollute the waters of the North Sea. The three major types of chemical waste are:

1 *Nutrients* (eg nitrogen, phosphorus) from sewage works and surplus artificial fertilisers which drain off farmland into rivers.
2 *Pesticides* (eg lindane, DDT, aldrin) used to protect crops (see page 157); these also drain from farmland into rivers.
3 *Metals* (eg mercury, cadmium, copper) from different industrial processes.

These are just a few of the chemicals that enter the North Sea. They are not all brought by rivers. Other sources of pollution include:

- dumping from ships
- fall-out from the burning of wastes at sea (in specially-designed ships)
- outfall pipes from the shore
- the atmosphere (see Figure M.1).

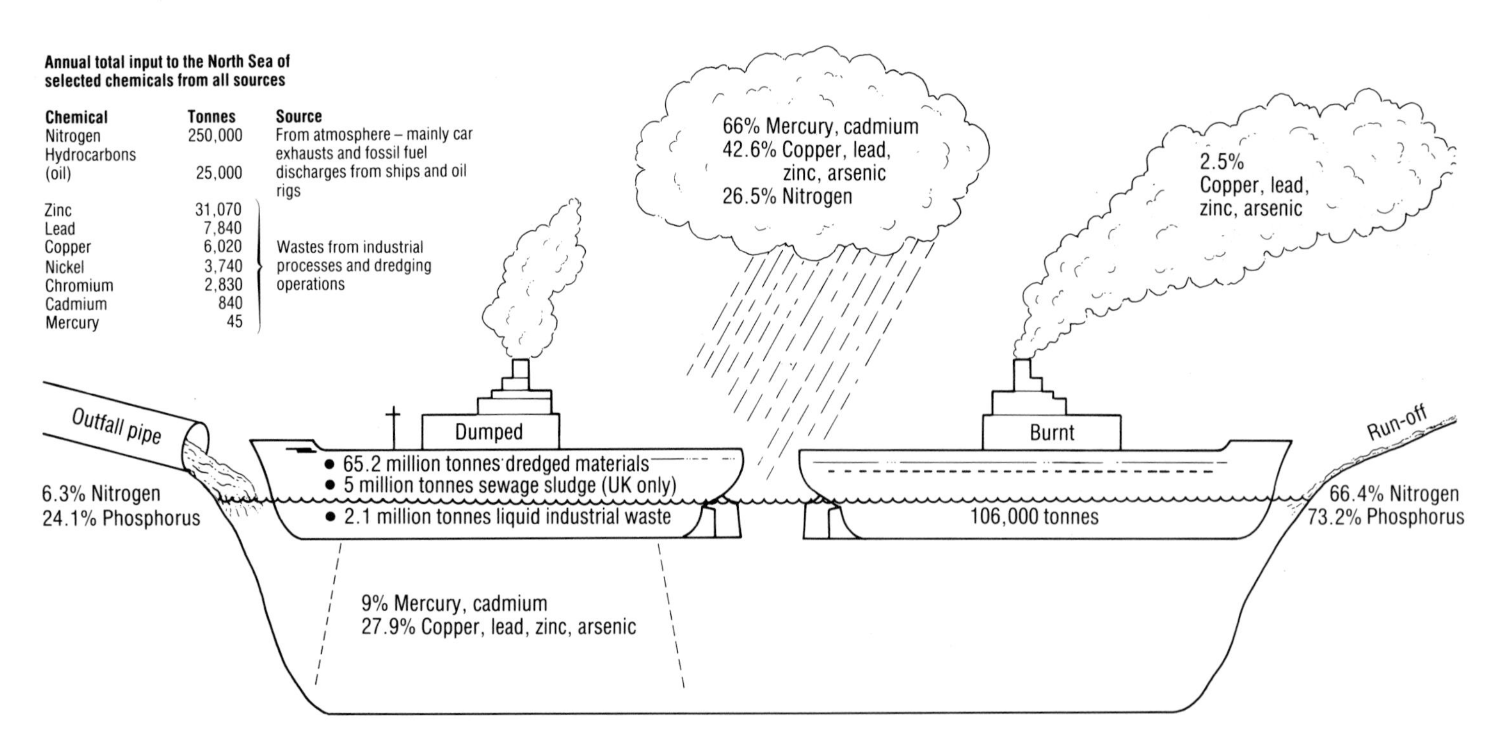

Figure M.1 Sources of chemical wastes in the North Sea. Figures are inputs from North Sea countries, with % contribution of selected chemicals from different sources

A recent report for a conference of European environment ministers called the North Sea 'one of the most polluted seas on Earth'. At the same time another report disagreed, arguing that with the exception of one or two 'hot spots' around Europe's coastline, the North Sea is a 'wholly healthy body of water'. Who is right? Both reports used the kinds of information in this unit; yet they came to opposite conclusions!

The water movements shown in Figure M.2 tend to carry wastes from Britain's shores across the North Sea towards the German Bight and Wadden Sea. The same water movements tend to trap the outflow of wastes from the European Continent in the same area. The result is that polluting chemicals concentrate around the eastern coast of the southern North Sea – one of the

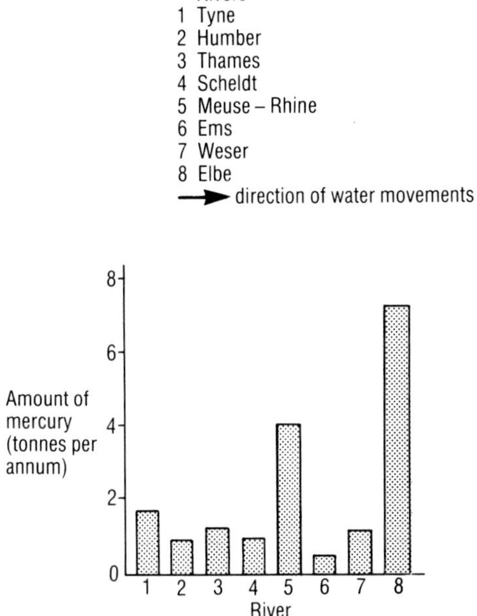

Different industrial processes use mercury (eg paper making), and seeds are treated with mercury compounds to prevent the growth of moulds

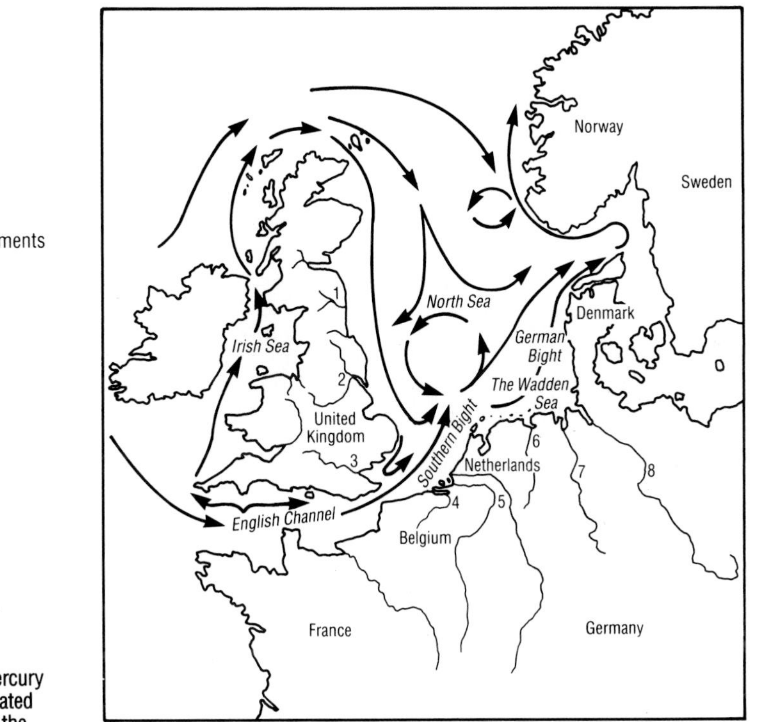

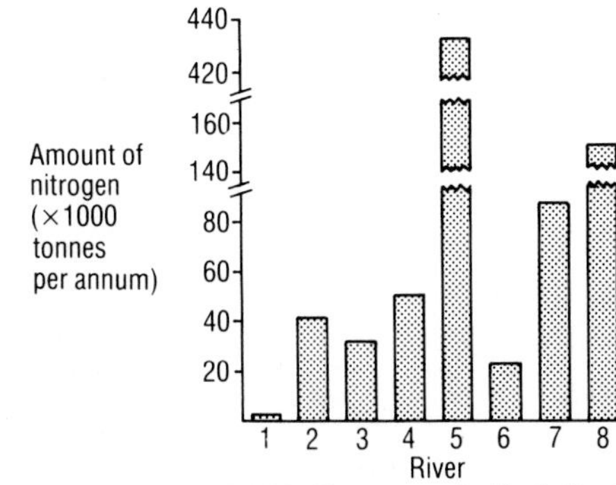

Sewage and artificial fertilisers enter the North Sea from the rivers named on the map

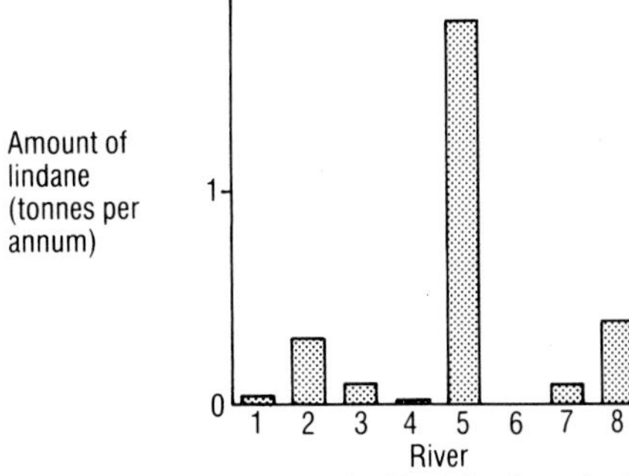

The pesticide lindane enters the North Sea from the rivers named on the map. Other pesticides entering the North Sea in run-off from farmland include aldrin, dieldrin and DDT

Figure M.2 Water movements in the North Sea – anticlockwise circulation pushes water east towards the European mainland and Scandinavia

'hot spots' mentioned earlier. Diseased fish are more common in the German Bight and Wadden Sea than elsewhere in the North Sea (Figure M.3).

Problems include skin cancer, ulcers, damage to the gills and abnormalities of the skeleton. Scientists have found metals in the water, particularly in the German Bight. So who is right? Can you arrive at an answer, based on the evidence in this unit? Give reasons for your decision.

Pollution in all its forms harms wildlife. For example, the common seals of the Wadden Sea are particularly hard hit. They carry in their bodies more PCBs (a substance like lindane) than any other population of sea mammals in the world. In the 1950s there were 3000 common seals living in the Wadden Sea; today there are less than 300 (Figure M.4).

Figure M.3 Diseased fish are becoming more common in the North Sea. Some scientists blame increased levels of pollution, others think the causes are natural

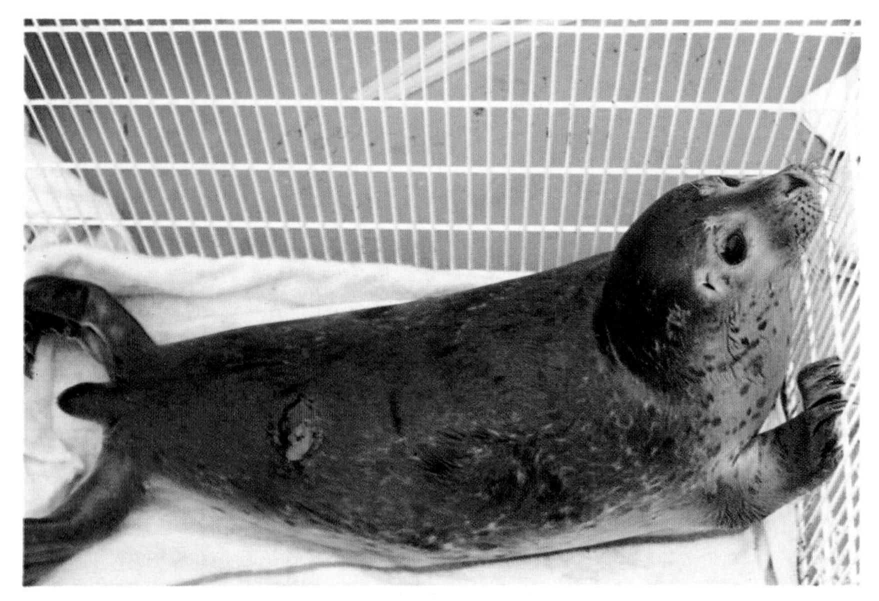

Figure M.4 A common seal from the North Sea. In 1988 common seal populations were badly affected by a virus that caused a disease like distemper in dogs. Many animals died before the epidemic faded away in the winter. The virus returned in 1989 but its effects were less severe. Again, opinion is divided over the link between pollution and the epidemics. Some scientists think there is no link, but others believe that pollution may have weakened the seals' immune system, making them vulnerable to the virus

The area is one of the most important breeding grounds for fish in the North Sea. For example, over 80% of the plaice caught in the North Sea for food begin life in the Wadden Sea. If we pollute it to the point where wildlife can no longer survive there, we may seriously damage the ecology of the North Sea as a whole.

Pollution can be dangerous for humans, too. Mercury, for example, accumulates up food chains. Fish species at the top of the chain have been found to contain large amounts. If these fish are caught for food, mercury enters the human food chain. An incident at Minamata Bay in Japan highlighted the tragic consequences of mercury pollution. A factory in the area had released mercury into the sea for at least 30 years. This gradually passed up the food chain to fish – an important part of the local diet. Mercury attacks the human central nervous system: some people died after eating the fish; others were brain damaged and pregnant women who had eaten the fish later gave birth to deformed babies.

Sewage pollutes many of the North Sea's most popular bathing beaches. Most of Britain's sewage systems were built over 100 years ago. The contents of sinks and lavatories emptied into pipes that took the untreated waste straight into the nearest stretch of sea. Here sunlight and salt water destroyed most of the harmful disease-causing organisms.

Figure M.5 A sewage outlet off the coast of Britain. Sea water is unfit for bathing if 100 cm^3 contains more than 2000 bacteria found in faeces. Bacteria causing diarrhoea, salmonella, hepatitis, cystitis, infections of the ear, nose and throat, typhoid and polio are found in sea water polluted with sewage

Figure M.6 Blackpool, crowded with holiday makers. How could you estimate the number of people on the beach?

	1981	1986	1990	1991	1992
Number of incidents	552	436	791	705	611
Spills requiring clean up	171	126	136	169	156
Costs (£1000s at 1992 prices)	219	190	1310	305	127

Table M.1 Offshore oil spills reported for the UK. The increase in the number of reported incidents is partly due to improved surveillance techniques

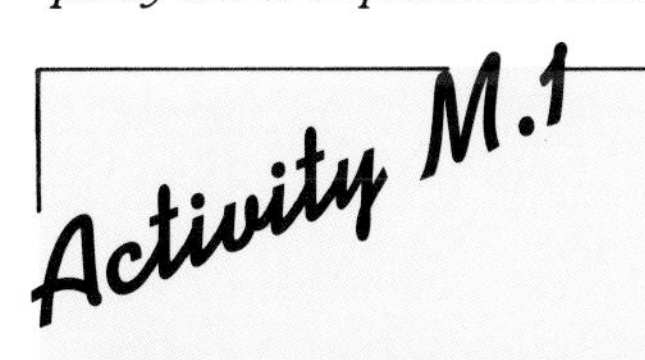

How to clean oil-soaked feathers

1. Put a few drops of water onto a feather. You should see that the feather naturally repels the water.
2. Dip five feathers into car engine oil (old car sump oil would do).
3. Dip each oiled feather into a boiling tube containing a different detergent, soap or solvent. Suggested solvents are ethanol (an alcohol) and propanone (also called acetone). Try the following soaps and detergents mixed with water: soap flakes, washing-up liquid and a biological washing powder. Move each feather about in its liquid. Leave for about ten minutes and then take out the feathers to dry.
4. Look at the feathers and assess them on cleanliness and power to repel water. Which treatment has been the best?
5. Find out how the RSPCA or other organisations treat the oiled feathers of birds. Do many birds survive this treatment?

The system worked quite well, but it was not designed to cope with the present size of seaside towns, especially when summer visitors swell the population (see Figure M.6). Today natural cleansing systems are overwhelmed and many of our beaches are unsafe for swimming.

Oil spills: a case study

Oil finds its way into the sea from accidents involving oil tankers, oil terminals or oil rigs. Some tankers clean out their tanks with sea water after delivery instead of using the proper facilities in dock. All of these incidents do a lot of damage to wildlife. The detergent used to disperse the oil often does as much harm to the marine life as the oil itself. The cost of the clean-up operations runs into many hundreds of thousands of pounds (see Table M.1).

Birds that swim or dive in the sea are killed in great numbers by the oil. Their feathers become coated in oil, which removes the insulating properties of the feathers against water and cold (Figure M.7). The birds take in the poisonous substances when they try to clean the oil off themselves. Millions of sea birds have been killed over the years in oil disasters (Figure M.8).

Figure M.7 An oily gannet. What is its chance of survival?

Figure M.8 Sea birds killed by oil spills

What can be done once the oil is spilt?

Detergents disperse the oil but can be pollutants themselves. They also spread the oil over a wider area, possibly increasing the threat to marine life. If oil reaches the shore it kills most of the life in shallow water. The coating of oil on the shore is smelly and difficult to remove. It causes great inconvenience to fishermen and local inhabitants. Attempts are concentrated on stopping the oil slick from reaching the shore.

One method uses bales of straw thrown into the oil slick. These absorb the oil and are rounded up using a line of booms. The booms are attached at one end to a barge and the other end to a motor launch. The oil-sodden straw is transferred to the barge.

Some of the other methods which have been tried include the use of huge vacuum cleaners to suck up the oil into floating tankers. The oil can then be recycled. Floating barriers can be used to keep oil out of particularly sensitive areas of the coastline, such as a wildlife sanctuary. BP have discovered a chemical that converts oil to a rubbery solid. If this can be developed, and it is not a pollutant itself, oil slicks can be made into an easily removed rubber sheet.

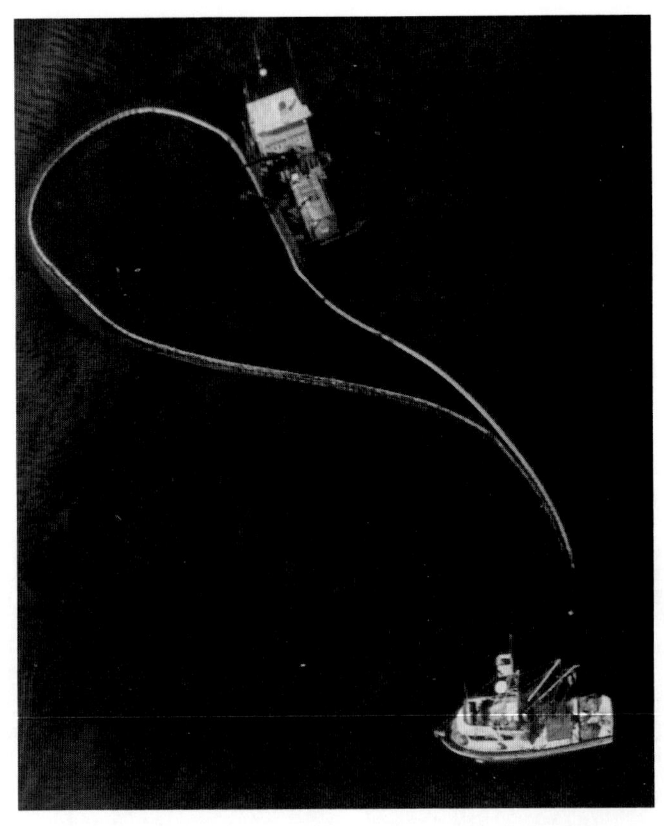

Figure M.9 Skimming a slick released from an oil tanker

The Great Lakes problem: a case study

During the 1960s problems were noticed with the lower Great Lakes of North America (see Figure M.10). A blue-green alga seemed to be taking over Lake Erie. The alga was being washed up on bathing beaches in such quantities that bulldozers were employed to remove it. The water supplies from the lake often had a poor taste and an unpleasant smell. There was hardly any dissolved oxygen in the deeper waters, so many insects and plants could not live in the polluted area. Lake Erie was dying. There were also problems in Lakes Ontario, Huron, Michigan and Superior.

The problem was traced to the elements nitrogen and phosphorus. The blue-green alga thrived on the high levels of these elements (a process known as eutrophication. It covered the lake's surface at the expense of other plants in the lake. As the huge masses of alga died, they fell to the bottom of the lake, where micro-organisms multiplied and began the process of decomposition (see page 141). Their activities used up all the dissolved oxygen in the water (Figure M.11). Without the dissolved oxygen, fish and other water-borne life died.

Both nitrogen and phosphorus are present as compounds in treated and untreated sewage. The lake was receiving increasing quantities of these elements from this source. Phosphates, present in detergents, also contain the element phosphorus. Large quantities of phosphates from this source

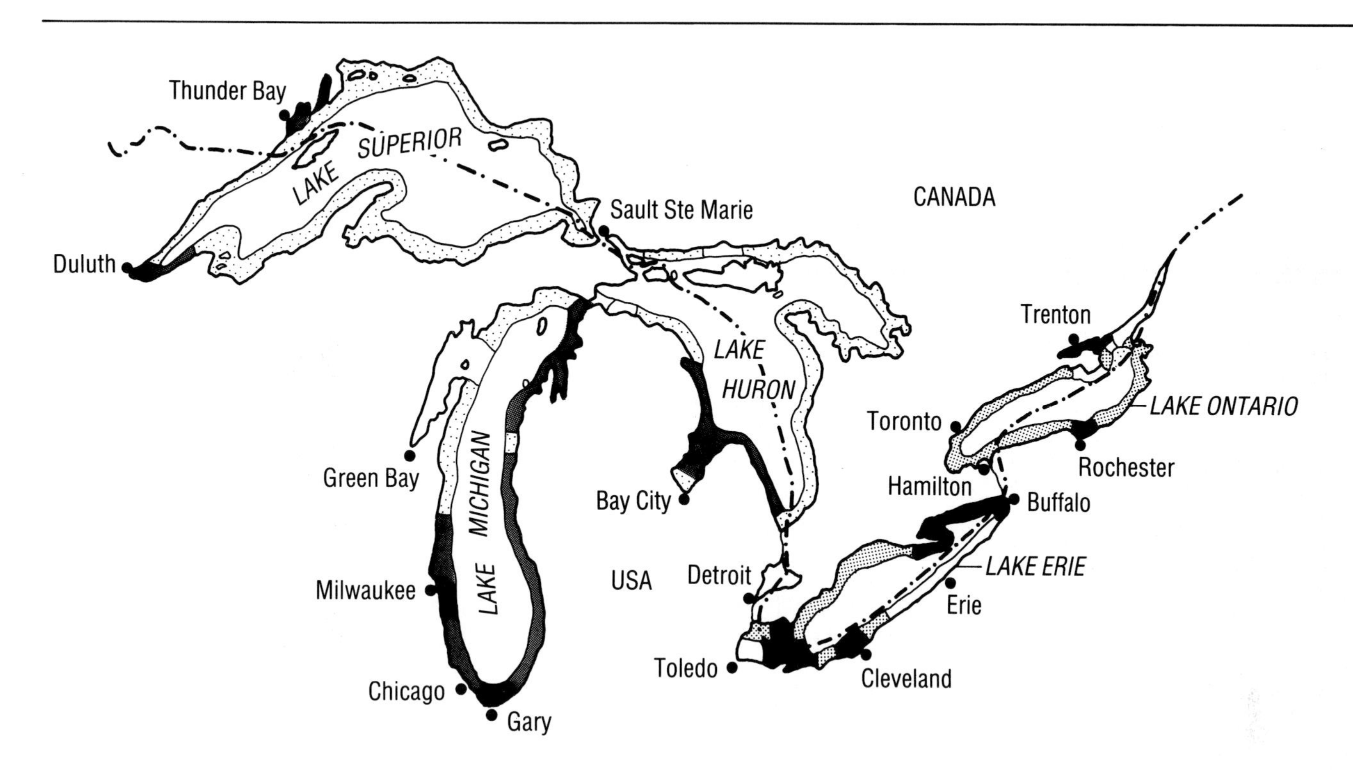

Figure M.10 The Great Lakes (the darker the shading, the more polluted the water) is the border between the USA (to the south) and Canada (to the north)

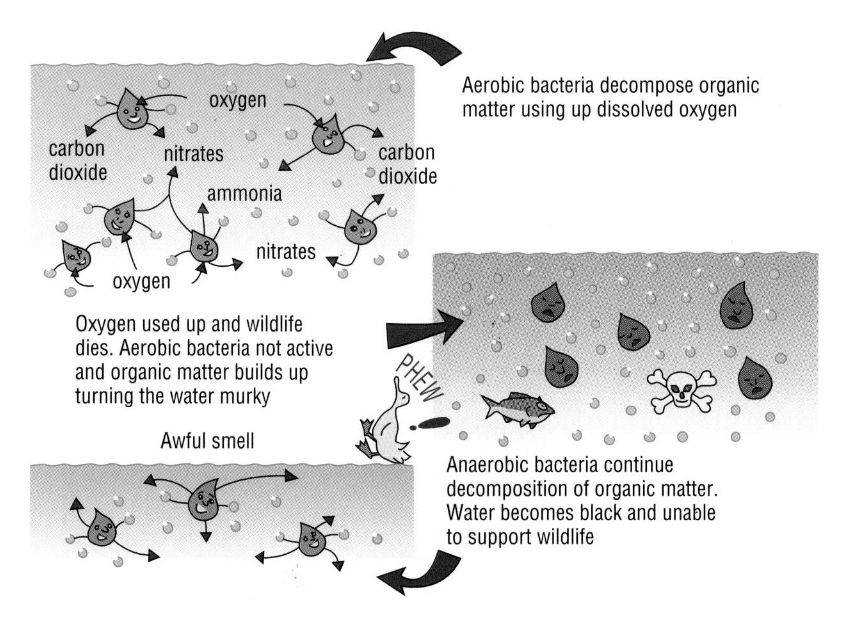

Figure M.11 Bacterial activity depletes water of its oxygen

were present in the waste water pouring into Lake Erie. It is estimated that in the 1960s phosphorus was pouring into Lake Erie at the rate of 62 000 kilograms per day.

In 1972 there was a Great Lakes Water Quality Agreement between Canada and the USA. They decided to limit the allowed phosphorus content in household detergent. The level of phosphorus in treated sewage was also limited to 1 milligram per litre.

The problem with the Great Lakes has not gone away. Levels of phosphorus are lower but levels of nitrogen are increasing. Some of the nitrogen comes from nitrates and nitrites washed into the lakes from agricultural land. The nitrates are used as fertilisers (see page 159). Nitrates also come from the atmosphere via acid rain (see page 174). Fortunately the levels of nitrogen are below the recommended limit of 10 milligrams per litre.

Activity M.2

Phosphorus levels in Lake Erie

The table shows how phosphorus levels in Lake Erie have changed since the Great Lakes Water Quality Agreement between Canada and the USA.

	Total phosphorus loadings (tonnes per year)								
Year	1975	1976	1977	1978	1979	1980	1981	1982	1983
USA	7898	6666	6478	5722	4040	3373	2642	2161	2247
Canada	232	262	259	228	234	212	218	252	253

1. Which country has contributed most of the phosphorus pollution over the period of years shown?
2. Plot two graphs, (i) the USA phosphorus figures against year, (ii) the Canadian phosphorus figures against year. Which country has dealt best with the phosphorus problem?
3. Compare the progress made by the two countries in cutting down on phosphorus levels.

Summary

- Oil spills harm wildlife. Clean-up operations can cause even more damage.
- Oil can be removed by huge vacuum cleaners or soaked up by absorbent materials.
- Floating barriers help to keep oil away from wildlife sanctuaries.
- Untreated sewage uses up oxygen dissolved in the water of rivers or lakes so animals and plants die.
- Eutrophication occurs when the process of decomposition removes the oxygen dissolved in water. The elements phosphorus and nitrogen contribute to the eutrophication of lakes and rivers.

Investigation M.3

Investigate the effects of nitrogen and phosphorus on water weed.

Figure M.12 Water weed can be affected by nitrogen and phosphorus

Questions

1. Look at Figure M.2 (page 163). Assess which river contributes the most pollution to the North Sea.
2. What damage can oil spills cause to marine life?
3. What laws should be passed to minimise the chances of spillage of oil at sea?
4. Describe ways to remove oil from a slick. Try to design your own method of stopping an oil slick reaching shore.

Extension questions

5. What damage can untreated sewage cause?
6. **a** Why did a blue-green alga take over Lake Erie?
   b What problems did it cause?
   c How is the problem being solved?
   d What is the latest threat to the Great Lakes?

Unit N
Air pollution

Polluting air: Dust and smoke

Smoke consists of particles of carbon and unburnt fuel. Dust and grit particles come from parts of fuels that will not burn. The average size of these particles increases in a regular pattern:

smoke 1 μm (millionth of a metre), dust 10 μm, grit 100 μm

All of these particles arise in the air as a result of burning fuels. Over a thousand tonnes of smoke are made in the UK each year, together with one and a half thousand tonnes of dust and grit.

Figure N.1 Smoke and grime ... Britain's city life pre-1960s

What damage can they cause?

- Smoke makes buildings and clothes dirty. Figure N.1 shows how smoky and dirty the air was in cities and towns before the 1960s
- The particles cut down the level of sunlight reaching the Earth's surface: dust from volcanoes has been known to upset weather patterns over a very wide area
- Plant growth is less in a smoky environment
- Grit and dust get into machinery and increase wear, so lowering the efficiency and lifetime of the machine
- The particles in dust and smoke get into the lungs and can cause breathing problems

Hay-fever

In the spring and summer pollen grains float around in the air and cause untold misery for countless hay-fever sufferers. The pollen count is the number of pollen grains in a cubic metre of air; when the count is above 50 the suffering starts. When it rains, pollen and other particles in the air fall with the droplets, so the pollen count is reduced in wet weather.

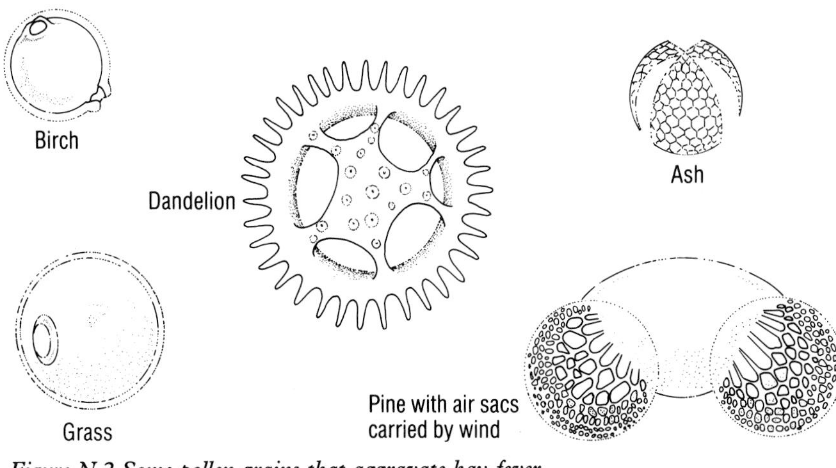

Figure N.2 Some pollen grains that aggravate hay-fever

Activity N.1

To collect pollen grains from rain water

Choose a day when the pollen count is high for this experiment.

1. Fold a filter paper into the usual cone and use a paper clip to hold it together.
2. Rest the cone in the top of a wide-necked bottle and stand it out in the rain for a few hours.
3. Open the filter cone by removing the paper clip. Look at the collected particles under a microscope. Are there any pollen grains on the filter paper? Some common pollen grains are shown in Figure N.2.

To collect dust from the air

1 Take off the protective backing from several sticky cards lined in the same way as graph paper (available from laboratory suppliers).
2 Place one card at each of the sites you wish to test with the sticky side upward. Leave for a few days for dust to stick to the surface. If you put any cards outside make sure they cannot blow away.
3 Look at the cards under a microscope. Can you see any pollen grains? Sketch the different types of particles seen. Count the number of particles in a few millimetre squares and find an average number for each card. Do the different sites show different numbers and types of dust?

Dust blowing in the wind can be collected using the sort of apparatus shown in Figure N.3. You could make a dust collector using plastic tubing and small jars for collecting the dust. Some rain water may be collected too. It would also show which direction delivers the most dust. Remember to label each jar with a number and note down its direction (geographically).

Figure N.3 A dust collector

Figure N.4 London's smoggy streets

London smog

Fog is a thick mist of water droplets. It may also contain particles of sulphur and polluting gases such as smoke. Smog is the name coined for smoke mixed with fog. This lethal mixture caused havoc during the winter of 1952 in London. The weather was calm, moist and cold for a few days. The cold weather meant plenty of coal fires with their polluting smoke. The smoke and the water droplets mixed, so smog was formed. The calm weather meant that the smog did not blow away. The traffic came to a halt because of the poor visibility (Figure N.4). Thousands of people had breathing problems and developed coughs. Elderly people were most at risk and many died.

The sulphur dioxide and dust produced by burning fuels was to blame for the deaths. The '*Clean Air Acts*' of 1956 and 1968 followed the public outcry caused by this tragedy. The acts prevented the burning of smoky fuels in certain areas. Smogs are a thing of the past in London today.

Activity N.3

To study graphs of the London smog

Look at the three-graphs-in-one in Figure N.5, showing human deaths per day and levels of smoke and sulphur dioxide in the air. Make sure that you are looking at the correct graph and vertical axis when you answer the questions:

1 Estimate the number of people who died in London on each of the following days: 4th, 7th and 10th of December 1952.
2 Which day showed the maximum number of human deaths?
3 Which day shows the maximum value for (**a**) smoke pollution, (**b**) sulphur dioxide pollution? Is there any connection between these answers and that to question **2**? Give the figures for these maximum values with their units.
4 The worst of the problem lasted about six days. Suggest the dates covered by this six-day problem. What changes in the weather are likely to have caused the problem to subside after the six days?

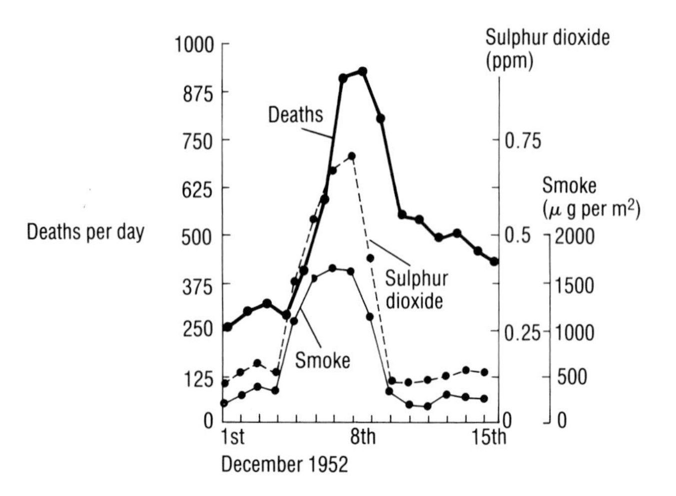

Figure N.5 Death rate and levels of sulphur dioxide and smoke in London's smog

The smokiest areas of Britain

European Community standards for smoke and sulphur dioxide levels came into force from the 1st of April 1983. In the UK 29 areas were found to be higher than the EC standards and were set targets to lower the pollution over a ten-year period. In 1985 six of these highly polluted areas were even unable to keep to the new adjusted targets! The six places with most smoke and sulphur dioxide pollution in the UK are numbered in Figure N.6.

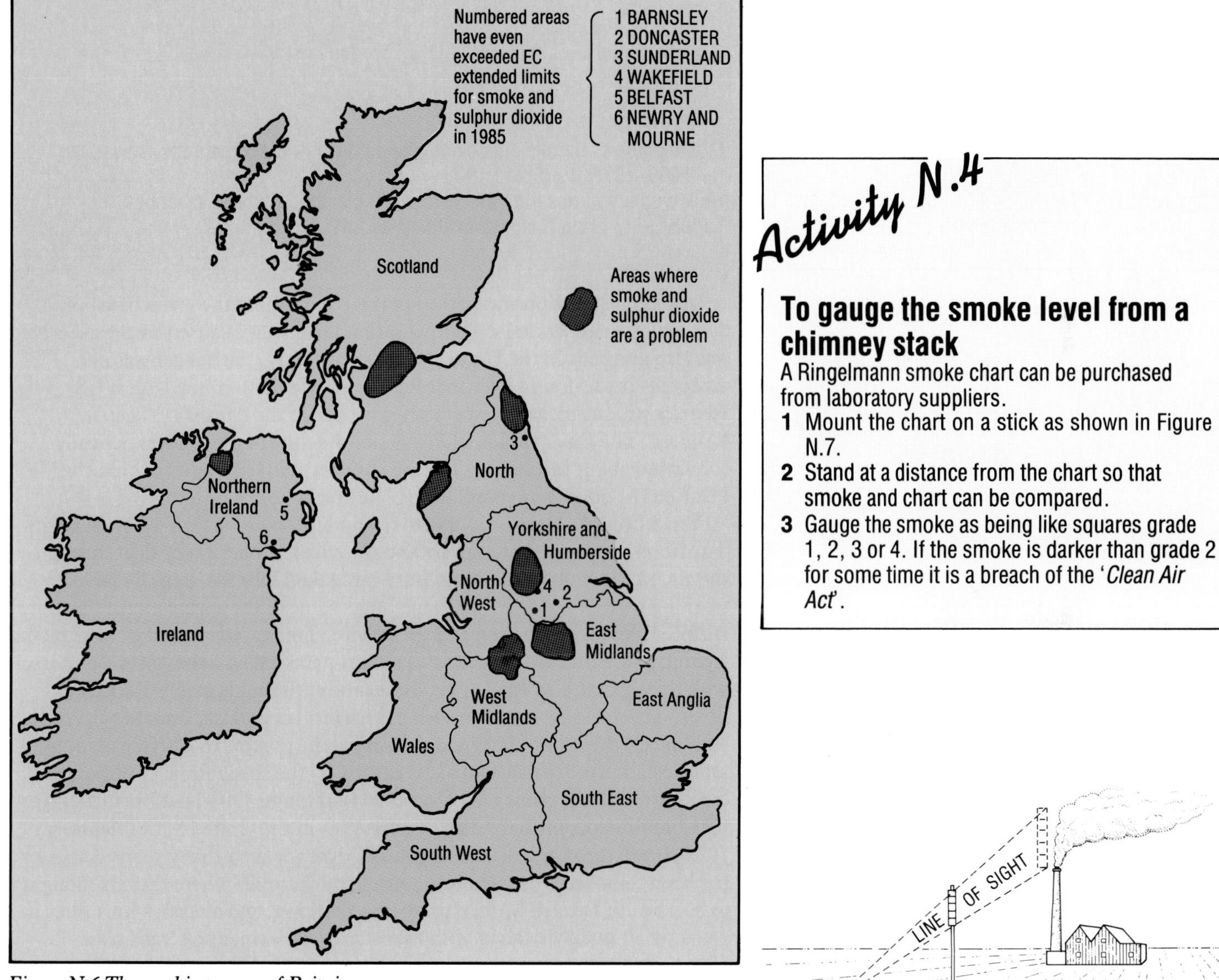

Figure N.6 The smokiest areas of Britain

Activity N.4

To gauge the smoke level from a chimney stack

A Ringelmann smoke chart can be purchased from laboratory suppliers.

1. Mount the chart on a stick as shown in Figure N.7.
2. Stand at a distance from the chart so that smoke and chart can be compared.
3. Gauge the smoke as being like squares grade 1, 2, 3 or 4. If the smoke is darker than grade 2 for some time it is a breach of the '*Clean Air Act*'.

Figure N.7 Using the Ringelmann smoke chart

Asbestos cover up

Asbestos was used in brake-linings, roof-insulation, oven-linings, fire-blankets, air-filters and fire-doors. It has a great power to resist heat while remaining unchanged itself. The first case of lung trouble due to asbestos was diagnosed in 1906 by a British doctor. He clearly did not think the discovery very important because he did not report it to any medical journal. It was not until 1924, when workers involved with asbestos were barred from getting life-insurance, that people sat up and noticed.

Activity N.5

To show how the quality of our air has changed with time

Look at the data in the table below, which shows the average smoke emissions from coal combustion in the UK for different years.

	1951	1956	1961	1966	1971	1976	1981	1986	1992
Smoke emissions from coal combustion (million tonnes)	2.42	2.32	1.56	1.06	0.61	0.37	0.27	0.27	0.26

1 Plot a graph of smoke emissions against year and estimate the emissions for 1965, 1975, 1985 and 1995.

2 How would you describe the success in cleaning the air? What has happened to bring this improvement about?

In 1933, legislation was passed in the UK limiting the levels of asbestos dust exposure to workers. Compensation for damage caused by the asbestos was also granted. In the United States, amazingly, the danger was not accepted. It was found later that Raybestos Manhatten, the largest US asbestos producer, had deliberately covered up the dangers to keep in business. The Environmental Protection Agency in the US was not fully convinced about the dangers of asbestos until 1964; 31 years behind the UK! Its often the other way round with pollution problems.

The fibres of asbestos can range from microscopic size to 30 cm in length. The fibres are shaped like spears and the smallest ones easily float through the air. In a day up to 16 billion fibres are taken into the lungs by people working with asbestos. Once in the lungs some of the spears pierce the inner surface (pleura) of the lung and stick there. This causes scarring of the pleura leading to breathlessness and severe chest pains called **asbestosis**. If a person with this disease also smokes cigarettes the suffering is magnified many times. The asbestos spears absorb the harmful chemicals from the cigarettes and 'inject' these into the blood stream via the pleura. In extreme cases of asbestosis a massive heart attack is caused by the sheer strain of pumping blood through the damaged blood vessels of lungs. Official estimates suggest that 2 million people have died of asbestosis in the United States. Some say 500 000 will die of the disease in the UK over the next thirty years. Asbestos has been removed from schools, houses and factories where there is thought to be a health hazard. Special protective clothing, and a mask with a filter is worn by all people dealing with asbestos for its remaining 'safe' uses (Figure N.8).

A problem of dust

Cement works generate huge quantities of dust which can cause lung irritation leading to breathing problems. To be kind to the environment cement works have had to devise ways of reducing the dust. A combination of two ideas makes for a modern almost dust-free process.

The machinery making the cement is surrounded by what is an almost dust-proof box. The air pressure inside the box is kept lower than that outside by using an extractor fan. This fan also extracts the dusty air. The

Figure N.8 Asbestos being removed from a building by a worker wearing protective clothing

low pressure inside encourages air to travel into the box rather than dusty air to travel out.

The extractor fan directs the dusty air to an electrostatic precipitator. This has one electrode with a voltage of 50 000 volts and another which is earthed (the collecting plate). The dust particles are charged in the electrostatic field between the electrodes. The charged dust particles stick to the earthed collecting plate and can be removed safely. The dust collected is not wasted but recycled back to the boxed machinery.

A similar scheme can be used for all dusty manufacturing processes.

Summary

- Smoke is made up of particles of carbon, and comes from burning fuels. It makes things dirty, cuts down sunlight and can harm breathing.
- The *Clean Air Acts* have resulted in much lower smoke levels in the UK.
- Asbestos fibres can cause a lung disease called asbestosis.

Questions

1 What factors combined to make the London smog so dangerous? How was the problem solved?
2 Suggest how petrol-driven vehicles can also make smog given the right conditions.
3 Explain how the effects of asbestos fibres and cigarette smoke combine with surprising effects.
4 Argue against the idea of having a very tall chimney stack to solve smoke problems.

Unit O
Acid rain

Polluting air: what has happened to rain?

Every year millions of tonnes of acid-forming gases are released from power stations, motor vehicles and heavy industries. These gases join the air and pollute it. Rain, mists and snow carry acids and these can cause great damage (Figure O.1).

Millions of acres of forests have died or are dying. Thousands of lakes, rivers and streams can no longer support any life such as fish and plants. Crops and buildings are damaged. The acid rain passes through soil and takes out metals such as aluminium. This aluminium pollutes rivers and lakes. There is some evidence that aluminium in our drinking water can cause a condition known as Alzheimer's disease. This disease results in loss of memory, changes in personality and eventually death.

We breathe in the acidic air and drink acidic water. The effects of this cannot be good. The environmental group Greenpeace say '*ACID RAIN KILLS PEOPLE TOO*'.

Figure O.1(b) Acid rain damages trees and kills them

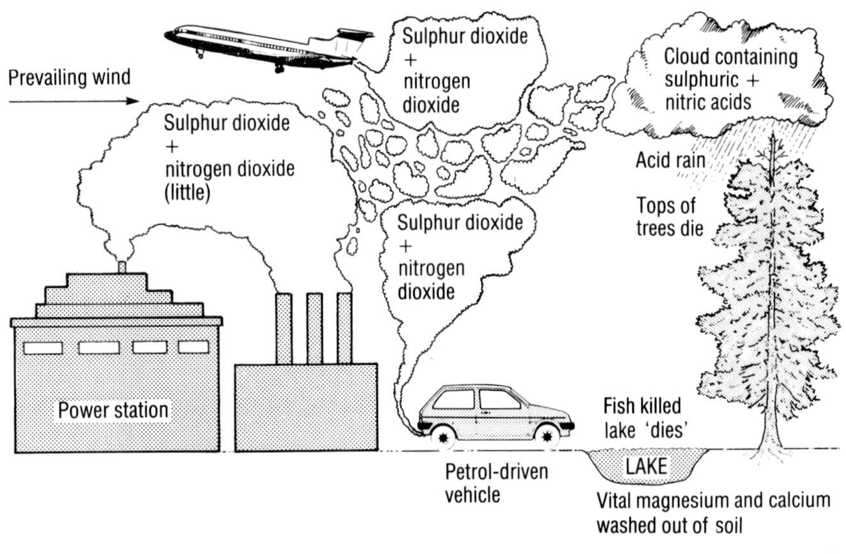

Figure O.1(a) The formation of acid rain

Where does the problem start?

Most of the power for industry is obtained by burning fossil fuels, such as coal and oil. These fuels contain some sulphur as an impurity. It is not easy to separate the sulphur and so the fuel is burnt as it is. The sulphur in the fuel becomes sulphur dioxide gas on burning.

sulphur (in fossil fuel) + oxygen (in air) → sulphur dioxide

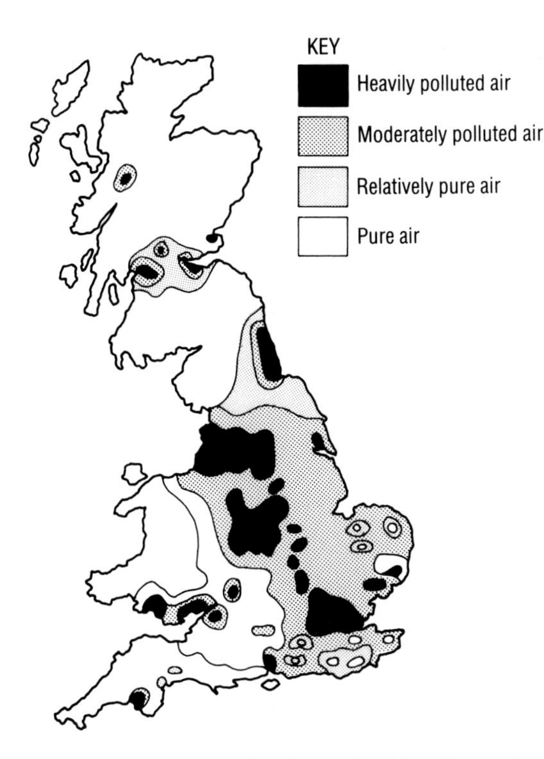

Figure O.2 How heavy is sulphur dioxide pollution in Britain? Where would you live?

The sulphur dioxide is released to the atmosphere with the smoke from the fuel. Our coal-fired electricity power stations belch out huge quantities of sulphur dioxide daily. Figure O.2 shows the areas in Britain which are most heavily polluted by sulphur dioxide.

The sulphur dioxide mixes with water droplets in the air and the oxygen in the air. A chemical reaction occurs in which sulphuric acid is formed.

sulphur dioxide + water + oxygen (from the air) → sulphuric acid

This results in sulphuric acid entering the atmosphere. It later falls as acid rain.

Acid rain is not only sulphuric acid however. Industrial processes are often at such high temperatures that the two most abundant gases in the air combine. These two gases are nitrogen and oxygen.

nitrogen + oxygen → nitrogen oxides

Petrol-driven engines also contribute greatly to the nitrogen oxides in the atmosphere. One of the oxides is nitrogen dioxide, which also combines with water droplets and oxygen from the air. The result is nitric acid, which is every bit as dangerous as sulphuric acid.

nitrogen dioxide + oxygen (from the air) + water → nitric acid

Once again the nitric acid is in the atmosphere ready to be deposited as rain. Acid rain is a mixture of mainly sulphuric acid and a little nitric acid.

The damage it causes

The devastating effect of acid rain was first observed in Sweden. In 1978 they found that 18 000 of their 96 000 lakes could no longer support fish life. Their measurements showed that the acidity of the lakes had increased alarmingly. This was killing all life in the lakes. They concluded that acid rain was responsible. The Swedish accused Britain of polluting their lakes. The sulphur dioxide and nitrogen oxides from our power stations and petrol vehicles were being blown to Sweden by the prevailing winds. They accused us of exporting 62 000 tonnes of sulphur dioxide, through acid rain, to their country in 1978. Not only lakes were affected. The tops of trees were dying off and acid rain was the chief suspect. Many trees were dying in the Swedish and Norwegian forests. Scientists found that acid rain washed out valuable chemicals from the soil and released poisonous ones. The acid rain washes out magnesium and calcium from the soil; vital elements for the healthy growth of trees. Once the calcium and magnesium have gone, aluminium is the next element to be released. Aluminium is poisonous to trees and clogs up the gills of fish, stopping them breathing. The acid rain was killing the trees on which the people's prosperity depended.

Another serious effect of acid rain is its attack on limestone and marble. Many ancient buildings and monuments are made of marble or limestone. These are both forms of the chemical calcium carbonate. Acid rain attacks the monuments, gradually pitting and crumbling them, as shown in Figure O.3.

Figure O.3 The effect of acid rain

calcium carbonate + sulphuric acid → calcium sulphate + water + carbon dioxide

What can be done?

Recent tests have shown that no country is escaping acid rain. Many of the lakes in Scotland and Wales are becoming increasingly acidic. Attempts have been made to convince all countries that sulphur dioxide levels must be cut.

Britain, one of the chief offenders, has been very slow in reducing its sulphur dioxide levels. In September 1986 the government announced a programme to reduce sulphur dioxide levels coming from three power stations. New power stations are fitted with the latest technology to remove this harmful gas. Many improvements are possible but they require huge amounts of money. The cost is stopping many countries from acting against this menace. They easily forget the cost of the damage caused by acid rain. Some companies generating electricity dispute the evidence that acid rain has very serious effects. They, together with others, say that there should be more research into acid rain. Others feel that action must be taken now if we value rivers, lakes and trees and their associated wildlife.

Investigation O.1

Investigate the effects of acid rain on plants.

Activity O.2 To measure rainfall acidity in a standard way

Making the collector: A 1 metre high post of 5 cm diameter is required. A wooden platform is screwed onto the top of the post as shown in Figure O.4. A plastic plant pot is then attached to the platform by a screw. A collecting vessel 20 cm deep and 15 cm wide is placed inside the plant pot; this can be the bottom half of a plastic squash bottle. Polythene bags 18 cm × 30 cm are used to contain the rain water.

Choosing the site: The collector must not be overhung by trees or buildings. It must be in an open space, perhaps very near to the school weather observatory. A detailed Ordnance Survey map should be used to make a sketch map of the area (about 2 mile radius from the collector). This map should include factories, buildings, power stations and also any woods or cultivated fields. Traffic conditions in the area should also be recorded.

Testing the water samples:

1. Place the polythene bag in the collecting vessel using a hand already covered with a plastic bag (or disposable plastic glove). Discard the bag that covered your hand.
2. The funnel must be washed with distilled water and left to dry before use.
3. Water samples should be taken once a week. This should always be on the same day, same time and preferably be done by the same person.
4. Place an Acilit (narrow range pH paper covering 0 to 6 in 0.5pH stages) indicator strip in the water sample. Leave it there for 10 minutes and then compare its colour with the chart on the box. Note this pH and also any observed soot, leaves, dust or bird-droppings in the water.
5. Note the temperature and other weather conditions at the time of collecting the sample.
6. Plot a graph of pH against weeks. Try to explain any variations in pH. Encourage other schools in your area to collect rainfall acidity data by the same method and compare results.

Note: a pH meter gives more accurate results than the Acilit paper.

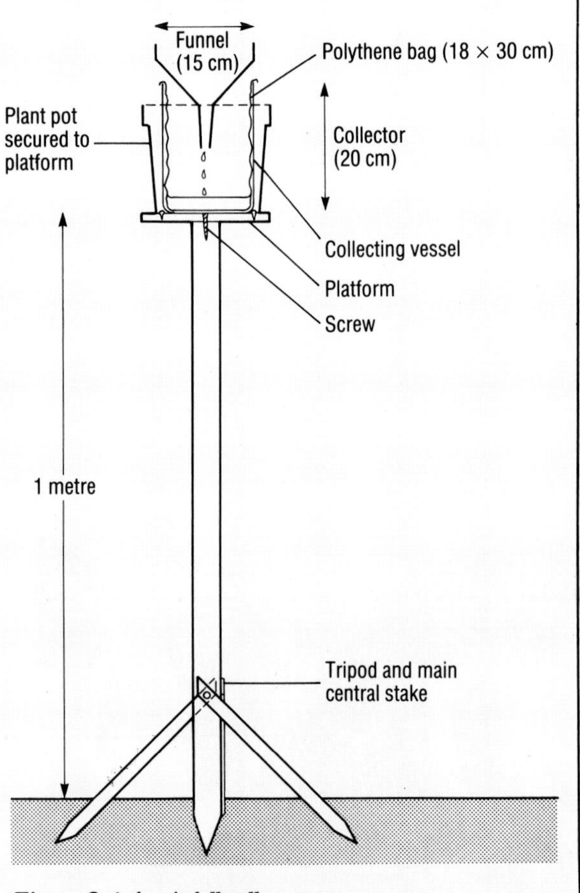

Figure O.4 A rainfall collector

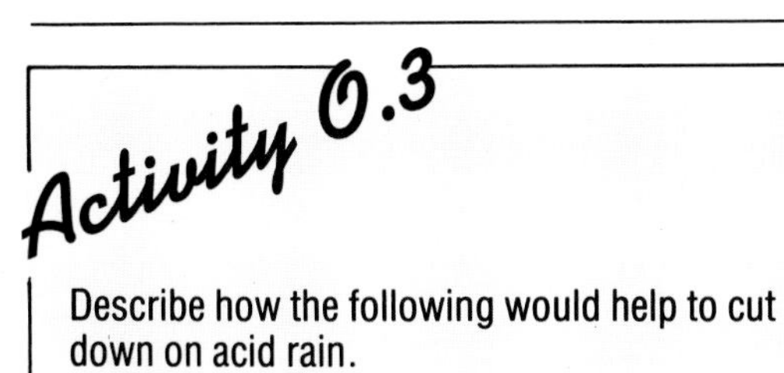

Describe how the following would help to cut down on acid rain.

(**a**) Filters to remove some gases in chimney stacks.
(**b**) Better public transport (trains and buses).
(**c**) Greater reliance on wind power and other **renewable** energy sources.
(**d**) Greater reliance on nuclear power.

Activity O.4

To find the pH of some water samples

1. Take a plastic funnel and a glass jar to collect samples of water. The samples could come from downspouts of drain pipes, garden water-butts or freshwater ponds.
2. Dip a strip of pH paper into the water sample.
3. Match the colour of the paper to the colour chart issued with the pH papers. If the pH is slightly below 7 this will be due to the carbon dioxide in the air. If it is in the region of 5.5 or below, this suggests sulphur dioxide and/or nitrogen dioxide are present.

Activity O.5

To study the pH of rainwater channels from trees to surrounding soil

1. Find a tree that has rainwater channels on the bark of its trunk. These are formed by water falling on the leaves, dripping off onto branches and then running towards the earth and then off into the soil. It is easy to spot the channels as nothing grows on them.
2. Make a small dam out of plasticine in a rain water channel when the bark is dry.
3. When it is raining, water collects behind the dam. This water can be tested with pH paper.
4. Also collect some of the soil from around where the water leaves the tree. Soak the soil in distilled water for a few minutes.
5. Filter and test the filtrate with pH paper.
   - Is the pH lower than for rain water that has not run down trees?
   - Can you explain the results?

Figure O.5 Is nuclear power one of the answers to the problems caused by acid rain?

Activity O.6

Biological detection of sulphur dioxide in the atmosphere

Some plants are very sensitive to the level of sulphur dioxide in the air. Lichens fit into this category. They are a combination of a fungus and a green alga. They grow on trees, stones and rooftops. Figure O.6 shows three types of lichen, the shrubby, the leafy and the slightly leafy types. The tolerance of each type to sulphur dioxide pollution is stated.

Figure O.6
(a) Shrubby: highly intolerant
(b) Leafy: medium tolerance
(c) Slightly leafy: high tolerance

1. Visit a town or city near to your home or school.
2. Use an Ordnance Survey map to work out where the town or city centre is.
3. Visit the town or city centre and look for lichens on buildings and graveyard headstones. Often the centre will be empty of lichens. If there are some, try to decide their type.
4. Mark lines on the map spreading out from the city or town centre. Walk outwards from the city or town centre using a compass to keep on course. Note the type and number of lichens at regular distances from the centre. Repeat for other directions from the centre.
5. Try to compose a chart on graph paper to show lines where the number and type of lichens are the same. Try to explain the shape of the chart.

Activity 0.7

Databank questions

The following questions refer to the tables and maps.

Give REASONS for your answers wherever possible.

Table A

1 What do you think are the sources of sulphur dioxide and nitrogen oxides in the home and on the road?
2 You have been given the job of drastically reducing acid rain. Where would your greatest chance lie in reducing sulphur dioxide and nitrogen dioxide?
3 You are the new chairman of a rail company. You wish to encourage people to travel by rail because it keeps the air clean. Write out a speech you would make to convince people that this is true.
4 Suggest measures we must take to rid the air of all acid rain. Do these measures have their own dangers? If so, what are the dangers?

Table B

1 Which European country makes the most sulphur dioxide or nitrogen dioxide?
2 Which European country makes the least sulphur dioxide or nitrogen dioxide?
3 Which European country exports the biggest percentage of its sulphur dioxide and nitrogen dioxide to other countries?
4 How many tonnes of sulphur dioxide and nitrogen dioxide are deposited in the following countries due to **their own** pollution: Germany, UK, Sweden?

Table C

1 What sort of damage do you think is being done to (a) buildings (b) people's health, and (c) crops, due to acid rain?
2 All of the estimated costs have a large range. Which has the largest range, and why?
3 These figures of damage due to acid rain do not seem to get through to the general public. How would you conduct a campaign to get the message across to the public?
4 Take the lowest figures for damage to the UK and find the total cost of acid rain to us.
5 Make a rough estimate of the total cost for all European countries.

Table D (next page)

1 Which types of fish are (a) most sensitive (b) least sensitive, to acid rain water?
2 You are a keen fisherman but note dead perch and pike on the surface of a lake. Would it still be worth fishing for eels or grayling?

Table A Percentage of air pollutants by source

Source	Sulphur dioxide	Nitrogen oxides
Road transport	2	51
Electricity supply	69	25
Home	3	3
Other	26	21

Table B

Country	Total sulphur dioxide produced	Sulphur dioxide sent to other countries
Belgium	810	638
France	3600	2012
Germany	3630	2338
Italy	3800	1804
Norway	150	102
Sweden	550	348
Switzerland	124	92
United Kingdom	5340	3750
ANNUAL FIGURES IN THOUSANDS OF TONNES		

Country	Total nitrogen dioxide produced	Nitrogen dioxide sent to other countries
Belgium	290	259
Denmark	181	172
Germany	1895	1487
France	2297	940
Netherlands	401	388
Norway	102	88
Sweden	247	183
United Kingdom	1838	1533
ANNUAL FIGURES IN THOUSANDS OF TONNES		

Table C
Building damage £80 million to £415 million
Health costs £50 million to £1.2 billion
Crop damage £55 million to £200 million
Estimated annual cost to the UK due to acid-rain losses (forest and lake losses would make the picture look even worse)

Map E

1. Which foreign country is being most '*unkind*' to Scandinavia? What factors combine to make the foreign country top of the '*unkindness list*'?
2. How much of this acid rain is self-inflicted by Scandinavia?
3. 27% of the acid rain comes from '*unknown*' sources. Suggest who these '*unknown*' contributors might be.
4. Given the choice, do you think Scandinavia should choose (a) financial compensation from other European countries, or (b) reduction in emissions of acid gases by other countries?

Table D

Acidity of the water	Fish that die
Below pH 6	Salmon Trout Roach
Below pH 5.5	White fish Grayling
Below pH 5	Perch Pike
Below pH 4.5	Eels

Remember the LOWER the pH the GREATER the ACIDITY

Map E

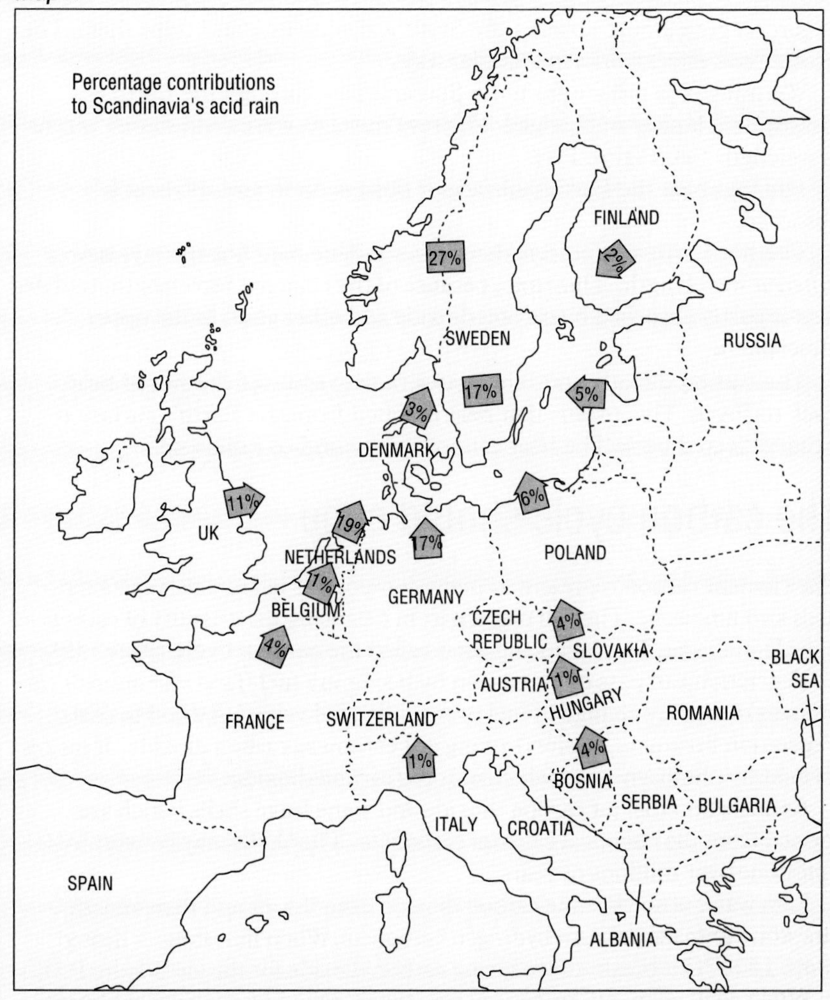

Summary

- Burning fossil fuels leads to accumulation of sulphur dioxide and nitrogen oxides in the atmosphere.
- These gases dissolve in water, forming acid rain, which contains sulphuric and nitric acids.
- Acid rain dissolves and crumbles building stone and therefore damages buildings. Wildlife is damaged because plants and animals cannot tolerate increasing acidity.

Questions

1. Name the acids in acid rain. Which gases make these acids when mixed with water (and sometimes more air)?
2. Make a list of the different types of damage inflicted by acid rain.
3. Why are Norway and Sweden particularly affected by acid rain?
4. Make a list of ways you could help reduce acid rain.
5. Imagine that the British Government wished to make big reductions in acid rain. What could they do, by way of encouragement and by passing laws, to achieve this aim?
6. **a** Name the two gases that combine with water and air to make acid rain.
   b Name two fuels that burn to form one or more of these gases.
   c Name the two acids formed in acid rain.
   d Acid rain is sometimes a problem for countries that produce little of the two acids themselves. Explain why.
   e Name a country that comes into category **d**.
   f Acid rain can dissolve some metals from the soil. Name one metal that might be connected with Alzheimer's disease.
   g List two other problems caused by acid rain.
   h Suggest three ways of reducing the acid rain problem, saying how you would do this.

Unit P
The greenhouse effect

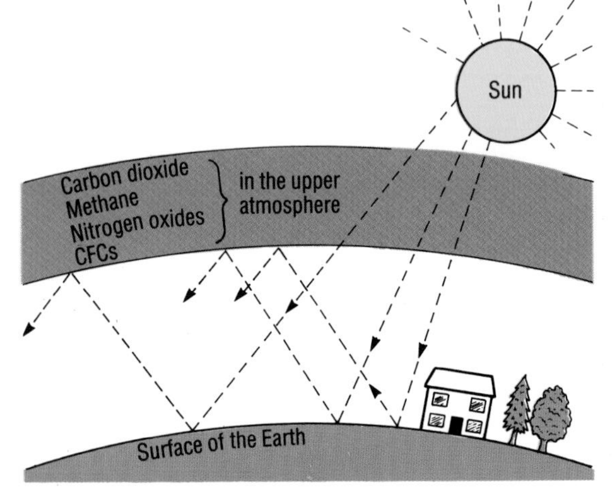

Figure P.1 The greenhouse effect

Polluting air: what is the greenhouse effect?

A garden greenhouse absorbs the Sun's warming rays and traps them. The greenhouse effect works in a similar way.

Warming rays radiate from the Sun and pass through the Earth's atmosphere largely unchanged. The rays travel as waves and have a certain wavelength (see Figure P.1).

The rays heat the Earth's surface or objects on it, and the heat is absorbed.

The hot Earth and objects themselves radiate heat but the rays have a ***different*** **wavelength. This time, because of the changed wavelength, radiated heat is partly absorbed by carbon dioxide and other gases in the upper atmosphere.**

The carbon dioxide and other gases radiate some of the stored heat back to Earth. This means that heat radiated from the Earth's surface is being reflected back. The heat cannot escape and so builds up.

The carbon cycle should cope

The element carbon is present as many compounds in animals, vegetation, fuels and limestone. They all play a part in balancing the quantity of carbon dioxide in the air. This balancing act is called the **carbon cycle** (page 142).

Vegetation can pass on its carbon by becoming fuel (food) for animals (or humans) or it may change to fuel over millions of years (eg wood to coal). Vegetation breathes (respires) giving out carbon as carbon dioxide. It makes its food by photosynthesis which absorbs carbon dioxide.

Animals breathe out carbon dioxide and some leave shells which are formed from the chemical calcium carbonate. The shells may convert to limestone over millions of years.

Rain water absorbs some carbon dioxide from the air and then attacks limestone to make calcium hydrogen carbonate. When limestone is heated above 1400 °C it breaks down giving carbon dioxide for the air (Figure P.2).

When coal, coke, oil, or natural gas (the so-called fossil fuels) are burnt, carbon dioxide is formed.

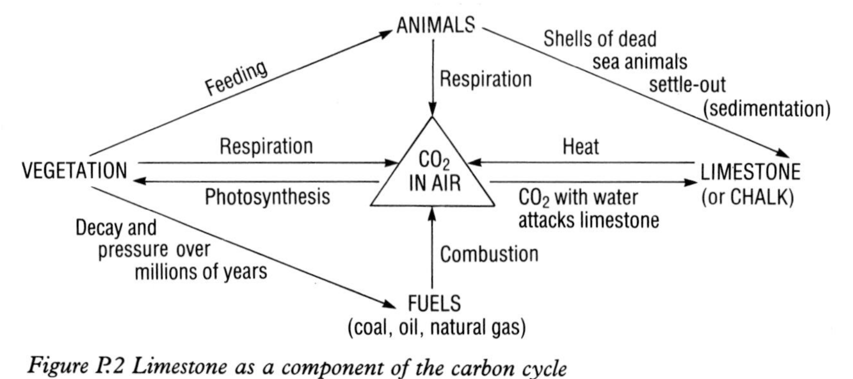

Figure P.2 Limestone as a component of the carbon cycle

Activity P.1

Understanding the carbon cycle

1. Which processes put carbon dioxide into the air?
2. Which processes directly take carbon dioxide out of the air?
3. Which processes merely change the carbon to something else rather than make carbon dioxide?
4. A morbid subject, but where does the carbon in animals (or humans) go when they die? Can the carbon ever end up as carbon dioxide in the air?

Why so much gas?

The carbon cycle is not coping because our fast increasing population requires more and more fuel. The fuel chosen is usually a fossil fuel. Eighteen thousand million (billion) tonnes of carbon dioxide are generated worldwide each year due to the burning of fossil fuels. The carbon dioxide concentration in the air has risen gradually over the years, averaging 2.5 – 4.5 % increase per year.

Carbon dioxide accounts for almost three quarters of the warming greenhouse effect. Methane, nitrogen oxides and CFCs also contribute to the warming of the Earth's atmosphere (Figure P.3). Methane is given off by ruminating cattle as a waste product, and by rice grown in paddy fields. As the sizes of herds, and the amounts of rice grown, increase to feed the fast growing population, so more methane is produced. Levels of nitrogen oxides, which are given off from petrol engines and high temperature industrial processes, are also on the increase. CFCs come from aerosols, foam and freezers. International agreements mean that levels of CFCs in the atmosphere should fall.

With the build up of greenhouse gases the temperature of the Earth has risen gradually. The rise is very small but if it continues major problems are predicted. For example, some scientists estimate that if carbon dioxide levels rise to double the natural level (which is 0.03% of gases in the atmosphere) then temperatures might rise by 1.5 – 3 °C.

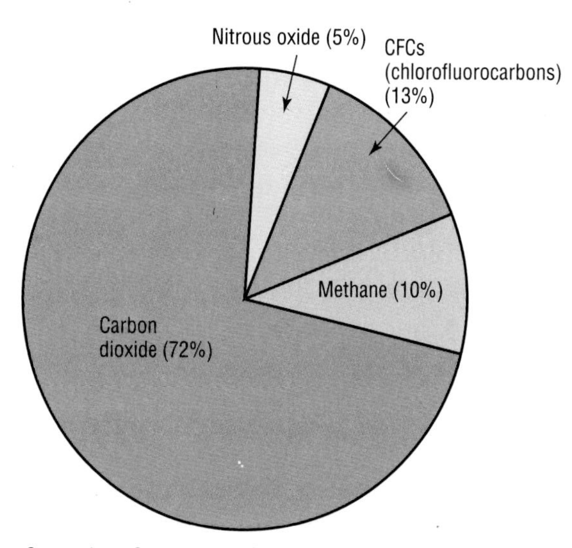

Source: Inter Governmental Panel on Climate Change

Figure P.3 Greenhouse gases: their % contribution to the greenhouse effect (1992)

The weather could change

The warming of the Earth could upset normal weather patterns. The amount of cloud cover, the force and direction of winds and the rainfall patterns could all change. One piece of hard evidence that climate has already been affected has been collected by scientists at the University of East Anglia. They fed annual rainfall figures for two large areas of the world into their computer. The one area had gained the rain from the other. More seriously, the dry desert area, which could not afford to lose rain, had lost what it had. The area already rich in rain had gained more (Figure P.4).

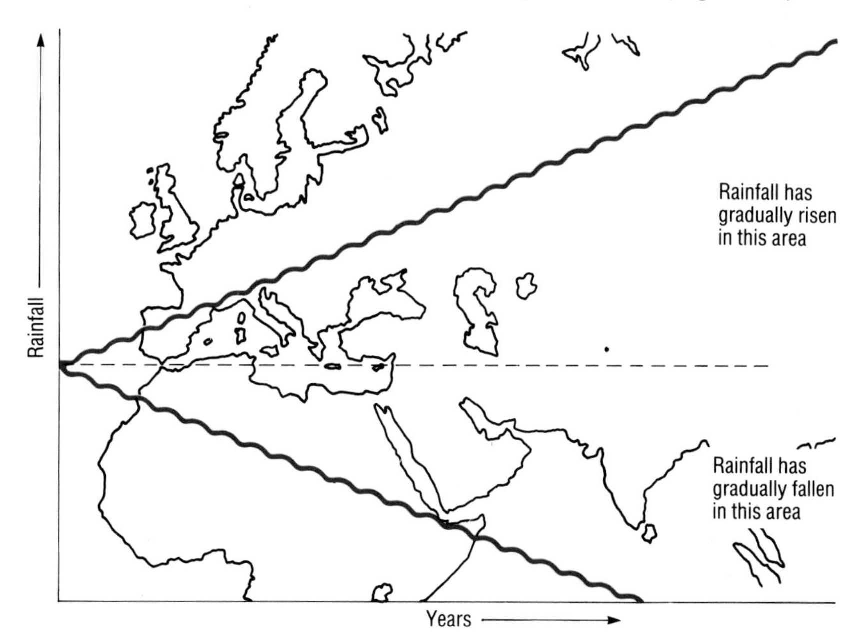

Figure P.4 Rainfall trends in two different (but adjoining) areas of the world

Activity P.2

What can be done about the greenhouse effect?

Look at the list of suggestions and write down how each would affect the level of 'greenhouse' gases.

1 A switch from fossil fuels to 'clean' fuels like solar, wind, wave and nuclear power.
2 Grow more forests.
3 Remove carbon dioxide before it leaves the chimney stack.
4 Have strict rules forbidding waste of fuels (30% of energy generated is wasted). Help people to save energy at home by loft insulation, wall-cavity insulation and double glazing.
5 Insist on the fitting of catalytic converters to all car exhausts (there is a snag to this one, can you spot it?).
6 Insist on birth control to restrict growth in the population.

Which of these suggested measures are most likely to work best?

Floods?

As the Earth warms there is a threat that the polar ice-caps might melt. This would raise all sea levels by at least 1.5 metres. The scale of flooding would be immense. Wealthy countries could plan for the floods and build up river banks and sea walls to prevent flooding. The poorer countries would suffer most because they cannot afford to take adequate precautions.

Plants

It was thought that plants would thrive on the extra carbon dioxide because it is used in their energy-building process (photosynthesis). However laboratory tests show that many cultivated plants do not grow better with higher concentrations of carbon dioxide, but some weeds seem to thrive on it! Will weeds become more of a threat to crops? Tests with high levels of carbon dioxide are now taking place. One such test involves 400 tonnes of the gas. Some pests seem to be more destructive with extra carbon dioxide.

Summary

- Increased levels of carbon dioxide, methane, nitrogen oxides and CFCs in the atmosphere cause the 'greenhouse effect'.
- The build-up of these 'greenhouse gases' traps heat on the Earth's surface and could change the weather, cause flooding and favour weeds and destructive insects rather than cultivated crops.

Extension questions

1 Try to link all the reasons for the greenhouse effect to the statement *'the greenhouse effect can be blamed on the huge growth in human population'*.
2 List the predicted problems that might arise from the greenhouse effect.
3 Explain how the greenhouse effect works. If rays radiated from the Earth did not change wavelength, what would happen to them?
4 In Vasteras, a city of 100 000 people in Sweden, waste heat from the nearby power station is used to heat every household. This is called **combined heat and power**. How does this project help in solving the greenhouse effect? Can this type of project be used elsewhere in the world?
5 **a** Explain how a gardener's greenhouse warms up.
b Some gases are accumulating in the atmosphere causing the Earth to warm up. Name three such gases.
c How do these gases contribute to the warming-up of the Earth?
d Why is this effect called the 'greenhouse effect'?
e How are people, plants and the weather likely to be affected by the greenhouse effect?
f Name two sources of energy that do not produce 'greenhouse' gases.
g The fitting of catalysts in car exhausts could increase the greenhouse effect. Explain how this could happen.
h One answer to the greenhouse effect is to plant many more trees. How does this help?

Unit Q
Disappearing forests

The tropical rain forests

Nowhere is the environmental impact of human activities better illustrated than in the tropical rain forests. Rain forests girdle the Equator (Figure Q.1) covering 14.5 million km^2 of land in some of the world's poorest countries.

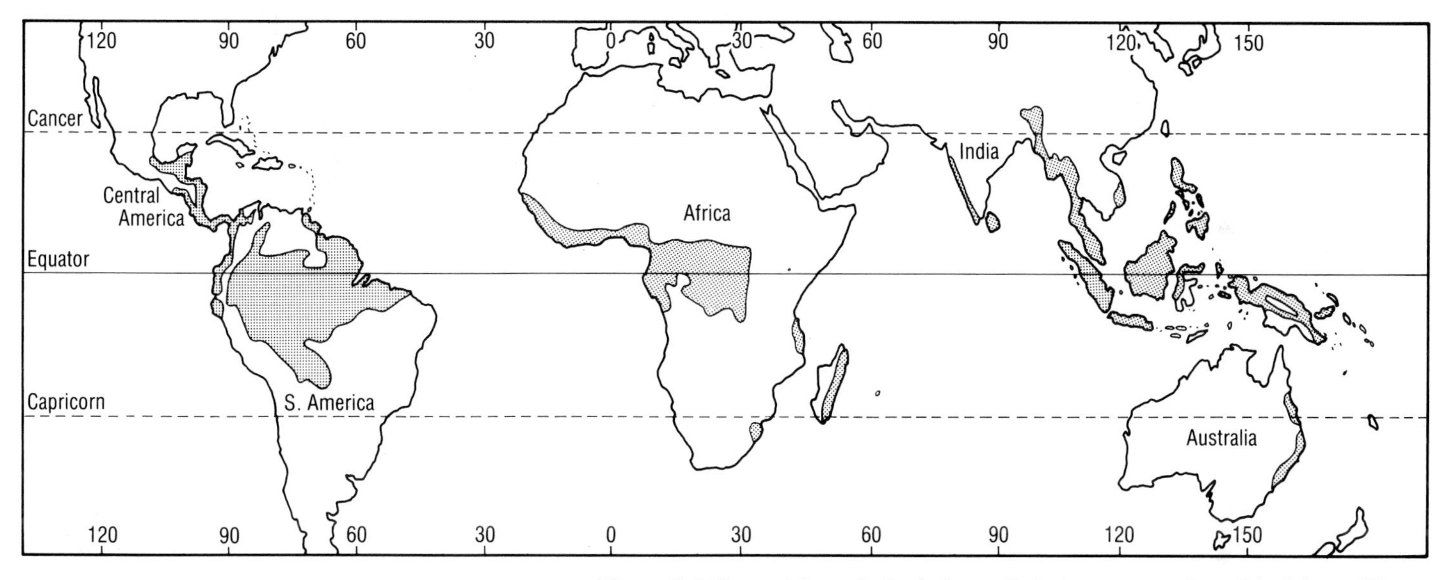

Figure Q.1 The world's tropical rain forests. Rain forest covers about 8% of the world's land with dense vegetation. The rain forests contain about 50% of all growing wood and the greatest variety of life of any ecosystem on Earth. It is estimated that in the rain forest of Peru more than 30 million species of insect alone live in the forest canopy

Rain forests are often described as the world's lungs. The forests' giant trees recycle carbon dioxide and oxygen through photosynthesis. Moisture absorbed by the forest evaporates back into the atmosphere, crossing oceans and continents to fall as rain thousands of miles away (Figure Q.2).

About 100 000 km^2 of rain forests (the size of Scotland and Wales together) are now being lost each year. Cutting down and burning the trees (**deforestation**) takes away the soil's protective covering of plants, leaving it open to fierce tropical rain storms. Erosion soon sets in as soil is washed into the rivers and carried out to sea (Figure Q.3). In the dry season the sun's heat soon bakes the bare soil dry. Chemical reactions in the drying soil produce a hard, brick-like layer called **laterite**.

Scientists estimate that burning the trees of the rain forests could release as much as six thousand million tonnes of carbon dioxide into the atmosphere each year. Deforestation also encourages the activities of soil micro-organisms. Their decomposition of the huge quantities of plant debris releases perhaps a further two thousand million tonnes of carbon dioxide each year. Reduction in tree cover also means that less carbon

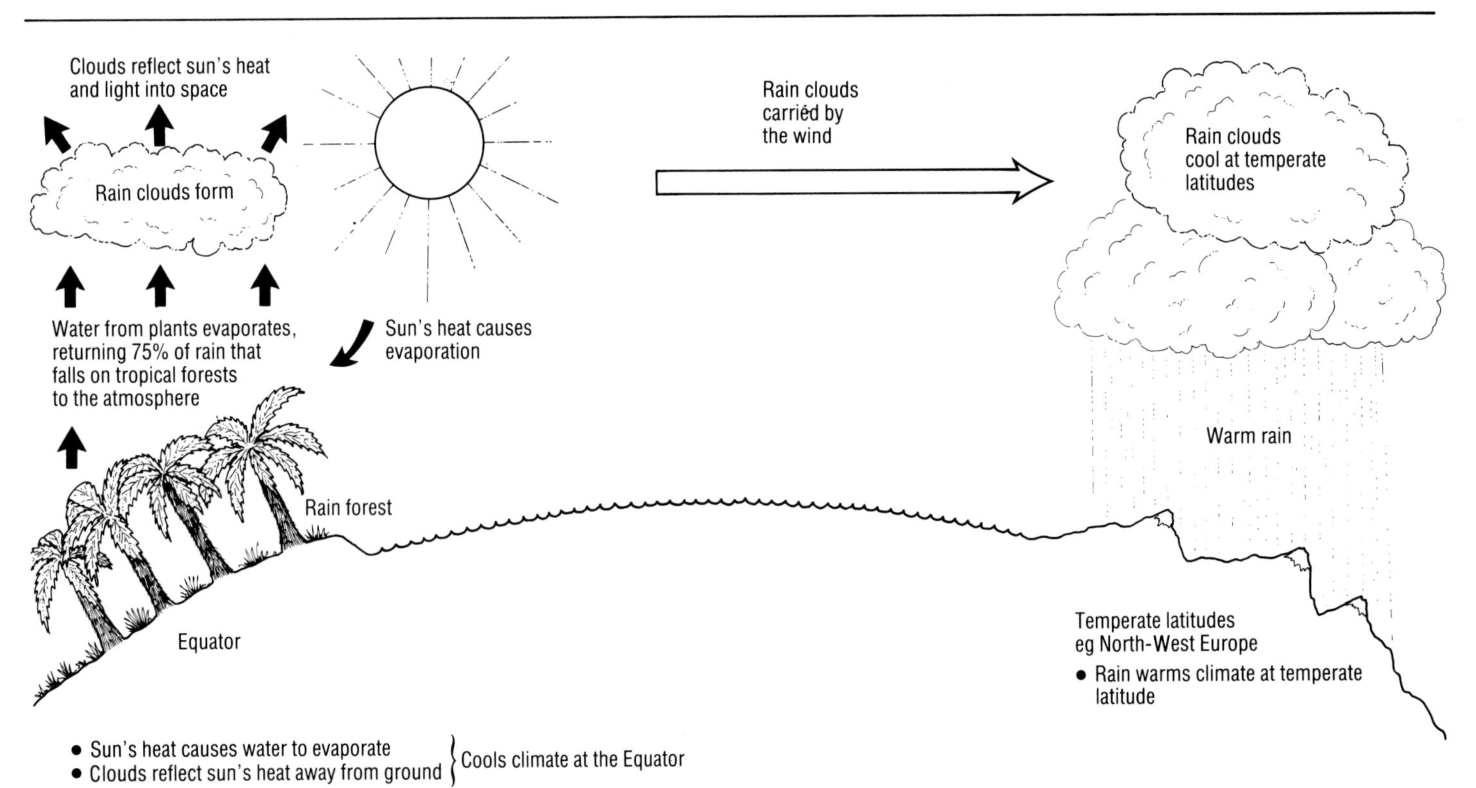

Figure Q.2 How rain forests affect the world's climate. Without them, the Equator would be warmer and temperate latitudes cooler. The pattern of rainfall would change worldwide, with widespread effects on agriculture

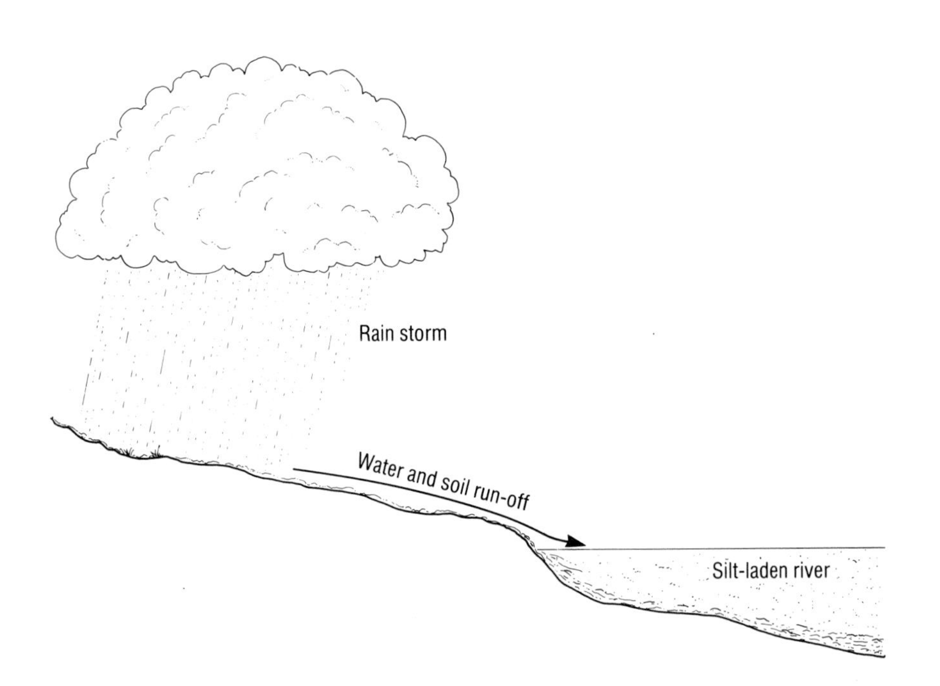

Figure Q.3 Rain forest soil covered by plants absorbs water, releasing it slowly into rivers. This means that rivers receive water even in the dry season. Clearing vegetation leads to soil erosion and flooding in heavy rains. However, in the dry season, rivers quickly dry up as the soil cannot hold water. Without trees to put water into the atmosphere, there will be regional and possibly global changes in the climate

dioxide is taken out of the atmosphere through photosynthesis. All of these effects of deforestation contribute to the enhanced greenhouse effect.

Why are we clearing the rain forests? 'We' is used deliberately because it is *our* demand for food and goods which use rain forest resources that is partly the reason for clearing. Other reasons are caught up in the social, political

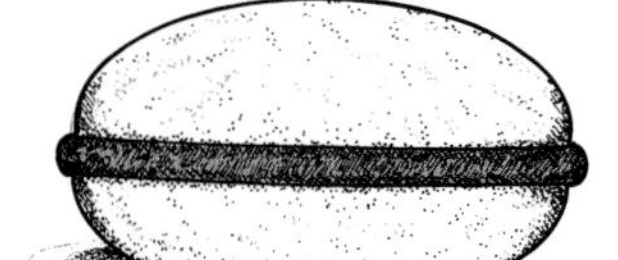

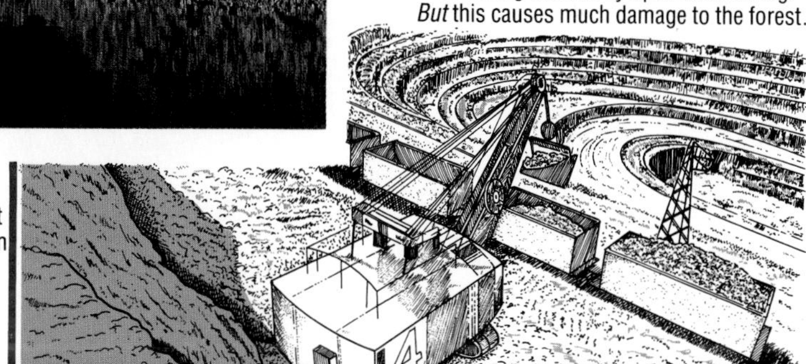

Figure Q.4 Exploiting rain forest resources

and economic development of the rain forest countries. The effects of this mix of national and international pressures on the rain forest environment is highlighted in Figure Q.4.

The future?

Many people believe that the over-use of energy and resources by developed countries together with rapidly growing populations in developing countries are straining the environment to breaking point. Ultimately, all economic development depends on natural resources from the environment. We cannot continue to use them extravagantly and without thought for their renewal. The rest of this unit looks at programmes of conservation (page 154).

The World Conservation Strategy

In 1980 a plan called the World Conservation Strategy was prepared by the United Nations Environment Programme, World Wildlife Fund and the International Union for the Conservation of Nature and Natural Resources. The plan set out an agenda for development that met human needs but which also protected the environment for future use. Figure Q.5 summarises its programme.

CONSERVING SOIL

Erosion

Soil is the lifeblood of farming. But more and more soil is lost each year through erosion caused in part by clearing forests and poor farming methods. Each year erosion removes:

- 6 billion tonnes of soil in India
- 4 billion tonnes of soil in the USA
- 2.5 billion tonnes of soil in the USSR
- 1.6 billion tonnes of soil from the highlands of Ethiopia.

As a result, crop yields decline. Evaporation accelerates the loss of soil water and land gradually turns to semi-desert. A cycle of drought and flood is established.

Ways of conserving soil

1 Reforestation – replanting trees and hedgerows. Plant roots hold soil in place; leaf fall adds humus which improves soil structure and water-holding properties. However, not all trees are suitable: quick-growing species like eucalyptus and conifers (used for making paper and furniture) may cause further damage to the soil.

2 Improved farming methods – increasing the variety of crops grown, rather than monocultures (see page 158).

CONSERVING WILDLIFE AND ECOSYSTEMS

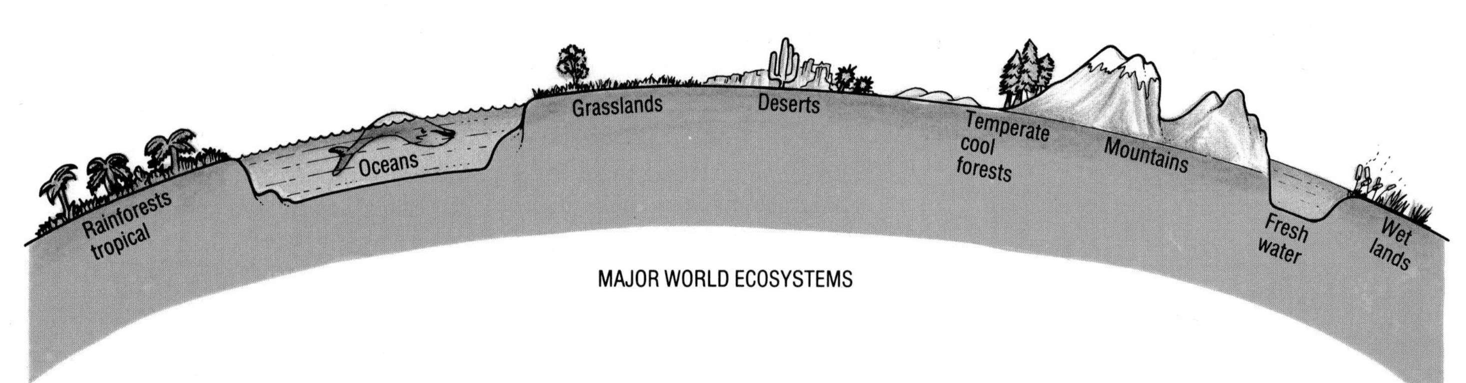

Extinction

More and more species are becoming extinct as a result of human activities. Since 1600 about 100 mammal species and 162 bird species have been lost.

6000 animal species and 25 000 plant species are now threatened with extinction. These are the *known* losses. It is likely that rainforest clearance wipes out many species which have not yet been identified.

Ways of conserving wildlife and ecosystems

1 Wildlife and ecosystems need to be managed in a way which avoids damage but which also benefits humans. This may mean looking for alternative ways of tackling problems. For example, the African savanna is infested with tsetse fly. The tsetse transmits a parasite which causes sleeping sickness in cattle and humans. If farmers wish to raise cattle, they have to wipe out the tsetse.This often means clearing the savanna of trees and scrub which are the tsetse's habitat, and killing wildlife which harbours the parasite. This clearance is costly, it damages the ecosystem, and it is not 100% successful – so that repeated clearance is necessary.

However, the wild antelope of the savanna are immune to tsetse fly. Properly managed, they provide more meat per hectare than cattle. The antelope offer the possibility of long-term development of food resources together with conservation of the savanna ecosystem.

2 Managing ecosystems for human benefit is difficult. However, rainforest products such as fruit, nuts, rubber latex and medicinal plants can be exploited while leaving the ecosystem intact. For example, vinblastine and vincristine are powerful anti-cancer drugs. They come from the rosy periwinkle plant that grows in the rainforests of Madagascar. Many more beneficial plants may remain undiscovered – it is vital to preserve areas of undisturbed rainforest if people are to benefit in the future.

Figure Q.5 Aims of the World Conservation Strategy

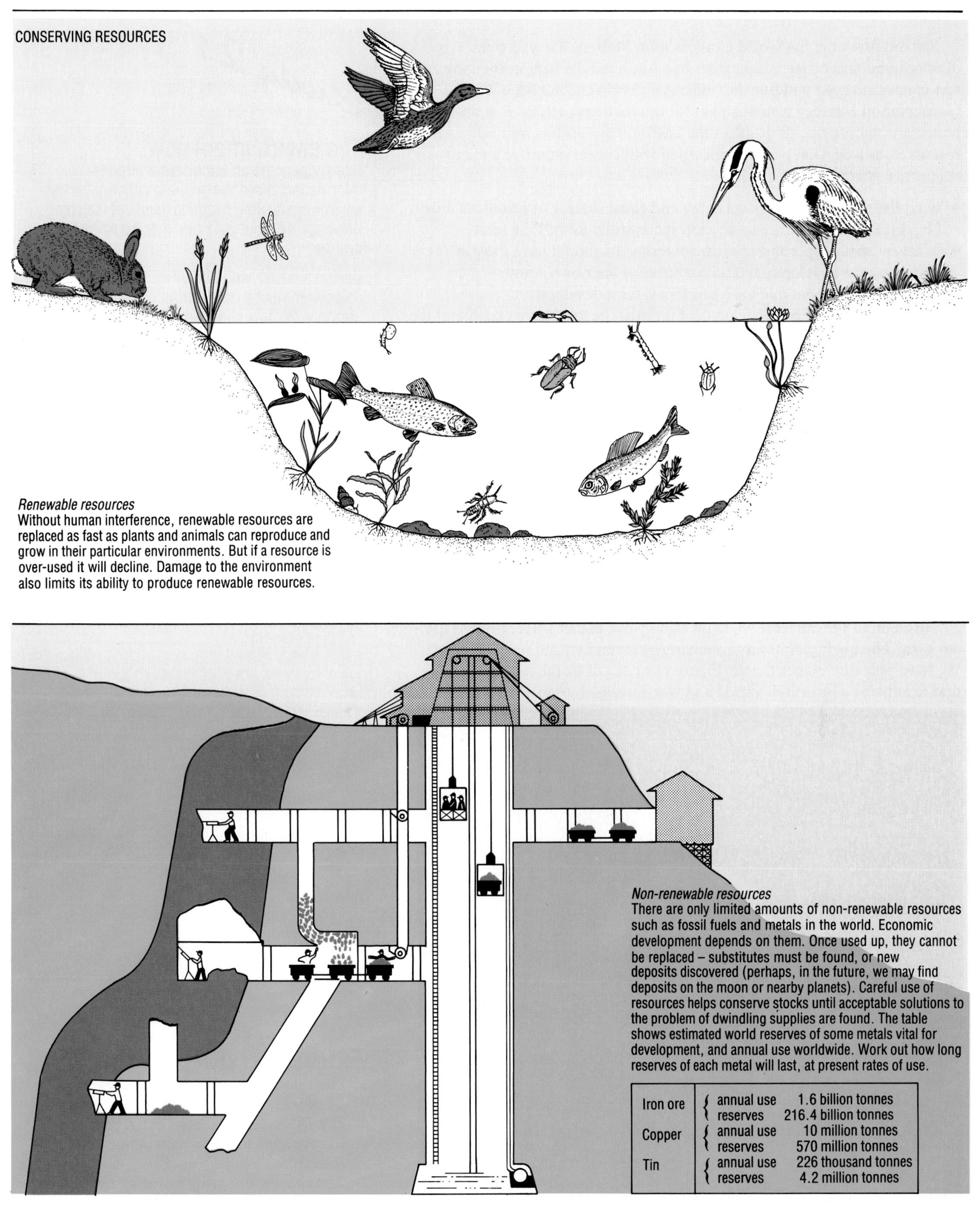

Renewable resources
Without human interference, renewable resources are replaced as fast as plants and animals can reproduce and grow in their particular environments. But if a resource is over-used it will decline. Damage to the environment also limits its ability to produce renewable resources.

Non-renewable resources
There are only limited amounts of non-renewable resources such as fossil fuels and metals in the world. Economic development depends on them. Once used up, they cannot be replaced – substitutes must be found, or new deposits discovered (perhaps, in the future, we may find deposits on the moon or nearby planets). Careful use of resources helps conserve stocks until acceptable solutions to the problem of dwindling supplies are found. The table shows estimated world reserves of some metals vital for development, and annual use worldwide. Work out how long reserves of each metal will last, at present rates of use.

Iron ore	annual use	1.6 billion tonnes
	reserves	216.4 billion tonnes
Copper	annual use	10 million tonnes
	reserves	570 million tonnes
Tin	annual use	226 thousand tonnes
	reserves	4.2 million tonnes

For the first time the World Conservation Strategy showed that development and conservation went hand-in-hand. In fact, in the long run one cannot succeed without the other. A second version of the World Conservation Strategy is now a plan for international action. It is aimed at politicians, whose decisions affect the environment and its natural resources, as well as at people concerned with conservation. It aims to encourage governments:

- to realise that ecology, the economy and equal sharing of resources must be partners if nations are to develop successfully in the long term
- to accept that people and not just governments should have a say in managing the development and resources of their own country
- to improve the economic analysis of long-term development
- to regularly check that development is not at the long-term expense of the environment
- to recognise that we share the planet Earth with wildlife which has a right to exist
- to control the harmful effects of industrial development, over-use of energy resources and growing populations increasingly concentrated in big cities
- to recognise the harmful effects of economic insecurity and warfare (including nuclear war) on the environment
- to put right ecosystems damaged or destroyed by human activities and upon which we depend for food, space and indeed life itself.

What we do now will affect the future of generations to come. We owe it to them to leave them an environment that meets *their* needs as it meets ours today and a world where resources are shared equally by rich and poor.

Throughout the universe we know of only one planet where there is life – our own. The quality of its environment is our concern and responsibility. We have only one planet Earth (Figure Q.6). All of us must make sure that it remains a beautiful, varied and worthwhile place in which to live.

Figure Q.6 What does the future hold for planet Earth?

Activity 2.1

The environment now

Newspapers are an important source of information about the growing concern for the environment. Bold headlines and eye-catching drawings help put over the facts contained in articles.

Design a poster with headlines and drawings on an environmental issue about which you feel strongly. Write a short article that uses your poster design.

Summary

- Cutting down and burning the rain forests releases huge quantities of carbon dioxide into the atmosphere. It also impoverishes soil.
- The World Conservation Strategy is an attempt to protect resources and use them sensibly.

Extension questions

1 How do rain forests affect the world's climate?
2 Briefly discuss the links between hamburgers and the destruction of rain forests.
3 Summarise the programme of the different versions of the World Conservation Strategy.

End-of-module questions

QUESTION ONE

The diagram shows a food web. Choose words from the list for each of the labels 1–4 on the diagram.

producer **prey** **predator** **fungus**

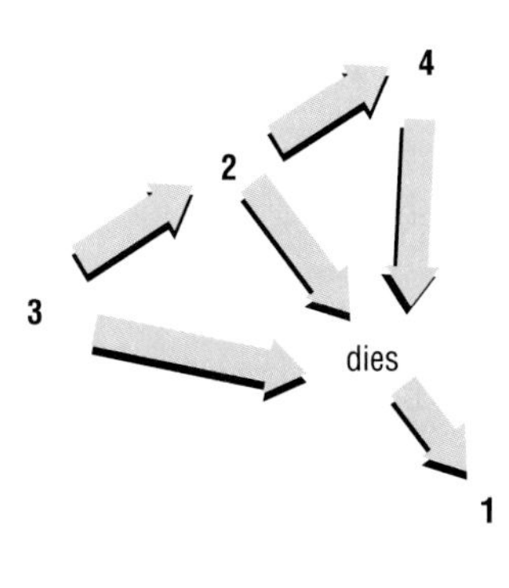

(C/D)

QUESTION TWO

The passage is about the carbon cycle. Choose words from the list for each of the items 1–4 in the sentences.

respire **photosynthesis** **fat** **microbes**

When an animal eats a plant, some carbon becomes part of ____1____.
Carbon dioxide returns to the atmosphere when plants ____2____.
When a plant dies, carbon is returned to the atmosphere by ____3____.
Carbon dioxide is removed from the atmosphere by ____4____.

(C/D)

QUESTION THREE

The diagram shows a woodland food web.

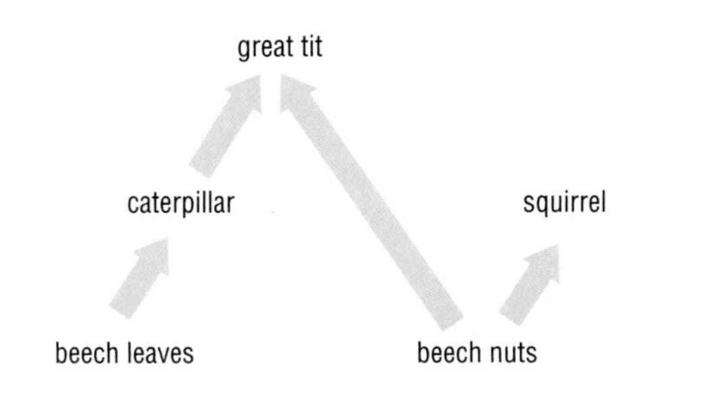

Choose from the list **two** producers.

beech leaves
squirrel
beech nuts
caterpillar
great tit

(E/F)

QUESTION FOUR

Choose from the list **two** factors that are necessary for plants to survive, but not so important for animals.

carbon dioxide
light
water
oxygen
nutrients

(E/F)

QUESTION FIVE

Choose from the list **two** gases that are produced when fossil fuels burn.

hydrogen
sulphur dioxide
nitrogen
nitrogen oxides
oxygen

(C/D)

QUESTION SIX

The diagram shows a food web.

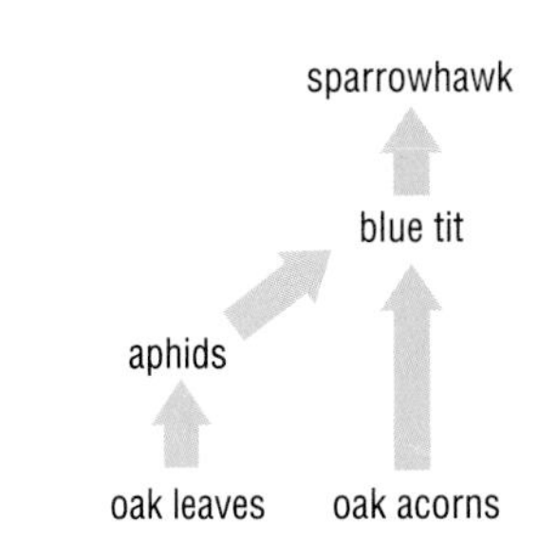

Choose from the list **two** organisms that are prey.

oak leaves
sparrowhawk
oak acorns
blue tit
aphids

(C/D)

QUESTION SEVEN

Food chains are usually very short. Choose from the list **two** statements that help to explain this.

movement increases the amount of energy available in the food chain
the biomass available increases along the food chain
some energy is lost as waste materials
respiration increases the amount of energy available along the food chain
a lot of energy is lost as heat

(A/B)

QUESTION EIGHT

1 Describe **two** uses that humans make of microbes. (C/D)

2 After an animal dies, how does the protein in its body eventually become nitrate? (A/B)

3 Fertilisers that farmers spray on the land can, eventually, get into rivers and lakes. How might this lead to the death of animals living in these waters? (A/B)

QUESTION NINE

Red deer live on the mountains of Scotland. The graph shows how the population in one area of Scotland varied over a number of years.

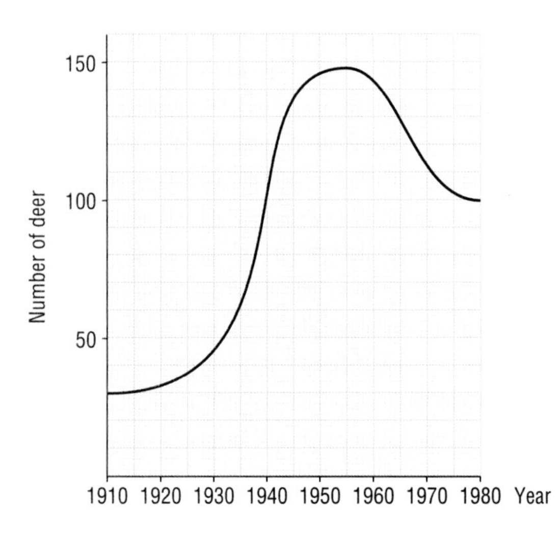

1 What was the population in 1920?
2 During which of the following years was the population increasing most rapidly?
1910 1930 1940 1950
3 Suggest **three** possible reasons for the population fall after 1960.

(C/D)

QUESTION TEN

The depth of leaf litter in a wood varies over a year. The graph shows this information.

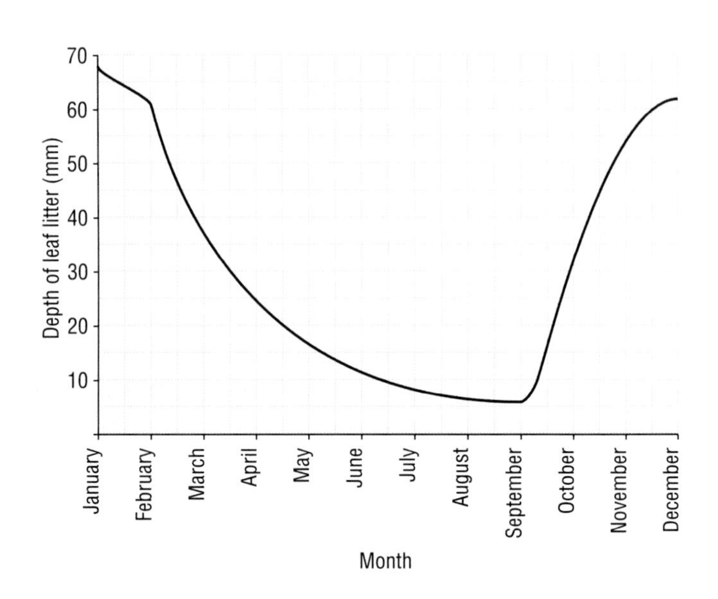

1 During which month is the leaf litter least deep?
2 What is the depth of leaf litter in April?
3 During which month does the leaf litter decrease in depth most rapidly?
4 **a** What causes the depth of the leaf litter to decrease?
b What conditions increase the rate at which the leaf litter disappears?

(C/D)

Inheritance and Selection

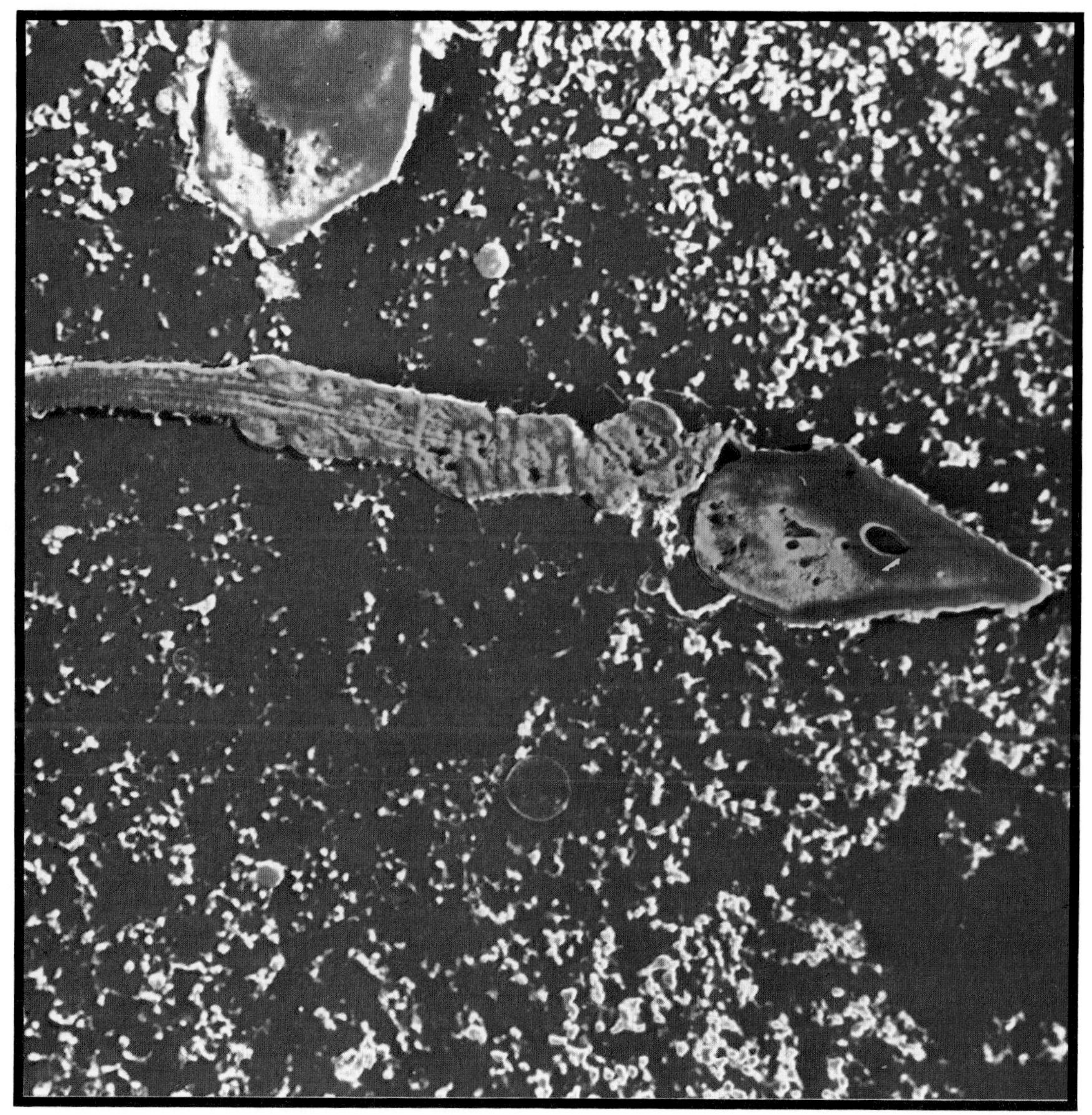

This module investigates the effect of heredity and environmental factors on the variation between individuals of the same species. It explains inheritance in terms of information carried on chromosomes, and explores the use of selective breeding, cloning and genetic engineering to produce plants and animals with preferred characteristics. The module considers how natural selection has led to the evolution of some species and the extinction of others, and examines the role of mutation in this process. It studies the determination of sex, and the inheritance of genetic diseases in humans, and finally examines and evaluates the hormonal control of fertility in women.

Inheritance and Selection

Learning Objectives

What you should know:

- Young plants and animals have similar characteristics to their parents because of information passed on to them in the sex cells (gametes) from which they developed. This information is carried by genes
- Differences in the characteristics of different individuals of the same kind may be due to the genes they have inherited, the conditions in which they developed or a combination of both
- Sexual reproduction involves the joining of male and female gametes resulting in individuals which have a mixture of genes from two parents
- Asexual reproduction gives rise to clones – individuals whose genetic information is identical with that of their parents
- New plants can be produced quickly and cheaply by taking cuttings from older plants
- We can use artificial selection to produce new varieties of organisms, some of which have led to increased yield in agriculture
- The ways in which fossils may be formed
- The theory of evolution – all species of living things which exist today have evolved from simple life forms
- New forms of genes result from changes (mutations) in existing genes
- Factors which increase the chance of mutations
- How sex is determined in humans by genes on the X and Y chromosomes
- The symptoms of Huntington's chorea and of cystic fibrosis
- The monthly release of eggs and changes in the thickness of the lining of the womb are controlled by hormones secreted by the pituitary gland and the ovaries
- How hormones can be used to control fertility in women

What you should be able to do:

- Investigate variation in tongue rolling, ability to taste PTC, hair texture and colour, eye colour, reflex time and height
- Explain how fossil evidence supports the theory of evolution
- Interpret evidence relating to evolutionary theory
- Evaluate the benefits of, and the problems which may arise from, the use of hormones to control fertility

EXTENSION

- How body cells divide by mitosis
- How gametes are formed by meiosis
- Why asexual reproduction results in identical offspring, but sexual reproduction gives rise to variation
- The drawbacks of selective breeding
- The principles involved in cloning and in genetic engineering
- How evolution occurs via natural selection
- Why mutations may be harmful
- How cystic fibrosis and Huntington's chorea are inherited
- How FSH, oestrogens and LH control the menstrual cycle
- How FSH and oestrogen can be used to control fertility

- Examine slides of dividing cells
- Clone cauliflowers
- Make informed judgements about the economic, social and ethical issues concerning genetic engineering
- Predict and explain the outcomes of genetic crosses
- Construct and interpret genetic diagrams

Unit A
Variety and variation

BACTERIA
- simple celled **micro-organisms**
- found everywhere: in water, soil and air; they also live on and in other living things
- some live in very harsh environments, where other living things could not survive
- a few bacteria cause disease
- most are **saprophytes**; they recycle elements through the environment

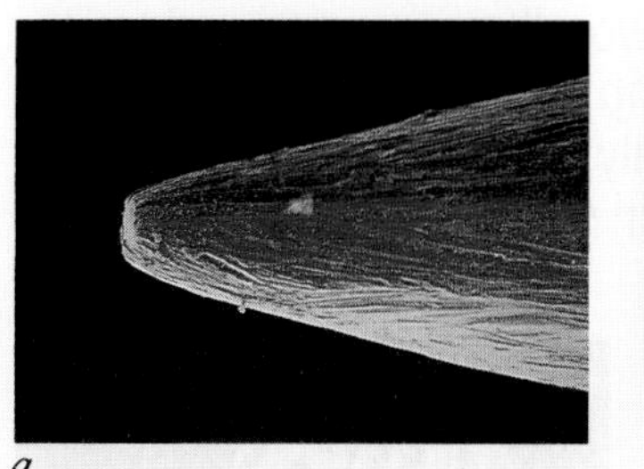
a

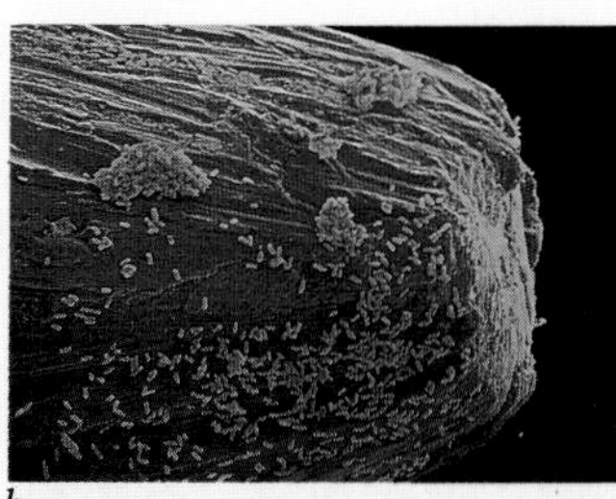
b

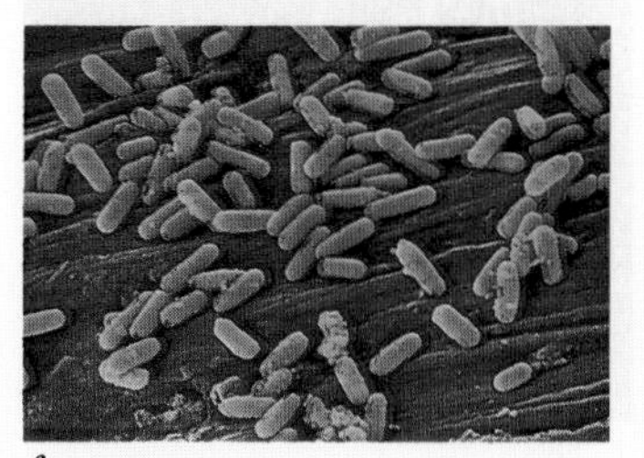
c

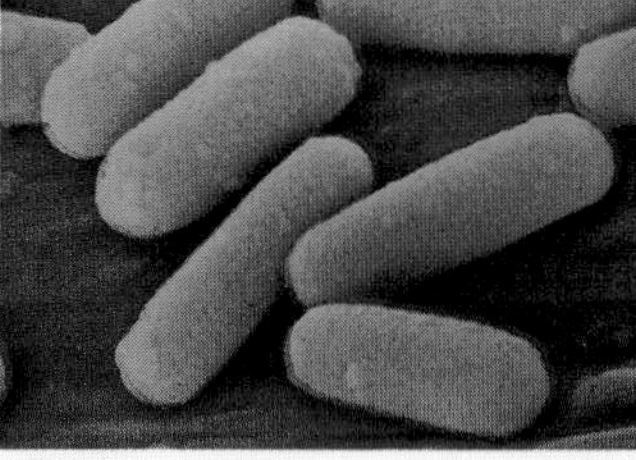
d

e

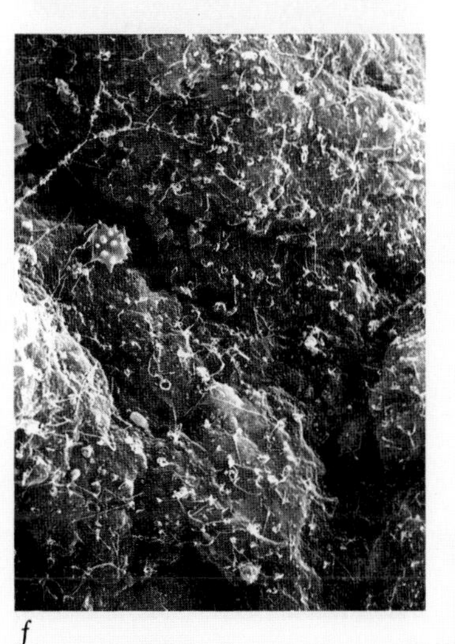
f

PLANTS
- release oxygen into the environment as a by-product of photosynthesis
  most algae live in water, although a few species (eg *Pleuroccocus* live in damp, shady places
- mosses and ferns are also restricted to damp, shady places – why?
- algae, mosses and ferns lose water easily; they need water for sperms to swim to eggs in reproduction
- over 80% of plants are flowering plants – why?
- flowering plants do not require free water for reproduction; a waterproof covering on leaves and stems cuts down water loss
- flowering plants are found in most land environments; few species live in water

g

h

i

j

k

l

Figure A.1 The living world

ANIMALS

- 95% of animals are invertebrate (without backbones)
- insects are the most numerous of all animals (10 billion per square km of land)
- insects, reptiles, birds and mammals dominate on land — why? The body's waterproof covering cuts down water loss; free water is not required for reproduction
- earthworms, snails and woodlice are restricted to living in soil or other damp, shady places because they lose body water easily
- earthworms help maintain soil structure and fertility

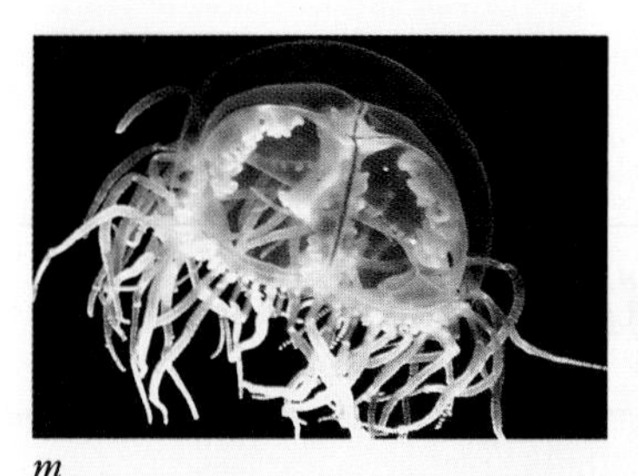
m

n

o

p

q

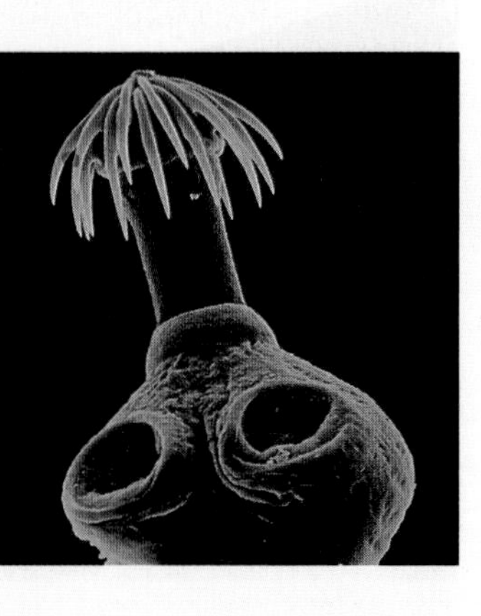
r

s

t

FUNGI

- most fungi consist of branching, thread-like filaments called hyphae
- together the **hyphae** make up the **mycelium**
- fungi live in damp, shady places since hyphae lose water easily
- yeast is a single-celled fungi
- some fungi are parasites, others are saprophytes which recycle elements
- a few fungi cause disease, some spoil food; others flavour food (eg blue cheese)
- some fungi are a source of antibiotic drugs (eg penicillin)

u

x

v

w

y

z

Figure A.2 The same species ... but no two people are exactly alike

Variation

One of the most fascinating things about living organisms is their enormous **variety**. They differ in appearance, behaviour and where they live. Garden snails, cockles, lions and zebras are all animals. They all differ from each other. However, garden snails and cockles belong together and lions and zebras belong together. But zebras and horses resemble each other more than either resemble lions.

Living organisms can be divided into groups, the members of which have similar features. Within each large group, smaller groups can be made. Let us consider ourselves and see how we are different from most other organisms, but how we are similar to some.

Figure A.1 illustrates some of the variety of life on Earth. It shows a way of grouping different sorts of living things. Notice that we humans are a distinct **species** of animal. A species is defined as a group of individuals which can interbreed to produce fertile offspring. In other words, their offspring must also be able to breed between themselves.

There may be a great deal of variation within species; for example, certain features differ between Asians, Africans, Scandinavians and American Indians. (Consider, for instance, the variation in skin pigmentation in Figure A.2.) However, they all belong to the same species because they can interbreed successfully to produce fertile offspring.

In humans there are hundreds of features which illustrate variation. Some can be measured and show a continuous trend from one extreme to another. An example of this is **continuous variation** in height of a population. Another type of variation shows two extremes without any intermediate types. Such a case is seen in the ability to roll one's tongue (see Figure A.3). This is **discontinuous variation**. You can either roll your tongue completely or you can't. There is no such thing as a 'half tongue-roller'.

Investigating variation

A great deal of variation can be observed by examining certain **characters** in members of your form at school. A character is a single feature of an organism which is under observation; for example, hair or eye colour. Variation means the distribution of these characters throughout the group. This can only be seen by collecting together all the results for the whole group. In this unit, you will investigate variation of certain characters for the students in your form.

Figure A.3 Tongue-rolling – you can either do it or you can't. This is an example of discontinuous variation

Continuous and discontinuous variation

We can use discontinuous variations to sort individuals into distinct groups which do not overlap. With continuous variation there are no distinct groups and individuals are spread over a range of measurements. By plotting histograms in the way suggested in the Activities, we obtain an idea of the limits within which most individuals fall. Also we can see how the more extreme cases are spread on each side. Before making measurements you should follow these rules.

1 Make sure that the character you are measuring does not depend on another factor. For example, if you are measuring height in a population of humans or length of leaves from a tree, the samples must all be of the same age. The age of an individual will certainly affect his or her height.

2 When measurements involve judgements of degrees of a character (for example, hair colour), everyone involved in measuring must agree with the judgement and use the same descriptions.

Activity A.2

Tongue-rolling

1 Try to roll your tongue as shown in Figure A.3.
2 Find the percentage of tongue-rollers and non-tongue-rollers in your form.
3 Construct a histogram from your results; it should resemble the one below. This example is similar to Activity A.1: both illustrate **discontinuous variation**
4 Observe and record whether the ability to tongue-roll is more clear cut than the ability to taste PTC.

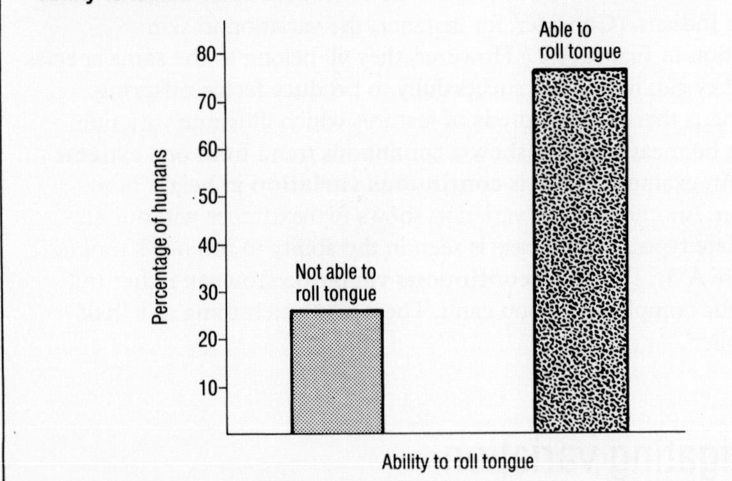

Activity A.3

Hair texture

Work in pairs.

1 Classify your partner's hair as straight or curly.
2 Find the percentages of people having straight hair and those having curly hair in your form.

Is there a distinct grouping or is there a continuous gradation in texture?

Hair colour

Hair colour is not at all easy to define. Hair can be black, dark brown, medium brown, light brown, blond or red.

Look at the hair colour of people in your form and modify this colour code if necessary.

Record the number of students in each group.

Are the differences within groups as great as those between groups? If so, what does this tell you about the variation of hair colour compared with the variation in tongue-rolling ability?

Activity A.1

Ability to taste phenylthiocarbamide (PTC)

PTC is a chemical which has a bitter taste to some people. It tastes just like the rind of a grapefruit. However, you won't know what this tastes like at all if you are a 'non-taster'!

1 Make a saturated solution of PTC in ethanol and soak pieces of filter paper with it. Now dry them.
2 Touch your tongue with the filter paper. Record whether you taste anything.
3 Calculate the percentage of 'tasters' and 'non-tasters' in your form.

Is this a clear-cut test? Are there any people in your form who cannot say definitely whether they are tasters or non-tasters?

Perhaps your form is not typical. Suggest a reason for this.

The table below shows data for the frequency of non-tasters in various populations.

Race	Number tested	Non-tasters (%)
Hindus	489	33.7
Danish	251	32.7
English	441	31.5
Spanish	203	25.6
Portuguese	454	24.0
Malays	237	16.0
Japanese	295	7.1
Lapps	140	6.4
West Africans	74	2.7
Chinese	50	2.0
S.American Indians	163	1.2

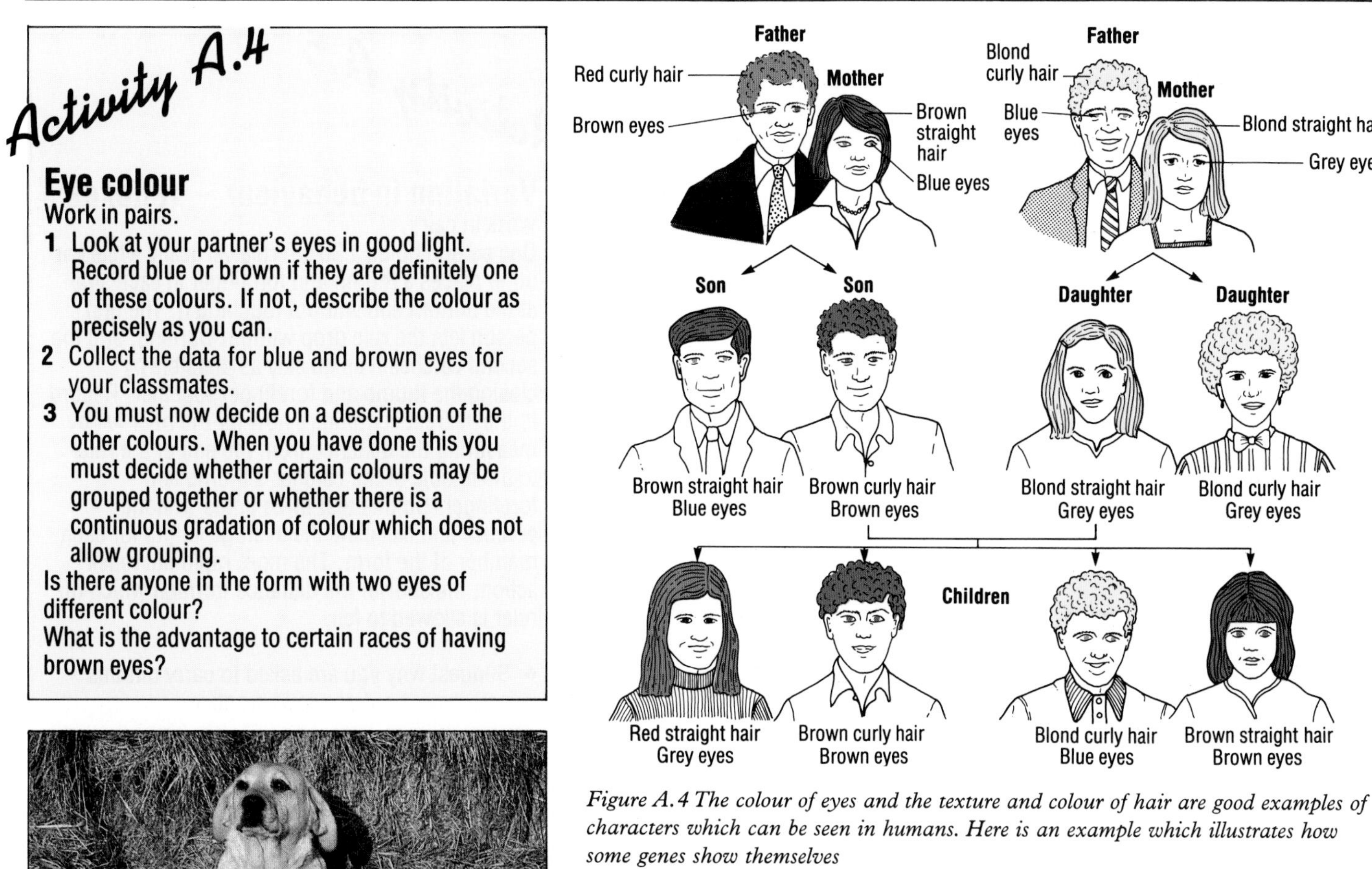

Activity A.4

Eye colour

Work in pairs.

1 Look at your partner's eyes in good light. Record blue or brown if they are definitely one of these colours. If not, describe the colour as precisely as you can.
2 Collect the data for blue and brown eyes for your classmates.
3 You must now decide on a description of the other colours. When you have done this you must decide whether certain colours may be grouped together or whether there is a continuous gradation of colour which does not allow grouping.

Is there anyone in the form with two eyes of different colour?

What is the advantage to certain races of having brown eyes?

Figure A.5 Different environments will affect some of these pups' characteristics (eg weight, size) ... but they will never grow up to be cats!

Figure A.6 These corgis will always be small, regardless of how well fed they are

Figure A.4 The colour of eyes and the texture and colour of hair are good examples of characters which can be seen in humans. Here is an example which illustrates how some genes show themselves

Causes of variation

The outward appearance of an organism (the **phenotype**) depends on a combination of inherited characters (the **genotype**) and characters due to the effects of the environment. We can say:

$$\text{phenotype} = \text{effects of genes} + \text{effects of the environment}$$

A simple example gives the idea. Imagine a litter of labrador puppies (Figure A.5). They all have **genes** (units of inheritance) from the same parents.

If we compare the growth of these puppies, we might see that one out of the litter becomes less strong and not so large as the rest. The reason could be that it has not been given the correct diet and exercise. In other words, the effects of the environment have influenced its phenotype. By altering the environment we can alter the phenotype.

On the other hand, a litter of corgis (Figure A.6) will grow up to be small dogs no matter how well fed they are. The genes determine the phenotype if the environment is kept the same.

Environmental factors

The **environment** of an organism means the type of place where it lives. However, the biological meaning of environment is every external influence which is acting on the organism. It therefore includes food and any other materials taken into the organism, temperature, degree of acidity, light and any physical forces acting on the organism from the outside.

Some examples will serve to show how important these factors are in shaping the phenotype.

Nutrients

The type of food eaten by animals and the types of mineral salts used by plants influence their growth. The graph in Figure A.7 shows the effects of improved diet on the growth of some Japanese children.

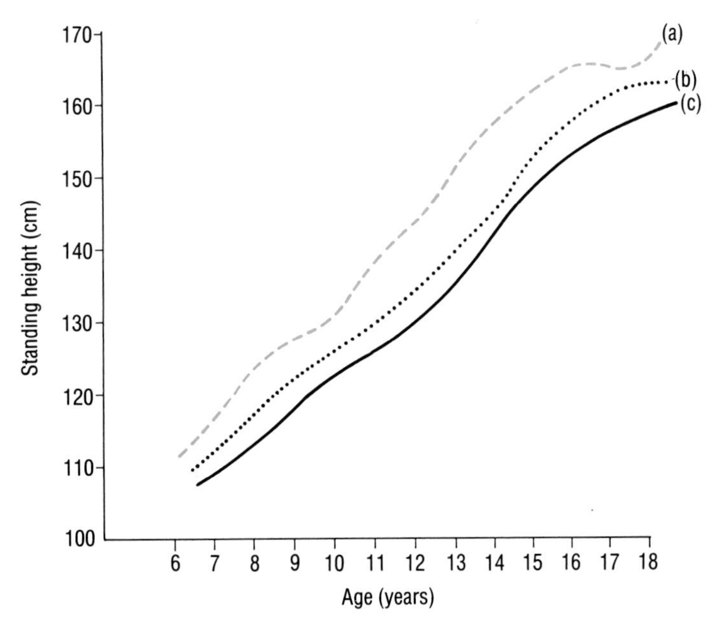

Figure A.7 The effects of improved diet on growth
(a) Height plotted against age for 898 Japanese boys aged 6–18, born in the USA and measured in 1957
(b) As (a), for 898 Japanese boys aged 6–18, born in Japan and measured in 1952
(c) As (a), for 898 Japanese boys, aged 6–18, born in Japan and measured in 1900

The children were brought up in different environments. Some were born and raised in the USA (graph (a)). The other data refer to children raised in Japan with and without an improved diet (graphs (b) and (c), respectively).

From the graphs you can see that the US-born children are the tallest at all ages. The improved diet and standard of living in Japan between 1900 and 1952 has also resulted in an increase in height.

The same trend has also occurred in Britain. Children are now taller and heavier, age for age, than they were 50 years ago.

Figure A.8 shows the effect of nutrition on the size of bones. One pig was fed on a normal diet, the other had the same food but in restricted amounts. Both pigs were from the same litter and were the same age when killed.

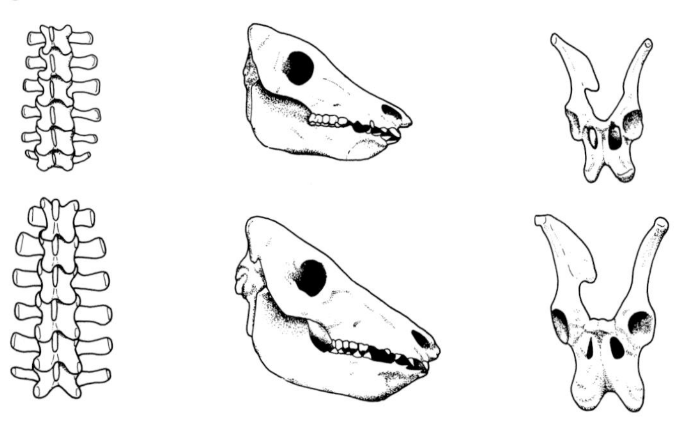

Figure A.8 The effect of nutrition on the size of bones

Activity A.5

Variation in behaviour – reflexes

Work in pairs.

One person holds a 30 cm rule vertically while the other places a thumb and forefinger at each side at the bottom end without touching it. The first person lets the rule drop without warning, and the second catches it as quickly as possible by closing the thumb and forefinger together. Record to the nearest centimetre how far the ruler fell by measuring the distance from the end of the ruler to the middle of the catcher's thumb and forefinger. Do this test three times without practice and calculate the average length for each member of the form. The more rapid the reflex action, the shorter the distance through which the ruler is allowed to fall.

- Suggest why you are asked to carry out this activity without previous practice.

Display the class results in a table like the one below and construct a histogram by plotting number of individuals against distance fallen.

Distance fallen (cm)	1	2	3	4	5	6	7	8	9
Number of individuals									

Activity A.6

Height

Record the height of each member of your class. Complete a table similar to the one shown below and construct a histogram as you did for Activity A.5.

Height (cm)	150–5	156–160	161–5
Number of individuals			

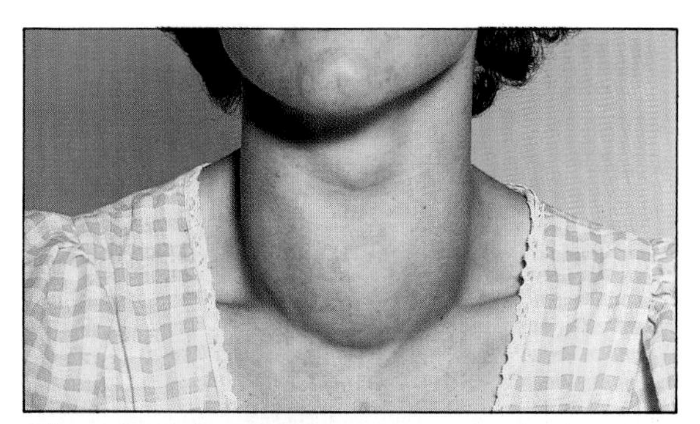

Figure A.9 Iodine deficiency causes goitre formation

Activity A.7

Weight of acorns

Collect 100 ripe acorns from under an oak tree. Do this randomly; that is, do not deliberately choose any particular ones. Clean and weigh each of them. Construct a histogram after tabulating the data as follows.

Weight (g)	0–0.5	0.6–1.0	1.1–1.5	1.6–2.0
Number of acorns				

Repeat this activity using the length of leaves fallen from the oak tree as the characteristic you wish to investigate.

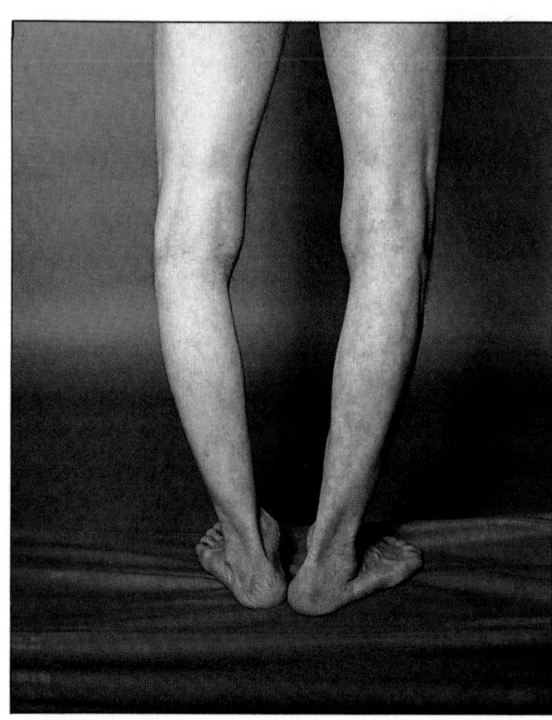

Figure A.10 Rickets is caused by a lack of vitamin D in the diet

An abnormal phenotype can also be seen in humans as a result of a deficiency of iodine. Lack of iodine in drinking water may be the cause of **goitre** formation. The swelling (Figure A.9) is caused by an enlarged thyroid gland which makes a hormone (**thyroxine**) that contains iodine.

Vitamin deficiency can also result in abnormal phenotypes. For example, lack of vitamin D causes **rickets** (see Figure A.10). If a person develops this disease, the bones become soft and are distorted under the weight of the body.

Chemicals

Some chemicals which cannot be called 'food' enter the body from the environment. They are not needed for growth or energy release and may affect the phenotype. One such chemical which had a disastrous effect was **thalidomide**. This drug was given to pregnant women to prevent sickness and to help them sleep: as a result, some of these women gave birth to deformed children, some of them without arms and legs. The foetus develops in an environment of amniotic fluid and the mother's blood. Thalidomide exerted its effects by entering the mother's blood and crossing the placenta.

Temperature

Changes in temperature affect the rates of enzyme-controlled reactions in all cells. These effects are shown in Figure A.11. The higher the temperature, the faster the enzymes work, up to a maximum rate. This explains why animals and plants generally grow more slowly at lower temperatures and why most organisms cannot survive at temperatures higher than 45 °C.

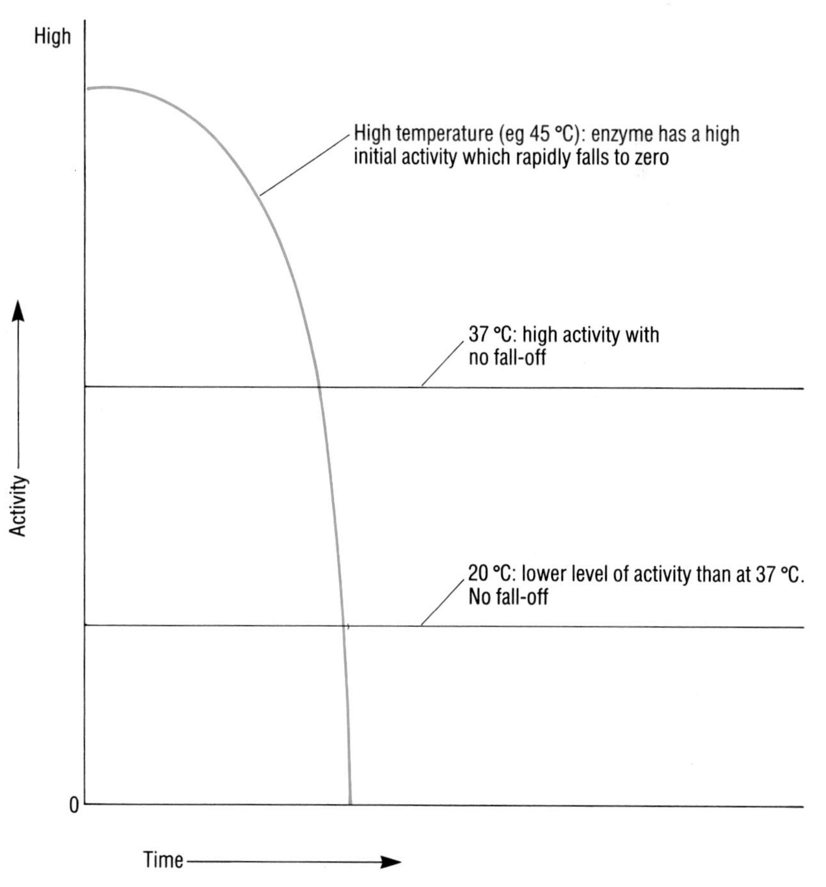

Figure A.11 The effect of temperature on enzyme activity

Light

The browning of human skin when exposed to light is an obvious change of phenotype due to an environmental change. More subtle changes may also occur. The photomicrographs (Figure A.12) show two leaves from the same plant. One has been grown in the shade (a) and the other in the light (b). Make a list of the differences between them.

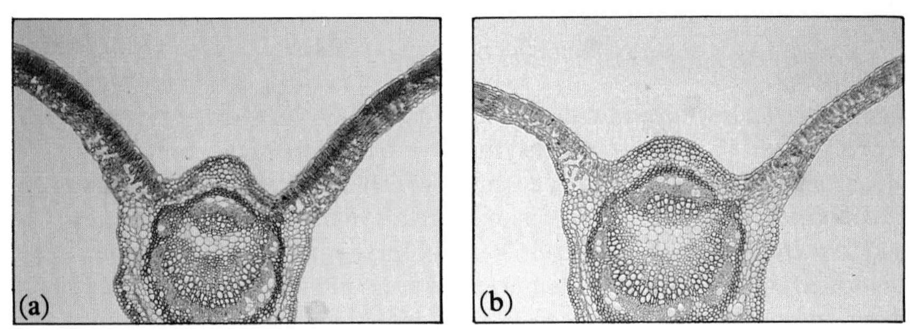

Figure A.12 Transverse sections of horse-chestnut leaves: (a) taken from a shady position; (b) taken from a well-lit position

Figure A.13 Muscles can be developed by exercise

Physical forces

Training by using muscles more than they are normally used can alter the phenotype. Muscle enlargement is obvious (see Figure A.13). Another effect of training is that the blood vessels to the muscles enlarge to supply more nutrients and oxygen.

Summary

- Plants and animals show continuous and discontinuous variations. These can be measured and expressed graphically.
- Causes of variation may be genetic or environmental.

Questions

1 Two hundred limpets of the same species were collected from a rocky shore. Their shell diameters were measured and then grouped by size in 5 mm ranges. The results are shown in the graph in Figure A.14.
  a What is the most common range?
  b How many limpets measured more than 30 mm in diameter?
  c What percentage of the animals had a shell less than 21 mm in diameter?
  d If this species of limpet grows 5 mm each year, how old were the largest limpets in the sample?

2 A seedsman wanted to find out the frequency of three flower colours in a mixed bed of Livingstone daisies. Random samples were counted where they grew. The results are as shown below.

Sample	Red	Pink	Orange
A (53 plants)	16	25	12
B (54 plants)	14	30	10
C (50 plants)	11	15	24
D(43 plants)	18	10	15
Total (200 plants)	59	80	61

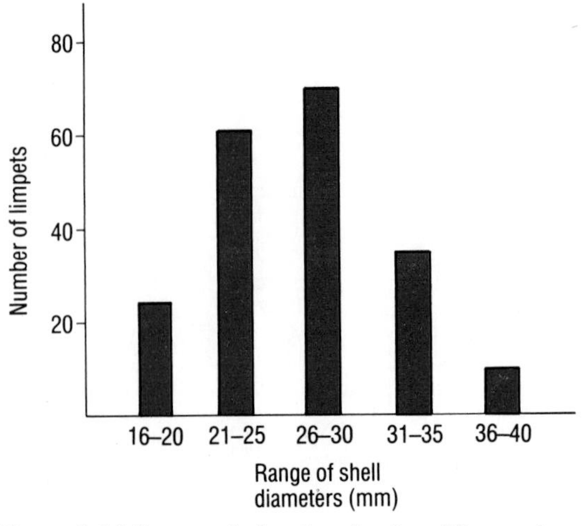

Figure A.14 Bar graph showing the size of limpets in a random sample

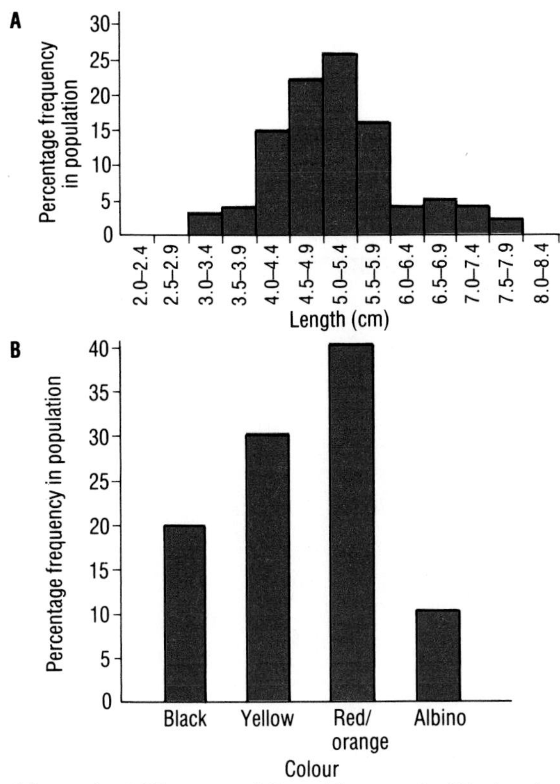

Figure A.15 Histogram (A) and bar graph (B) showing the length and colour of goldfish in a random sample

a Draw a suitable bar graph to show the total numbers of flowers of each colour.

b Suggest why more than one sample was taken.

3 The histogram and bar graph in Figure A.15 show variations in two characteristics in a population of goldfish. The population contained 250 fish. Histogram A shows variation in length. Bar graph B shows variation in colour.

a What type of variation is shown by the histogram and bar graph?

b From bar graph B, calculate the number of red/orange goldfish present in the sample.

4 Two girls studying a small wood near their school decided to measure the thickness of some leaves on a lime tree in the middle of the wood. They carried out their survey four weeks after the tree came into leaf. They took just five similar-sized leaves from three different heights above the ground. Some of their results are shown in the table below.

Height above ground (m) / Leaf	Thickness (mm)				
	1	2	3	4	5
2.25	0.8	0.9	0.8	0.8	0.7
6.0	0.5	0.7	0.5	0.6	0.7
8.5	0.3	0.3	0.2	0.3	0.4

a What is the most common thickness of leaves at the 8.5 m mark?

b Calculate the average thickness of leaves at a height of 2.25 m.

c Calculate the difference in thickness between the thinnest and thickest leaf found.

d Explain why this variation is more likely to be due to the effects of the environment than to genetic factors.

e Explain how you would test this hypothesis.

f The girls then used a lamp and light meter to see how much light came through some of the leaves. They found that some leaves let 80% of the light through while others let only 15% through. Which leaves do you think were found higher up the tree? Why?

Unit B
Genetics and variation

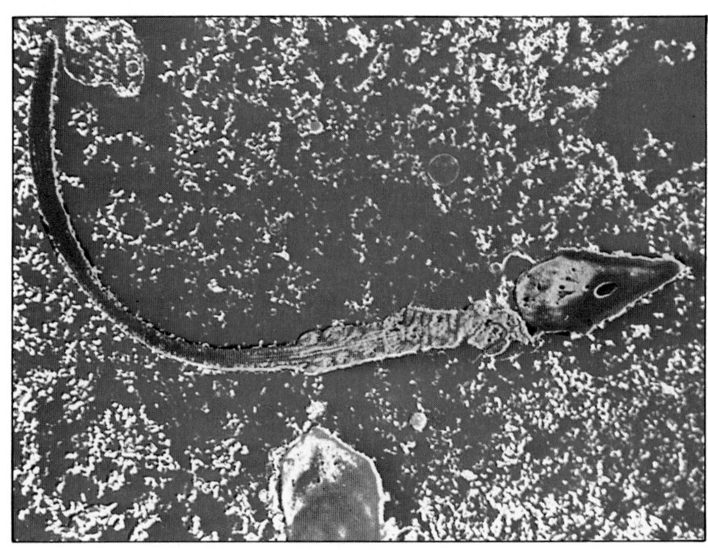

Figure B.1 A sperm as seen through a scanning electron microscope

Figure B.2 The female gamete – a human ovum

It would be ideal if we could look down a microscope and see the cause of variation revealed. Alas, this is not possible. Variations are handed down from one generation to the next through genetic material. Those cells which form a link between generations, the gametes, must contain genetic material and are therefore the best starting point for our study.

Figure B.1 shows the male gamete (**sperm**). In common with other cells, it has a nucleus and a membrane. Unlike female gametes (**eggs**) (Figure B.2), it has very little cytoplasm. Male gametes contain all the genetic material in the cell nucleus, which is in the so-called 'head' of the sperm. Sperms are **motile** (they can move). When fertilisation takes place the nucleus enters the female gamete and fuses with the female nucleus (Figure B.3).

The nucleus holds all the secrets of genetics. It contains **chromosomes**. A chromosome consists of the nucleic acid **deoxyribonucleic acid** (DNA), wound around a core of protein and folded tightly into a compact structure. The genes are the genetic material that living things inherit from their parents. They are lengths of DNA. Each chromosome carries a large number of genes.

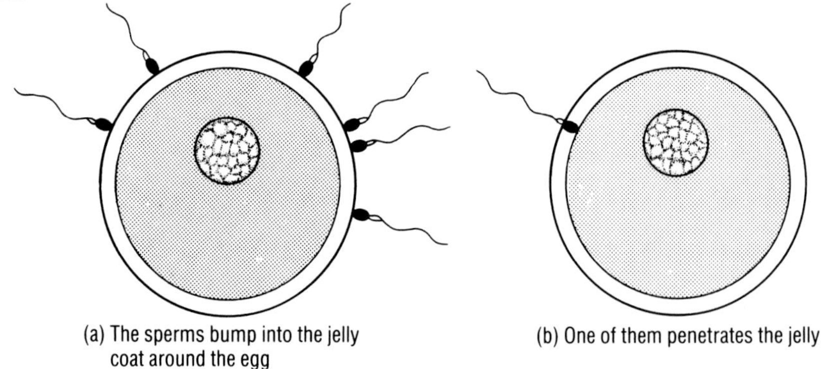

(a) The sperms bump into the jelly coat around the egg

(b) One of them penetrates the jelly

(c) Its head passes into the egg and the nuclei combine

Figure B.3 Fertilisation – the male and female gametes join to form a zygote

Activity B.1

A closer look at the nucleus

The best material to use is the growing region of a root tip of garlic (Figure B.4). This plant is available throughout the year and is easy to grow. Simply place the garlic bulb on the top of a test tube containing water. In a few days, roots will appear.

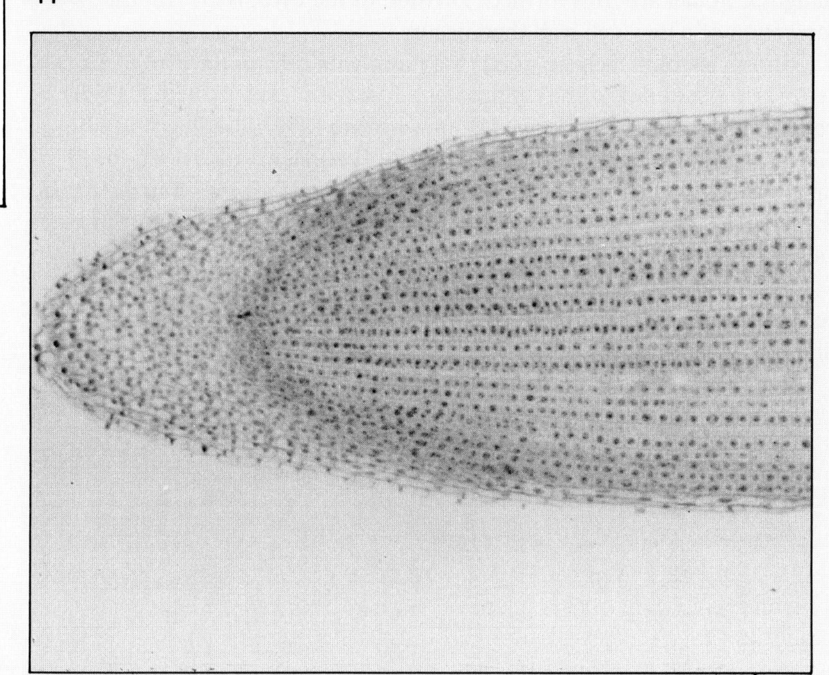

Figure B.4 Photomicrograph of a longitudinal section through a root tip

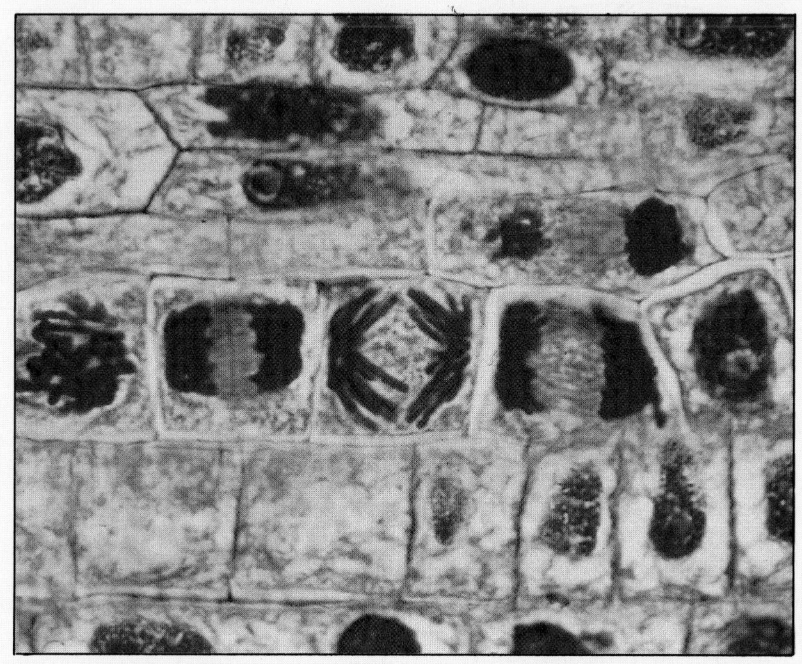

Figure B.5 From the left, the central four cells are in different stages of mitosis

Procedure

You will be given root tips which have been stained with Feulgen's stain – this reacts with chemicals in the nucleus to produce a pink colour.

1. Cut 2 mm off the end of the root tip using a sharp scalpel.
2. Place the tip on a microscope slide in a drop of 45% acetic acid. Use two mounted (dissecting) needles to pull the root tip apart. Try to break down the tissue into very small parts – this spreads the cells out.
3. Place a cover slip on the material. There should be just enough acetic acid to fill out the cover slip.
4. With the slide resting on some filter paper, put a few layers of filter paper on top of it. Press your thumb straight down on the region of the cover slip. Avoid any sideways pressure. This should flatten the cells and separate the chromosomes.
5. After about five seconds, peel off the blotting paper. The acetic acid should fill the space under the cover slip.
6. Examine the slide under the microscope. Use the high-power objective after focusing under low power. Draw the cells.

You should be able to make out the chromosomes.
Try to find the earliest stage at which the chromosomes are visible as double structures (see the left-hand cell in Figure B.5).
Look at several cells and compare them to those in Figure B.5.

As a result of your observations you should be aware of the following.

a The nucleus is the most likely site of the genetic material.
b The nucleus contains chromosomes.
c Chromosomes are only visible in cells which are actively dividing.

Cell division

The cell division studied in Activity B.1 is called **mitosis**. Apart from the special case of gamete formation, most cell divisions involve mitotic division of chromosomes. Figure B.5 shows a cross section of a root tip; different cells at different stages in mitosis can clearly be seen. The stages in mitosis are shown diagrammatically in Figure B.6. Prior to mitosis, each chromosome makes a copy of itself. At the start of mitosis, the original and copy separate and migrate to opposite poles of the cell. Two daughter nuclei are thus formed. Division of the cytoplasm and the formation of a new cell wall then occurs.

Mitosis enables each new cell (the **daughter** cell) to have an identical set of chromosomes to the original cell. It is the means by which all the cells of an organism are derived from a **zygote** (a fertilised egg). It also allows organisms to reproduce **asexually**. Thus all asexually-produced offspring from one organism have the same genetic information as that of their parent. Because the individuals are genetically the same they are called **clones.**

Sexually-produced offspring arise from the fusion of male and female cells. The production of these cells takes place by a special form of cell division called **meiosis**.

Importance of mitosis

Mitosis keeps the genes constant from generation to generation of dividing cells. Therefore when cells wear out and die, mitosis replaces them with new cells that are genetically identical to them. For example, old skin cells are always replaced by new skin cells, not by cells that are different.

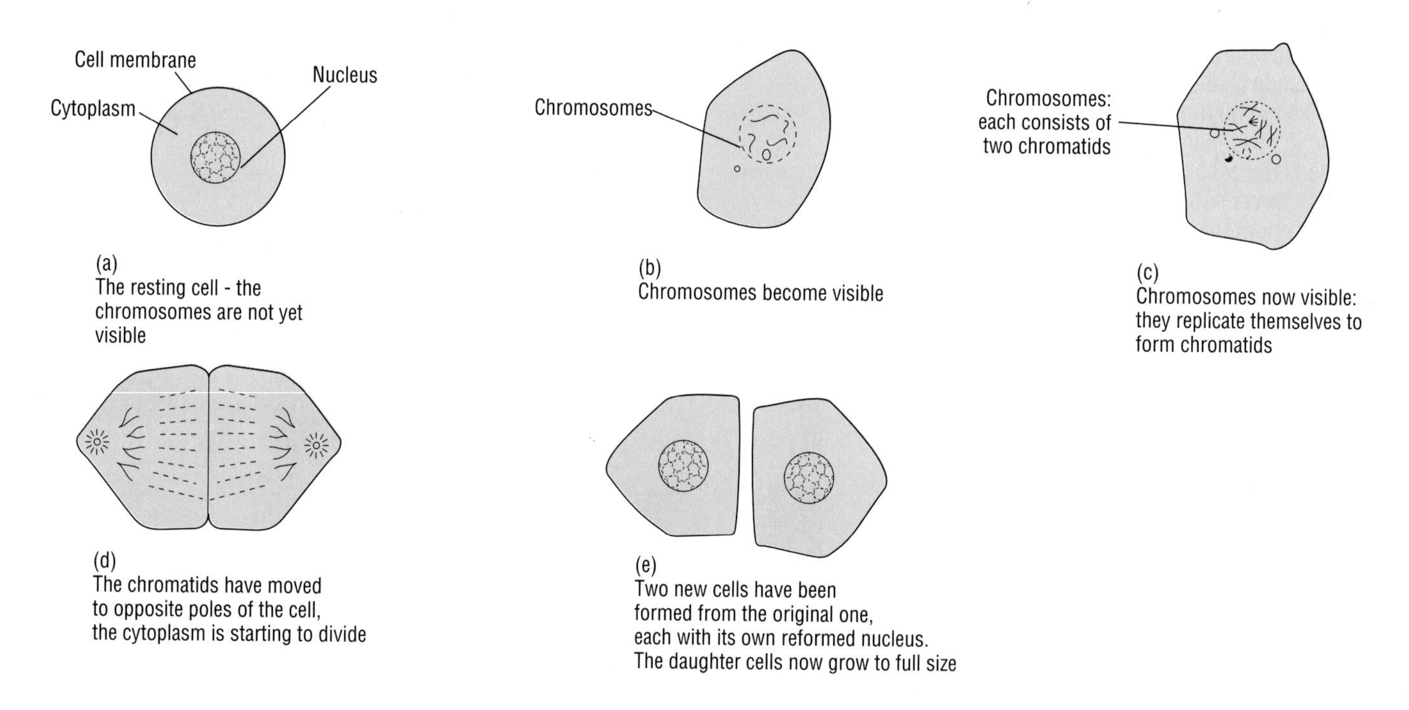

Figure B.6 A diagram of the stages in mitosis

Sex cells and fertilisation

The largest known cells are eggs (see Figure B.2). They contain little but cytoplasm; their bulk is mostly water and stored food. The important part of the egg is the nucleus. Sperm cells are hundreds or even thousands of times smaller than eggs. They are little more than a nucleus attached to a vigorous 'tail' (Figure B.1). When a sperm finds its goal, its nucleus joins the egg nucleus to form the nucleus of a new and separate cell.

The nuclei of both kinds of sex cells generally have fewer chromosomes than other cells in the body of the same organism. The reason for this becomes clear when we see what would happen if they did not. Cells in the human body have 46 chromosomes. If each matured sex cell, or gamete, also had 46, then a baby would have 92, and its children would have 184. Yet all normal cells in human bodies have 46. Human cells normally contain 23 pairs of chromosomes. One set comes from the mother's egg and the other set comes from the father's sperm. Mitosis ensures that each new cell gets a full set of chromosome pairs.

Microscopic studies of developing human sperms and eggs show that they have only 23 chromosomes, one from each pair. How does an organism produce cells with half the number of chromosome pairs?

Of all the countless millions of cells in our bodies – and those of other organisms – only egg- and sperm-producing cells divide in a way to split up the chromosome pairs. Logically enough, the process is called **reduction division** or **meiosis**. To see how it works we can follow meiosis as it occurs in an insect that has only four pairs of chromosomes.

The first step in meiosis is similar in some ways to mitosis. Each chromosome pairs off with its opposite number across the middle of the nucleus, and each chromosome duplicates itself. There is now a double chromosome for each original, the halves of which are called **chromatids**. Just as in mitosis the nuclear membrane breaks down, but unlike mitosis each double member of a pair goes to the new cells. At this point we have two cells, each containing four double chromosomes – one from each original pair.

A brief resting period follows, then begins a new wave of nuclear events, during which the double chromosomes break apart and each chromatid becomes a separate chromosome. Then the cells divide again. There are now four cells, each with four chromosomes, one of each pair of chromosomes in the original. The process of meiosis is illustrated diagrammatically in Figure B.7.

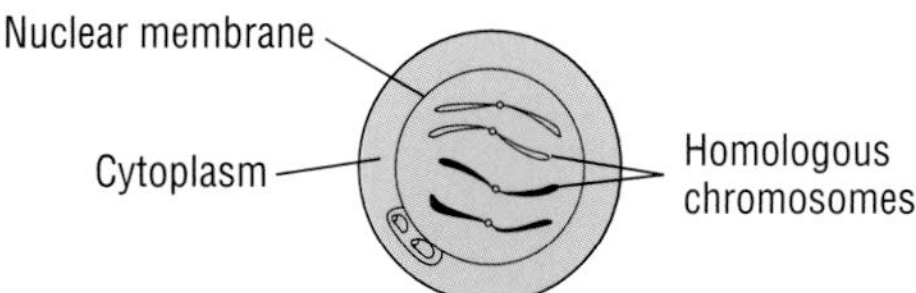

(a)
Chromosomes become visible, each chromosome moves towards and begins to pair up with its partner, or **homologous** chromosome

(b)
Chromosomes duplicate into two chromatids

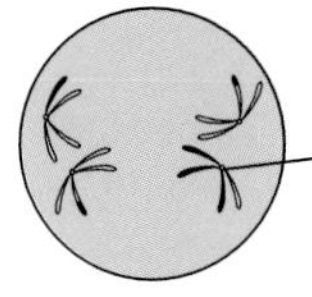

(c)
Chromosomes move to opposite poles. Cytoplasm divides

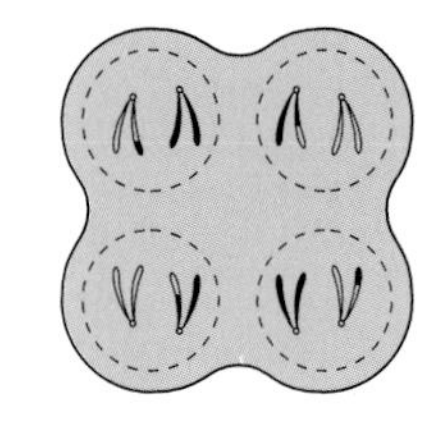

(d)
The chromatids now separate from each other and the cell starts to split again

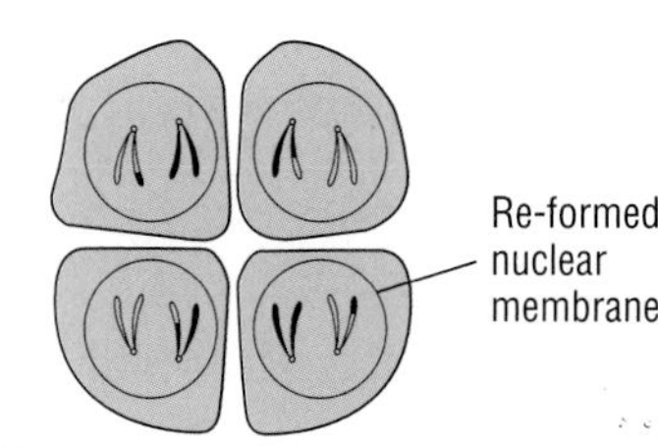

(e)
The cells have finished splitting. The result is four daughter cells, each of which contains half the original number of chromosomes

Figure B.7 The stages in meiosis

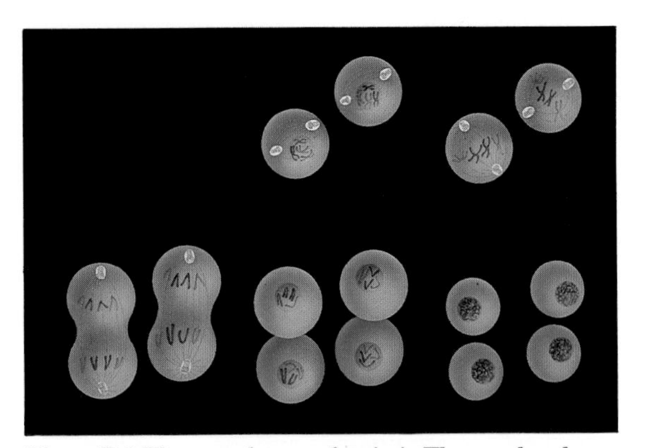

Figure B.8 The second stage of meiosis. The two daughter cells arising from the first division of the mother cell divide. Each single chromosome splits, giving rise to chromatids that migrate to opposite poles of the spindle. Overall, a single diploid mother cell gives rise to four haploid gametes (bottom right)

Both sperm and egg cells undergo meiosis. When the sperm unites with the egg, each provides half the chromosomes for the new individual.

A cell with half the normal number of chromosomes is called **haploid**; a cell with the full set of chromosomes is called **diploid**.

Since it is highly unlikely that there ever existed a cell with this identical set of chromosomes, each fertilised egg cell is unique. The advantage of sexual reproduction lies in the fertilised egg which makes new life similar but just a little different from either of its parents. This variation among offspring may produce one that may be able to adapt to changing conditions in the environment.

Once the egg has been fertilised, all further cell divisions produce cells with the full number of chromosomes (the fertilised egg undergoes *mitotic* cell division).

Eventually the new organism will reach maturity, and its time will come to reproduce. Its reproductive organs will then produce sperm or eggs and the cycle of life will have come full circle.

Sources of genetic variation

Offspring reproduced asexually are genetically identical to each other and to their parent. (They are clones – see page 204.) Mutation (page 237) is the only source of genetic variation. Sexually reproduced offspring inherit genetic material from both parents. They are genetically different from each other (except in the case of identical twins) and from their parents.

Since sexually reproduced offspring receive the same number of genes from each parent, any particular characteristic must be controlled by a pair of genes. Paired genes controlling a particular characteristic are called **alleles**.

Summary

- The nucleus of every cell contains chromosomes, each carrying a large number of genes.
- A gene consists of a length of DNA.
- Pairs of genes control particular characteristics. The genes of a pair are called alleles.
- Sex cells are called gametes. Male gametes are sperms; female gametes are eggs.
- Sperms and eggs are produced in the testes and ovaries respectively, by meiosis — a type of cell division that halves the number of chromosomes in the nucleus of each gamete. The full set of chromosomes is restored in an egg fertilised by a sperm.
- All other body cells divide by mitosis — a type of cell division that maintains a cell's full set of chromosomes.

Questions

1 Explain the relationship between:
a DNA and a gene
b genes and a chromosome.

2 What is the difference between asexual reproduction and sexual reproduction?

3 What is a clone?

4 Explain why sexual reproduction produces more genetic variation than asexual reproduction.

Extension questions

5 Distinguish between:
a meiosis and mitosis
b haploid and diploid.

6 **a** Why do the cells of a tissue need to undergo mitosis?
b What effect does mitosis have on the chromosomes of the parent cell?
c What is the relationship between the chromosomes of the parent cell and the chromosomes of the daughter cells?
d What is the importance of this relationship for the health of the tissue?

7 **a** In everyday language, what is a 'replica'?
b What is formed by the replication of DNA?

Unit C
Asexual reproduction, and selection

Cuttings

Many different plants can reproduce asexually. The root, leaf or more often the stem may grow into new plants. These parts are called **vegetative parts** and asexual reproduction in flowering plants is sometimes called **vegetative propagation**. Since new plants come from one parent plant, they are genetically the same (clones – see page 204). This means that the desirable qualities of the parent are preserved in the offspring.

Gardeners and farmers need to produce fresh stocks of plants with desirable characteristics like disease-resistance, colour of fruit or shape of flower. They exploit vegetative reproduction to guarantee plant quality from one generation to the next.

Taking **cuttings** is a method of vegetative propagation often used. New plants are produced quickly and cheaply from older plants with characteristics that we wish to preserve. Figure C.1 shows a geranium cutting. Short pieces of stem are cut just below the point (called the **node)** where a leaf joins the stem. Most of the leaves are stripped from the stem. New shoots grow at the points where the old leaves are stripped off. If the cutting is planted in a damp mixture of sand and peat then roots will sprout from the cut surface of the stem. The cutting then becomes a complete, independent plant. Cuttings are more likely to grow successfully if they are first kept in a damp atmosphere until roots develop.

Roots that grow from part of a shoot are called **adventitious** roots. Activity G.3 on page 79 investigates different conditions that promote the growth of adventitious roots from cuttings. Figure G.11(b) on page 81 gives you more information.

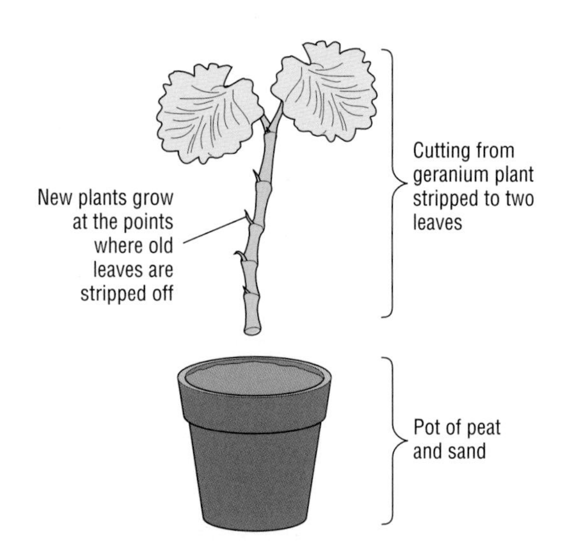

Figure C.1 A cutting from a geranium plant

Artificial selection

For thousands of years the selected pedigree animal has been prized. Such animals have a recorded ancestry for many generations. The dairyman wants selected pedigree cattle from which to breed his milking cows (Figure C.2). The hunter wants a pedigree dog. Similarly, pure lines of cultivated plants are needed by the farmer. Characteristics such as grain yield (Figure C.3), fruit yield and disease resistance are desirable qualities that can be maintained through selective breeding.

Figure C.4 shows two very different breeds of dog. Centuries of selecting dogs with particular characteristics such as size, colour, length of coat and shape of ear have resulted in many different breeds. Dog breeders have taken advantage of the large amount of variation (see page 195) in the characteristics of the members of generations of dog and chosen the characteristics which they want to be passed on to the next generation. The dogs with these characteristics have been allowed to reproduce and so the genes controlling the selected characteristics have been passed to their offspring. This process of choosing which characteristics should pass to the next generation and which should not is called **artificial selection**.

Figure C.5 shows the variety of vegetables which have been produced through artificial selection from one species of plant, *Brassica oleracea* (a relative of the mustard plant). By choosing different characteristics which make good eating, different vegetables have been developed to suit different tastes.

Figure C.2 A pedigree herd of cows

Figure C.3 Wheat yield has been increased by selective breeding

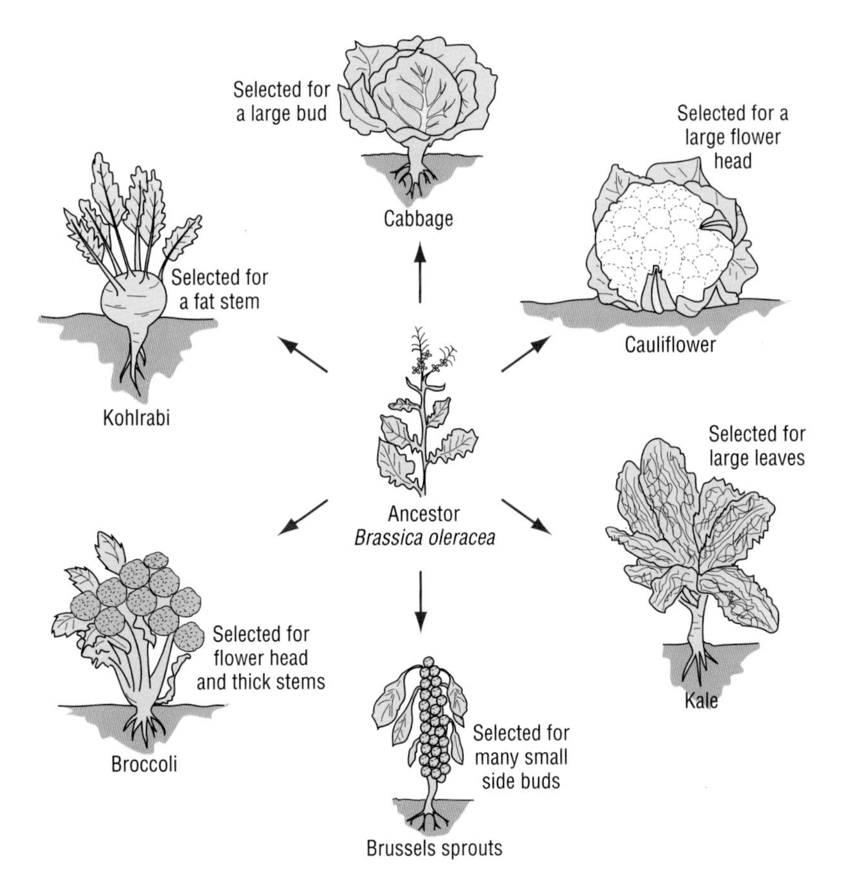

Figure C.5 A variety of vegetables artificially selected and bred from a single ancestral species

Figure C.4 They are the same species!

More help for the farmer

There is little point in sowing seeds if their yield will be poor or if there is a high chance of pests and diseases destroying them. For many years farmers have aimed at producing the 'perfect' crop. The grasses of the world provide the staple diet for most of its population. We know them as cereal crops. Humans have cultivated cereals since the earliest records of history. More people eat rice than any other single food. Wheat, maize, oats, barley, rye and millet make up the bulk of the rest of the world's cereals (Figure C.6).

Today, an enormous variety of cereals are available to the farmer as a result of artificial selection over many generations. Planned **hybridisation** (crossing two different varieties) forms the basis of the development of new varieties. The aim of hybridisation is to concentrate as many desirable features as possible in a single variety. One variety of wheat, for instance, may have good milling properties but poor resistance to fungal disease. Another may resist fungi but provide a low yield of grain. A third might not thrive in dry conditions but may give good baking flour when well irrigated. New varieties are constantly being developed by plant breeders which retain the favourable properties and eliminate the bad ones. To the basic properties already mentioned are added others that vary according to the climate and soil in which the wheat is to be grown. Also, the season of planting may differ: some varieties are planted in autumn, others in spring. An assortment of wheat species is shown in Figure C.7.

The improvement in varieties of cereal crops is largely responsible for the enormous increase in world production of wheat during the last 50 years. Figure C.8 shows a crop of high-yielding wheat being inspected in Nepal. Poorer farmers now have this type of crop available to them: a boost to their national economy.

Hybridisation and selection of varieties of rice has boosted yields by 25%. It is a slower process than wheat breeding but promises to give even higher yields in the future.

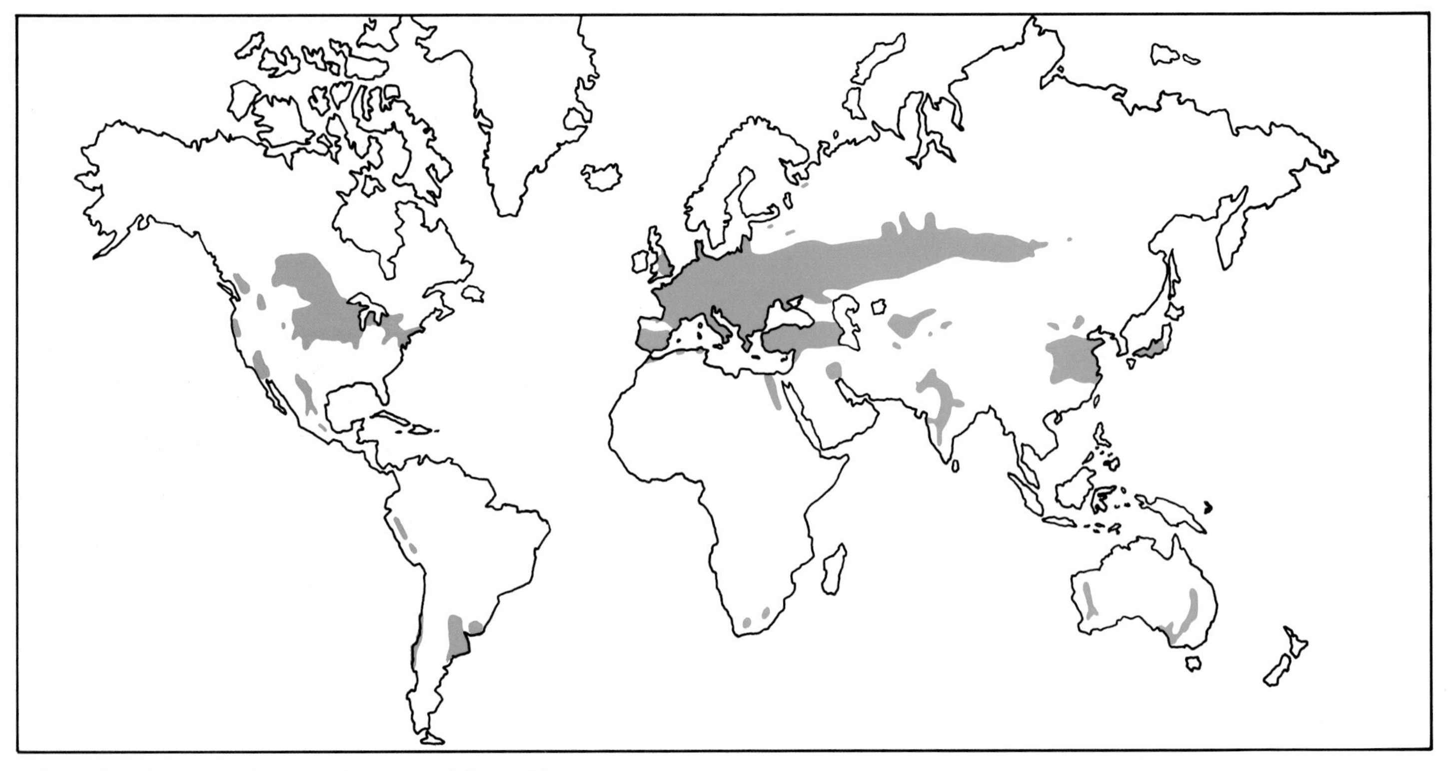

Figure C.6 The main wheat growing areas of the world

Figure C.8 A crop of high-yielding wheat being inspected by two members of the Small Farmer Development Programme in Nepal

Figure C.7 Most authorities agree that there are 13 wheat species, in addition to the wheat used for bread. The six shown here are (left to right): 7-chromosome einkorn, 14-chromosome emmer, macaroni wheat, Polish wheat, 21-chromosome club wheat and spelt. All are cultivated today

Potatoes and bananas

Most of the plants we grow for food are propagated from seed. However, potatoes are usually grown vegetatively from tubers. The tubers grow at the ends of underground shoots (Figure C.9). Soil is usually banked up around potato plants, so that plenty of shoots grow into the soil. The swollen tubers are the potatoes we eat. Each bud on a potato is capable of growing into a new plant. Confusingly, the potato tuber is often called a 'seed potato'. All the plants (Figure C.10) grown from potatoes collected from one plant will have the same genes. If the parent plant has good qualities such as disease resistance and high yield, the grower can be certain that the same qualities will be passed on to the crop plants. Of course, if you were to leave the potatoes in the soil year after year, the new plants which grew up would be very overcrowded.

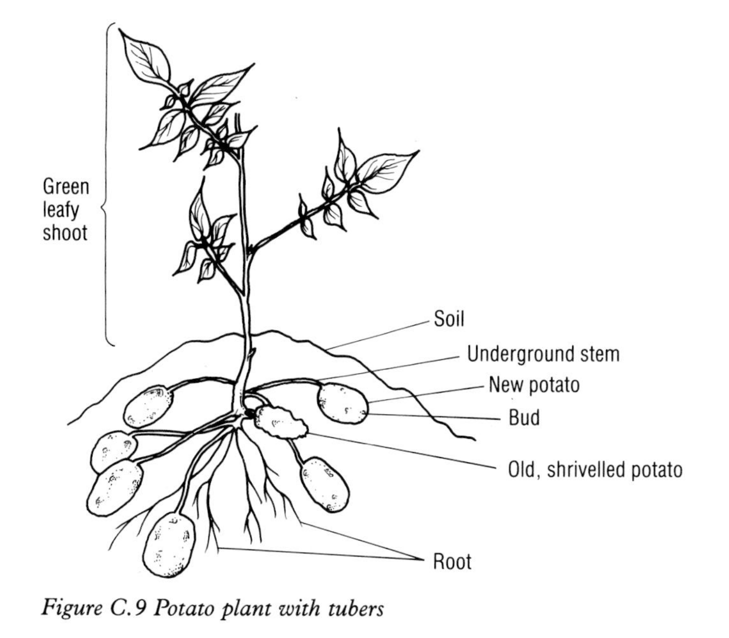

Figure C.9 Potato plant with tubers

Figure C.10 Potato plants in flower

Potato plants do produce fruits (Figure C.11) which are poisonous, although they look very much like tomatoes. The seeds in these fruits can be dispersed away from the parent plant. Seeds are produced by sexual reproduction. At fertilisation the genes in a male nucleus and those in a female nucleus come together. All the new plants growing from seeds will have different combinations of genes. The qualities of the plants which grow from the seeds will be unpredictable. Some of them may be very desirable, some of them may be very poor plants. Other seeds may not germinate at all. For potato breeders, the production of seeds provides a valuable way of producing new varieties, some of which will be worthwhile. The gardener, however, wants to be certain that all his plants are successful, and so he uses the reliable tubers.

Wild bananas reproduce both by seeds and vegetatively. But, the seeds in most bananas which are grown for sale cannot germinate. The commercial plants can only be propagated vegetatively from side shoots or suckers. All the plants grown will be genetically identical, except for any mutations (genetic changes) which take place spontaneously.

Figure C.11 Potato fruits

Summary

- Asexually reproduced plants are genetically identical to one another and to their parent. Desirable characteristics of the parent, therefore, are preserved from one generation to the next.
- Cuttings are a method of reproducing plants asexually.
- We can choose which individuals (plant or animal) will be allowed to breed the next generation. Our choice is for the desirable characteristics in parents that we wish to preserve and develop in future generations. In this way plants and animals important for agriculture have been selected to increase yields.

Questions

1 Grafting is a method of vegetative propagation (see page 208). It is often used to reproduce roses and fruit trees. Find out how grafting is carried out. What are the advantages of reproducing a particular variety of apple tree by grafting?

2 Make a list of as many breeds of dog as you can think of. Choose one breed and describe the characteristics which define the breed.

3 Some barley plants can be attacked by a fungus, while others are resistant to it. In an investigation, a plant breeder found that plants that could be attacked produced offspring that could all be attacked when self-fertilised. However, resistant plants produced a mixture of resistant plants and plants that could be attacked by the fungus when self-fertilised.

a How could the plant breeder obtain a stock of barley plants that are all resistant to the fungus?

b Assuming that resistance to fungus is controlled by a single gene, what must be the genotype (see page 197) of the resistant stock?

Unit D Cloning

Cloning

Once a plant breeder has produced a 'tailor-made' hybrid plant with all the required desirable properties – what next?

First, the breeder will want to satisfy the market demands for the product. This means the production of vast numbers of plants. Even more important, the plants must be identical with regard to the desirable genes that they carry. The genetic material possessed by an organism is called its genotype. All organisms produced asexually (no fusing of gametes) from the same parent have the same genotype. They are called **clones**. Mitosis gives rise to two cells with the same genotype. Asexual reproduction helps us to see the results of mitosis on a large scale.

The plant breeder makes use of this to produce identical plants. The basis of all plant cloning is the production of whole plants from parts of plants by asexual means. This eliminates the possibility of outside genes reaching the egg cell of the plant by sexual means. The outside genes may not be the ones that are required in the 'tailor-made' variety. Many varieties of fruits and flowers are maintained for generations in this way. Desirable characters can be genetically controlled precisely.

More about potatoes

Although cloning maintains desirable characteristics, there are disadvantages. Widespread use of clones reduces the number of alleles (see page 206) available for selective breeding. More about potatoes illustrates the point.

Table D.1 compares varieties of potato which are used in the manufacture of chips and crisps. *Dunbar Standard* developed in the 1930s makes excellent chips and *Record*, grown since 1944, is a favourite with crisp makers. Both varieties have particular characteristics that make them suitable for either 'chipping' or 'crisping'. The characteristics are the result of selective breeding.

The European potato industry was founded on varieties from tubers (see page 211) that were brought from Peru to Spain in the 1570s and to England around 1590. Since then cross breeding has mixed up the genes contained in the original introductions and recombined them into new varieties. Breeders and growers have then selected those with the most promising characteristics. However, even modern varieties are all rather similar as they all rely on the limited number of original genes.

Characteristic	**Variety**	
	Dunbar Standard	**Record**
Main use	Chip manufacture	Crisp manufacture
Parentage (cross)	Herald × Kerr's Pink	Trenctria × Energie
Shape	Long oval	Round
Skin	White	Rough yellow
Flesh	White	Yellow and floury
Cooking	Excellent	Processing only
Main asset	Keeps and cooks well	High food content
Main genetic flaw	Blackens after cooking	Vulnerable to virus Y

Table D.1 Characteristics of two varieties of potato

In 1845 the most popular variety of potato grown in Ireland was the *Lumper*. Similar in appearance and genetic make-up to the original varieties from Peru, the *Lumper* was partly responsible for the increase in Ireland's population during the first part of the nineteenth century. It produces excellent yields and provided a staple part of the diet for many years. However, the *Lumper* has a fatal genetic flaw which no-one at the time was aware of. It is a late maincrop variety which is particularly susceptible to blight – a fungus (*Phytophthera infestans*) – which was then unknown in Ireland (Figure D.1).

Cool wet weather in July 1845 provided the ideal conditions for the spread of potato blight spores. Blight devastated the crop of *Lumper* potatoes, and by 1847 the Irish potato harvest had failed completely. The rapidly growing population depended on this one type of food, so the blight resulted in widespread famine. More than a million people died of starvation (Figure D.2). A further million escaped famine by emigrating, mainly to the USA. Within a few years the population of Ireland was halved.

Figure D.1 Potato blight fungus feeds on the internal tissues of leaves, destroying them. The parts of the plant above ground die. As a result little starch is produced through photosynthesis for storage by the potato tubers underground. The tubers fail to develop

Figure D.2 More than a million people died in the Irish potato famine of the 1840s. Potato crops were devastated by potato blight fungus

Today millions of hectares of potatoes are grown throughout the world. Even so, inspection of supermarket shelves shows that there is still a reliance on a restricted number of varieties, despite the development of new breeds by crossing existing breeds. Also cloning the new breeds produces large numbers of potato plants in a short period of time for intensive production. Yields go up and up but the increase still depends on a relatively restricted number of genes.

In order to retain and improve the genetic basis for new varieties, gene banks of potato plants are maintained. The most extensive collection in the world is held in Peru. Many hundreds of different varieties of potatoes are duplicated at two separate centres (Figure D.3). Should one of the centres be destroyed, the potato genes are still protected at the other. Different countries including the Irish government help finance these potato gene banks in Peru.

Figure D.3 Gene banks help protect the genetic diversity of plants important to human welfare. This photo shows samples of potato species ready for storage in a bank in Huancayo, Peru

Intensive farming (see page 125) relies on far too few varieties of crops. For example, millions of hectares of grain crops, comprising only a few varieties, are grown on the prairies of the USA. If a new strain of virus or fungus attacked one of the main varieties (as was the Irish *Lumper* potato by *Phytophthera* in 1845), chaos would grip the grain markets and the world would face a major food disaster.

It is essential that the genes of rare varieties are conserved. The *Lumper* potato is a case in point. Despite vulnerability to blight, it may still contain genes that could be useful in the future. For example, the levels of ultraviolet (UV) light reaching the Earth's surface are rising because of depletion of the atmosphere's ozone layer. The *Lumper* might be more resistant to the harmful effects of UV than other more popular varieties of potato. Fortunately the *Lumper* is still grown, unlike other minor varieties of potato such as *Kemps*, *Buffs* and *Cups* which were grown in 1845 but are now extinct. Lost forever, their genes are not available to plant breeders aiming to produce new varieties suitable for an ever-changing environment.

Tissue culture

Nowadays it is possible to grow plants of many kinds from small pieces using **tissue culture** (Figure D.4). All the plants grown from pieces from one parent plant will be genetically alike and will grow in the same way if they are treated identically. The small fragments begin life growing in a liquid or gel which provides all necessary chemicals. The conditions will be sterile, so the new plants will be free of disease. This is an ideal way of propagating plants for glasshouse owners who want to produce many similar, healthy plants.

If tissue culture methods are used, special facilities are needed. Technicians produce the sterile growing medium. The fragments of plant are handled with sterile instruments, too. Finally, the correct temperature and other conditions have to be provided. The requirements for successful tissue culture are very precise.

Figure D.4 Growing plants in tissue culture

Figure D.6 The oil palm has many uses; both its leaves and fruit are used in the tropics

Cloning from cultures

Recently some commercially important plants have been mass produced from tissue cultures using a cloning technique. Between the 1960s and 1980s, research into production of oil palms took place. Today, many people in the tropics are benefiting from this form of biotechnology. Large numbers of people rely on the oil palm for food and detergents either directly or because its products are exported. Once the desired hybrid variety of oil palm was produced, clones could be made from it.

A small part of the growing point from the parent plant is collected. It is then grown in sterile conditions on agar jelly. It is supplied with the best conditions of minerals, hormones and temperature for growth. Mitosis produces identical cells. These are then separated. Each can develop into a new plant with the same desirable genes found in the parent plant (Figure D.5).

The process can be carried out in laboratories anywhere in the world and then the young plants can be exported to the tropics for growth in plantations (Figure D.6).

At present, most plants produced in this way tend to be decorative flowering plants for the horticultural industry. However, some fruits such as peaches and raspberries are also grown by this method. Perhaps the future points to the production of 'tailor-made' cereal crops. In this way the problem of world food shortages could be tackled by producing hybrids which could survive where, at present, no crops grow.

Selection
of hybrid variety of oil palm. Its fruit produces a lot of oil. A hybrid is the cross between two genetically unlike individuals

Tissue culture
Root tissue from the selected tree is placed in a medium which contains all the nutrients needed for growth. Auxin (see page 78) is used to stimulate growth

Clone formation
Plants develop from the root tissue. They are genetically the same (clones) since they come from the same parent. All of the clones have fruit which produces a lot of oil

Figure D.5 Breeding programme for the oil palm. Oil palms are produced that yield up to six times more oil than palms bred by traditional methods

Activity D.1

Cloning cauliflowers

Cauliflower florets (the flowers which form the white 'heart' we eat) are ideal material for plant tissue culture. Growth occurs within 10 days and plantlets are ready for transplantation within 12 weeks. CAUTION – you must make sure that your work is 'clean'. Aseptic working methods are essential to prevent micro-organisms from contaminating the cultures.

You will need: sterile petri dish, non-absorbent cotton wool and aluminium foil, cauliflower florets, 100 cm^3 of distilled water, 50 cm^3 of 70% ethanol, 100 cm^3 of bleach solution (chlorate 1 solution with added detergent), 1 dm^3 of plant tissue growth medium made up as follows:

20 g of sugar
10 g of agar
4.7 g of Murashige and Skoog (M & S) medium
25 cm^3 of kinetin stock solution

Making up kinetin stock solution
Dissolve 0.1 g of kinetin in 1 dm^3 of distilled water. The addition of two pellets of sodium hydroxide helps the kinetin to dissolve. Store the stock solution in a refrigerator at 4 °C.

Making up the growth medium
Dissolve the sugar, M & S medium and agar in 725 cm^3 of distilled water. Add the kinetin stock solution. Pour 2–3 cm^3 of the mixture into each of three boiling tubes. Plug the tubes with non-absorbent cotton wool and cap them with aluminium foil. Sterilise at 120 °C or more (in a pressure cooker) for 15 minutes, cool and then store in a refrigerator until needed.

Setting up the experiment
Clean all working surfaces by swabbing down with 70% ethanol. Cut out 3 small pieces of cauliflower floret on a clean petri dish (each piece of floret is called an explant). Sterilise the explants by flooding them with bleach solution for 10 minutes.

FROM NOW ON ASEPTIC CONDITIONS ARE ESSENTIAL.

- Dip forceps in ethanol and then flame them in a Bunsen burner. When cool use the flamed forceps to transfer the explants through three washes of sterile distilled water.
- Remove the plug and briefly flame the neck of one of the boiling tubes containing growth medium.
- Use flamed, cooled forceps to transfer an explant from the beaker of sterile distilled water to the boiling tube. Flame the neck of the tube once more and re-plug the neck.
- Repeat the procedure using the other two boiling tubes containing growth medium.
- Place the tubes in a warm, light place.

1 Why are all working surfaces swabbed with 70% ethanol?
2 Why are the forceps and neck of each boiling tube flamed in a Bunsen burner?
3 In the growth medium explain the role of:
a sugar
b kinetin.
4 Follow the events in each boiling tube on a daily basis. How many days pass before growth of each explant is visible?
5 Keep a diary charting the growth pattern of each explant.

Improving animals

A fertilised egg soon begins to divide by mitosis to produce a ball of cells forming the young embryo (see page 204). Splitting apart the embryo before the cells become specialised is the basis of different techniques which manipulate embryos to human advantage. For example, such techniques allow farmers to increase the numbers of their livestock and develop new breeds.

In simple terms, eggs from donor animals are fertilised in the laboratory and the young embryos that develop are split up into individual cells. The cells are transplanted into different regions of the uterus of a **surrogate** mother – so-called because the transplanted cells are not her own. The cells divide and embryos develop normally. Eventually the surrogate mother gives birth to a number of youngsters where one or perhaps two offspring would have been the more normal reproductive rate. In this way rare breeds of a particular type of animal can be conserved, or new breeds with desirable characteristics developed, reliably and quickly using surrogate mothers of a common breed.

A variation of the technique allows surrogate mothers to carry embryos of a different species (Figure D.7).The technique is useful because the fertilised eggs of rare species can be cooled and preserved for years after the original parents have died. When a suitable surrogate is available the fertile eggs can be transplanted.

Figure D.7 Eland (a species of antelope) has given birth to a baby bongo (another species of antelope)

These methods provide an increasingly important source of rare species. Zoos are developing the ideas to help conservation. Their breeding programmes build up numbers of rare stock and arrange for their release in the wild.

Embryo cloning produces many genetically identical copies of an animal. For example, it is possible to clone calves. A young embryo is split into two and each part is transplanted into separate cows. The calves that the cows produce will be exactly alike (clones).

Imagine that a mutation occurs in a cow that makes her an exceptional milk producer. Conventional breeding techniques would reshuffle her genes with the risk that the desirable mutation would be lost. Cloning to produce identical copies of the cow will conserve her unique milk-producing abilities in future generations

Unit E
Genetic engineering

The principle of genetic engineering relies on isolating a gene from one organism and putting it into another of a different species. But why should we want to do this in the first place?

The answer is because the gene might be responsible for carrying out a very useful function. For example, scientists often isolate genes from human chromosomes which control the production of hormones. They transfer these useful genes to bacteria or yeasts.

The idea of transferring human genes to bacteria is to increase production of useful substances (proteins). The micro-organisms multiply very rapidly and can be cultured relatively cheaply. In fact, they can provide almost unlimited amounts of substances that are practically unobtainable in bulk in any other way.

The process starts with biological 'scissors' called **restriction enzymes**. These cut chunks of DNA at points on chromosomes where useful genes are known to exist. The restriction enzymes therefore enable the scientist to cut out very precisely the gene that is needed. It may be one out of hundreds on a particular chromosome. The next stage is to put the gene into a bacterium. The gene is not put directly into the bacterial chromosome. Instead, genetic engineers use a loop of bacterial DNA called a **plasmid**. These are normally present in bacteria but are largely independent of the rest of the cell. Plasmids, like chromosomes, carry bacterial genes which control the micro-organism's metabolism.

The plasmid is cut open with restriction enzymes and the foreign gene is inserted. The break is sealed with another enzyme called a **ligase** (an enzyme which binds chemicals together). This process makes a mixed molecule called **recombinant** DNA.

The altered (infective) plasmids are then mixed in a test tube with bacteria which do not have any plasmids. Some plasmids move inside these bacteria. The infective plasmids carry the foreign gene inside the cell where it can instruct it to make the required protein (see Figure E.1).

The process is regarded as a success only when the gene is expressed. This means that the host cell has obeyed the instructions carried by the foreign gene and has made protein.

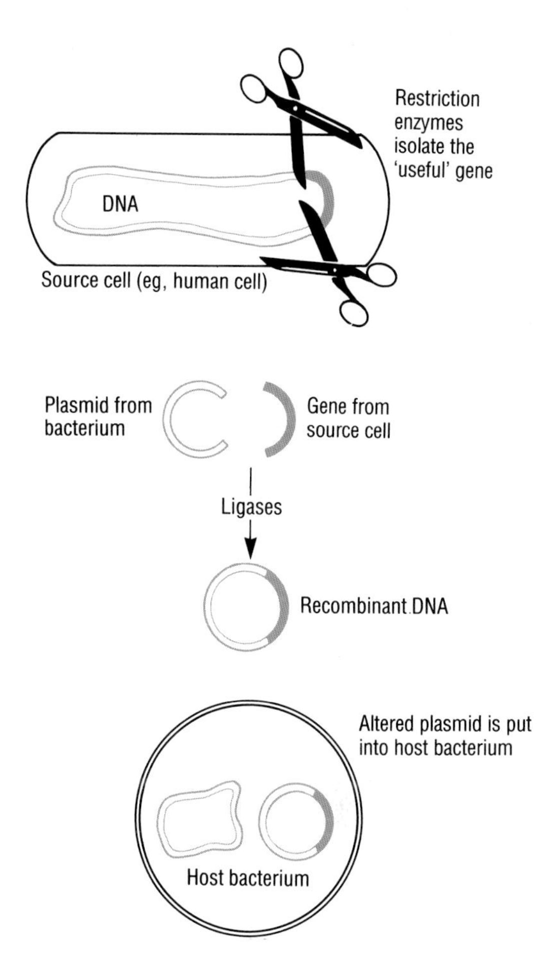

Figure E.1 Genetic engineering in a nutshell: a gene (length of DNA) from the source cell is combined with a plasmid to form recombinant DNA. The plasmid is then put into the host bacterium, where it moves into cells and controls their activity

Life-saving genetics

There is a hormone called **insulin** which is made by certain cells in the pancreas. Its function is to keep the concentration of glucose sugar in the blood at a constant level (0.1 g per 100 cm^3 of blood). If the glucose level falls much below this, the body does not have enough fuel to function properly. A glucose level above the normal concentration also disturbs body functions. In particular, the kidneys fail to cope and glucose is lost in the urine.

Without enough insulin people cannot control their blood glucose levels, and suffer from **diabetes**. Fortunately, diabetics can control their condition by injecting insulin (Figure E.2).

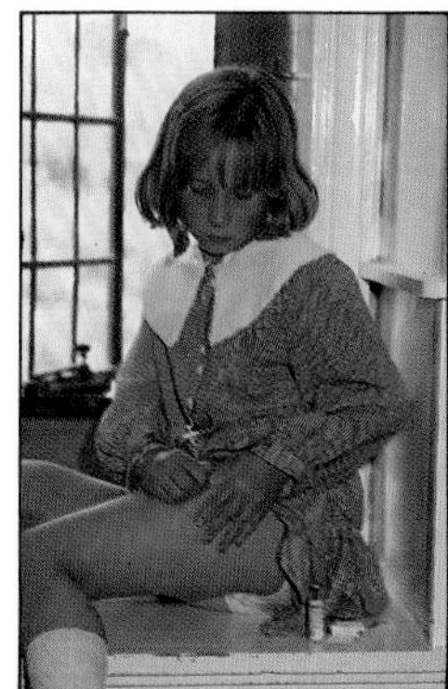

Figure E.2 Some people have to inject themselves with insulin every day

Activity E.1

Built-in pesticides

Biologists have grown a tobacco plant that makes its own pesticide. Genes of a tobacco plant have been modified so that the plant makes a poison that kills insects.

The gene which controls the production of the toxin has been isolated from a bacterium. It has been inserted into the tobacco plant in two ways:

(a) The cellulose cell walls of some of the leaf cells were broken down so that the plasmid could pass freely into the cytoplasm.

(b) Fragments of the stem were wounded with preparations containing the genetically altered plasmids.

1. Explain how the cellulose cell walls of the tobacco cells might have been broken down.
2. Explain why there might be possible dangers to the environment as a result of this technique.

Until recently the insulin used for this treatment was taken from the pancreas of various mammals. However, it was difficult to produce enough insulin in this way. Genetic engineers have solved the problem of mass production by using bacteria. The method is summarised in Figure E.3.

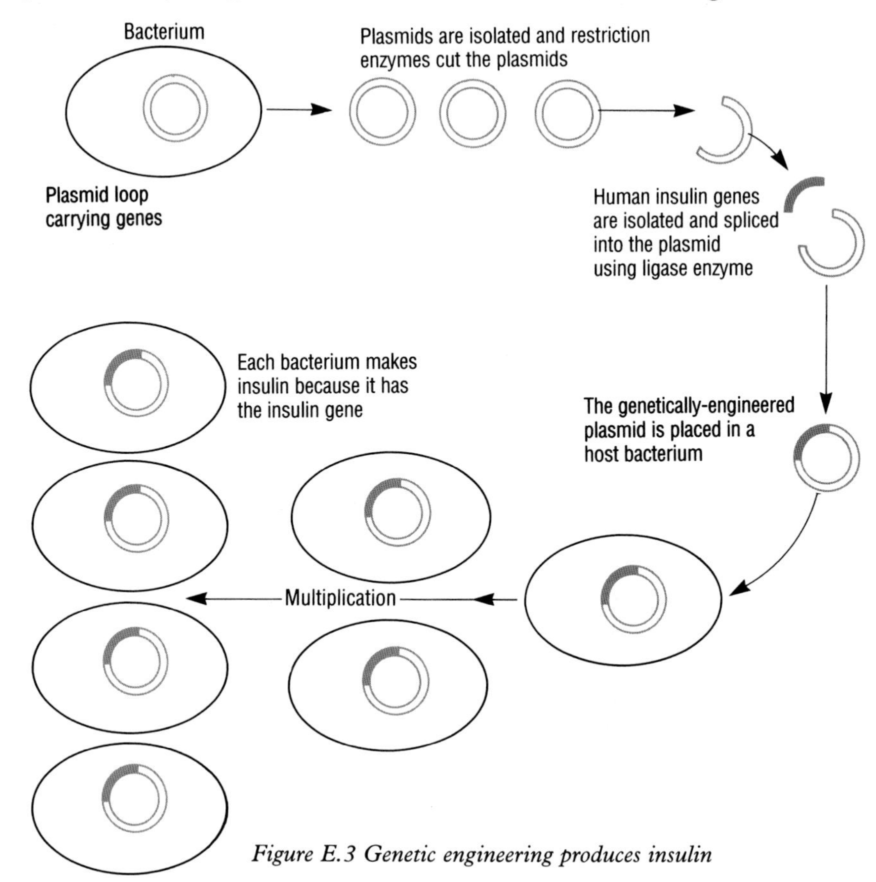

Figure E.3 Genetic engineering produces insulin

Activity E.2

Mutants down under!

The practice of releasing genetically engineered micro-organisms into the environment to help plant production has been introduced in Australia. An Australian scientist soaked the roots of almond seedlings in a solution of genetically altered bacteria to treat crown gall disease in roots (a disease caused by a bacterium). After completing the four-month experiment, the affected soil was sterilised and disposed of.

1. Explain why the scientist sterilised the soil before disposing of it.
2. Suggest how sterilisation could be carried out.
3. Discuss why environmentalists might oppose the introduction of mutant bacteria into the surroundings.
4. Suggest how bacteria would be able to combat crown gall disease in plant roots.

Other genetically engineered hormones, including **human growth hormone** and **calcitonin** (the hormone that controls the absorption of calcium into bones), are being produced by methods similar to those for producing insulin. Using human genes to produce substances like hormones helps to prevent the harmful side-effects that can come from products obtained from animal tissues. It also reduces the use of animals for medical research.

Genetic engineering is used to produce other health-care products. For example, **interferon** is a protein produced by virus-infected cells. It helps healthy cells break down viruses and prevents them from multiplying. Originally interferon was extracted from white blood cells. Thousands of litres of blood had to be processed to produce minute yields of interferon. At a cost of £10 million per gram, interferon was extremely expensive, and research designed to find out how it worked was slow. Then, in 1980, the gene for one type of human interferon was isolated and inserted into the bacterium *Escherichia coli*. Prices fell sharply as genetically engineered interferon became much more widely available at a fraction of the previous production costs.

It was hoped that interferon would be as effective a treatment for viral diseases as penicillin has proved to be for bacterial diseases. However, turning interferon into such a treatment is proving difficult.

Genetically engineered plants

Scientists use genetic engineering to isolate desirable genes and insert them into crop plants.

The bacterium *Agrobacterium tumefasciens* is an ideal cell for introducing desirable genes into host cells. Figure E.4 shows how *Agrobacterium tumefasciens* can be used to introduce a gene for herbicide resistance into crop plants.

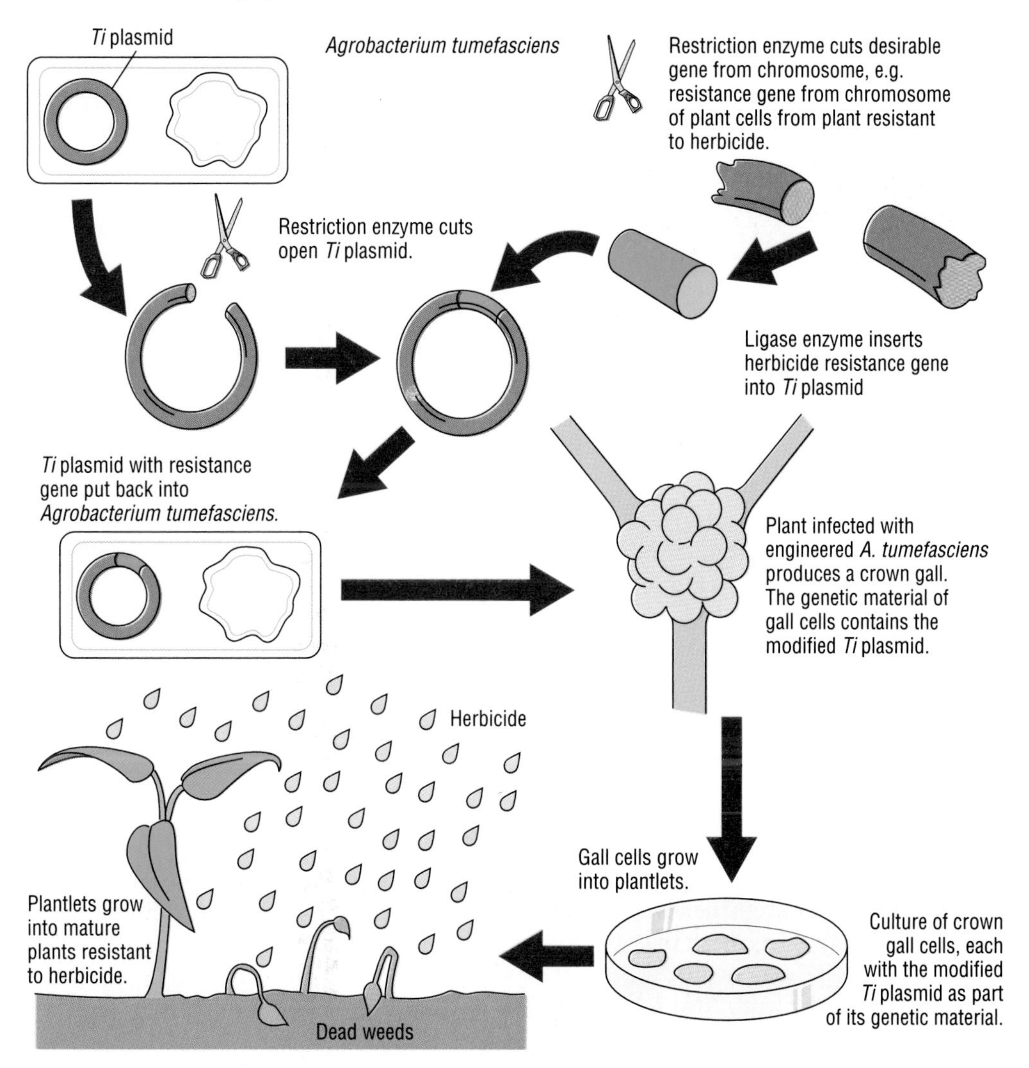

Figure E.4 Engineering Agrobacterium tumefasciens *by inserting a desirable gene (for example, a gene for resistance to herbicide) into its* Ti *plasmid*

The plant is infected with engineered bacteria and, because of the *Ti* plasmid, produces a cancerous growth (tumour) called a **crown gall**. Cells in the gall each contain the *Ti* plasmid with the desirable gene in place. Plantlets can be cultured from small pieces of tissue cut out of the gall. The plantlets are genetically identical clones (see page 204), each carrying the *Ti* plasmid with its desirable gene. The plantlets are transferred to soil where they grow into a crop. Herbicide used to control weeds does not affect them because they have the *Ti* plasmid genetically engineered for herbicide resistance, so the plants grow healthily in the absence of competing weeds.

Could genetic engineering be dangerous?

The answer to the question is 'probably'. Harmful genes concentrated in a population bred in the laboratory may 'escape'. They could then affect natural populations in the wild. A well-publicised example of the potential problem is given below.

Fish farmers have designed a 'super' salmon by artificially selecting genes that control desirable qualities. In doing this, the genes responsible for homing behaviour have been suppressed. If these salmon escape into the natural population, as is often the case, they could introduce mutated genes into the 'wild' salmon population and destroy their normal homing behaviour.

Figure E.5 Fish-farming can encourage the spread of mutant fish

Figure E.6 Mutant genes could prevent salmon from reaching their natural breeding grounds

Figure E.7 Leguminous fruits

Many of these potential problems are largely hypothetical. For example, there is always the fear of genetically altered disease-causing organisms escaping from laboratories. In fact, great care is taken to stop such accidents happening. For example, biotechnologists don't just transfer useful genes into a plasmid. They also include 'safety genes'. The safety genes mean that the altered cells can only grow in the laboratory; eg, the cells may require a certain type of food that would not normally be present in the wild. Also scientists usually build into the piece of DNA being carried, a gene that only allows the cell to grow in the presence of a certain antibiotic. The antibiotic is not found in the wild, so the gene and the organism carrying it would not survive.

Plants of the future

Crops of the legume family (most plants which have pods as fruits), such as peas or beans, are exceptional. They have their own built-in 'fertiliser factories'. These are colonies of bacteria in the roots that can change nitrogen from the atmosphere into a form that can be used by the plants.

If genetic engineers could transfer the ability of the legumes to fix nitrogen into other crops, such as wheat or rice, the economic implications would be enormous. The environment would benefit, too. The 'home-grown' bacterial fertilisers, unlike artificial fertilisers, would not leach out of the soil to pollute ground water and rivers.

Figure E.8 Root nodules of a broad bean plant. The nodules contain nitrogen-fixing bacteria

Before scientists can transfer nitrogen-fixing ability to other crops, they need to understand the relationship between the bacteria and their plant hosts. When the bacteria infect roots of legumes, the plants respond by developing small outgrowths, or **nodules** (Figure E.8). Within these, the bacteria enjoy an **anaerobic** (oxygen-free) environment and are supplied with food by the plant. Figure E.9 shows bacteria from a root nodule of a pea plant.

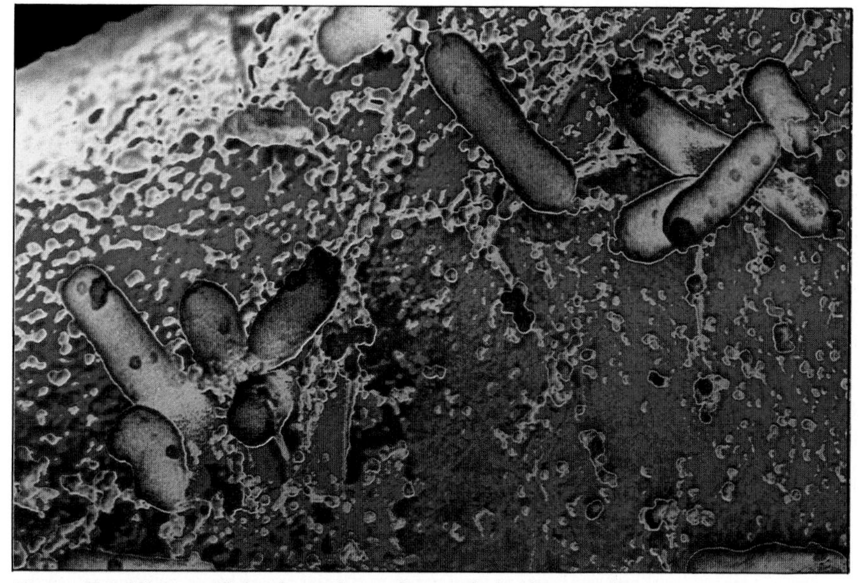

Figure E.9 Nitrogen-fixing bacteria on the root hair of a pea plant

The relationship is very specific. A given variety of bacteria can infect only certain groups of legumes. These are called **cross-inoculation** groups. Legumes in the same cross-inoculation group make the same types of chemicals. For example, peas all belong to one cross-inoculation group which also includes vetches and lentils. Clovers belong to a separate group. So clovers cannot be infected by the same strain of bacterium as peas.

However, genetic engineers have succeeded in altering the bacteria which normally infect peas so that they will now infect clover. The next stage perhaps will be to alter the bacteria so they will be able to infect cereal crops. A truly giant step forward.

Summary

- Useful genes can be taken from one organism and put into another, where they are used to the advantage of humans. This is called genetic engineering.
- Genetically engineered organisms can be used to produce large quantities of chemicals made by cells.
- Plants and animals can be altered by genetic engineering.
- Methods of mass production of plants by various forms of cultures have been developed.

Extension questions

1. How does recombinant DNA differ from other types of DNA?
2. Why is the bacterium *Agrobacterium tumefasciens* an ideal cell for introducing desirable genes into host cells?
3. Explain why a legume may be described as a 'fertiliser factory'.

Unit F
Evolution and extinction

Figure F.1 is a photograph of the night sky taken by the Hubble Space Telescope circling the Earth. Light from the dimmest of the galaxies emitted at around the time of the origins of the Universe has taken between 12 000 million and 15 000 million years to reach us. By comparison, the planets of the solar system are relative newcomers. Interstellar dust and gases circling a star we call the Sun condensed into the Earth and its companions around 4500 million years ago. Evidence suggests that soon after, the first stirrings of life on Earth occurred.

It seems likely that the variety of life on Earth today owes its origins around 4000 million years ago to a scenario similar to the one shown in Figure F.2. Present-day living things are descended from ancestors that have gradually changed through thousands of generations. We call this process of change **evolution**.

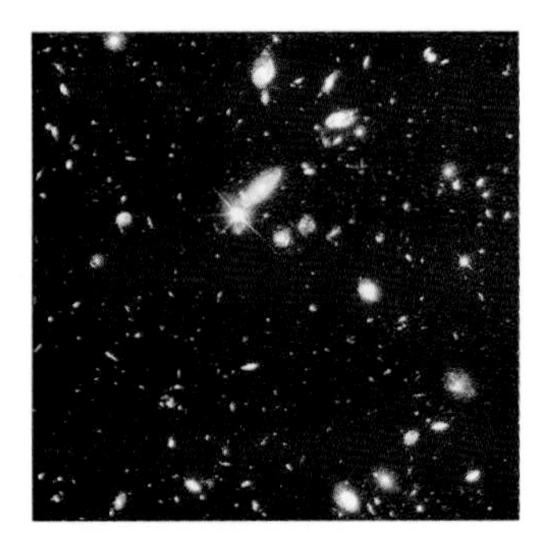

Figure F.1 Some of the galaxies of stars formed at the birth of the Universe. Each galaxy is seen here as a speck or swirl of light, but contains hundreds of billions of stars. The shutter of the camera on the Hubble telescope was left open for 10 days to capture the image

Figure F.2 Is this how life began on Earth?

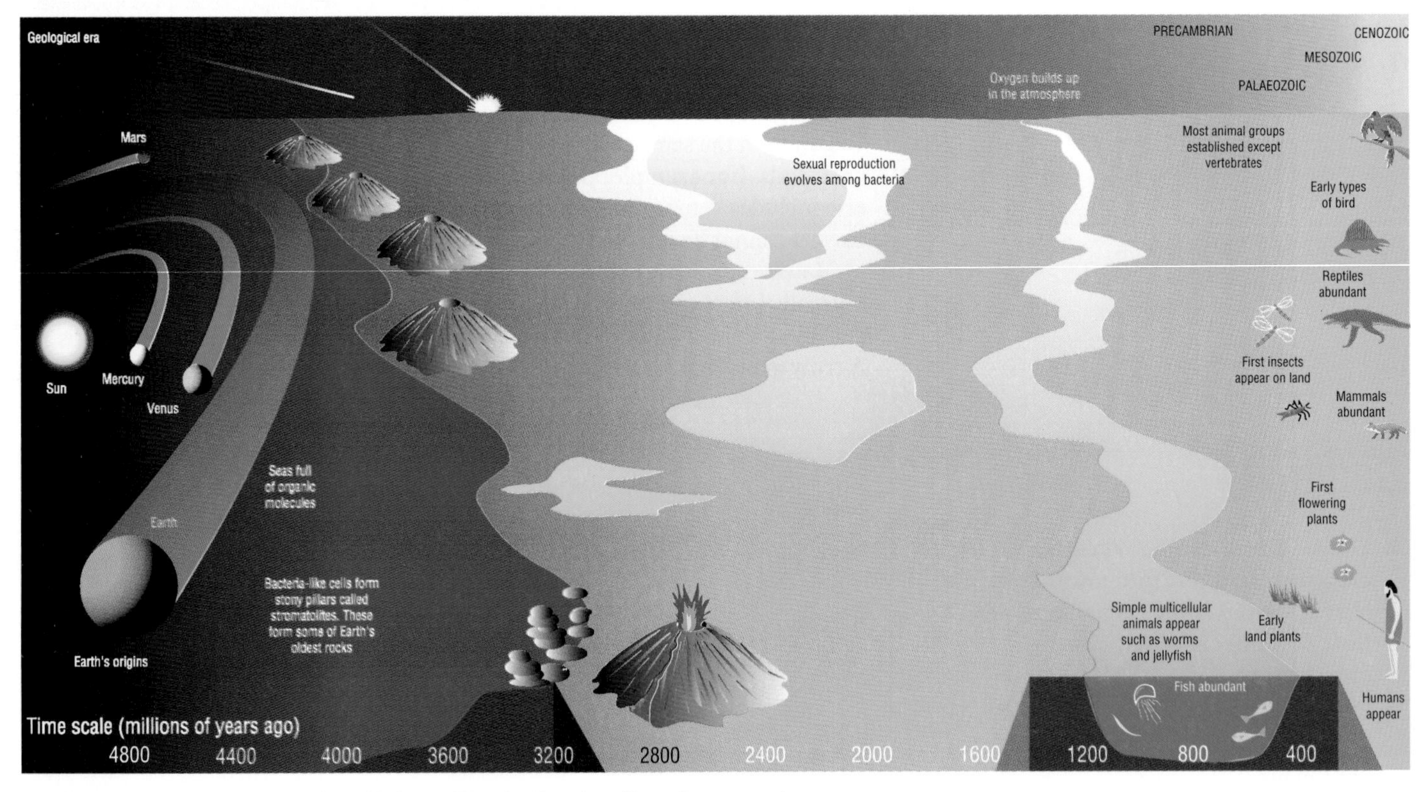

Figure F.2 is a time-table of life on Earth. Its timeline shows major events during the evolution of life from its earliest beginning. Notice that we humans are relatively recent. Our nearest ancestors first gazed at the night sky 2.5 million years ago. Think of it like this. If we condense all of time since the origins of the Universe until the present into one calendar year, humans appeared on Earth just a few seconds ago.

Fossils

Fossils are the remains or impressions made by dead organisms. They are usually preserved in **sedimentary** rocks which are formed layer on layer by the deposition of mud, sand, silt and calcium carbonate over millions of years. Much of the Earth's surface consists of layers of sedimentary rock. Provided it remains undisturbed, the more recently formed the layer, the nearer it is to the Earth's surface. Undisturbed rock layers, therefore, follow on like the chapters of a book. The fossils in each rock layer therefore are a record of life on Earth at the time when the layer was formed. The sequence of fossils traces the history of life on Earth.

The rock layers of the Grand Canyon have lain undisturbed for hundreds of millions of years (Figure F.3). They have been exposed by the Colorado River eroding the soft rock. Today the river flows through a gorge 1.6 kilometres deep. The layers of rock at the bottom of the canyon are 2000 million years old. To travel down one of the trails from the rim to the bottom is to travel back in time. The fossils in each layer represent a particular stage in the evolution of life (Figure F.4).

Figure F.3 The Grand Canyon – the rocks are mostly sandstones and limestones deposited in shallow seas that once covered western USA

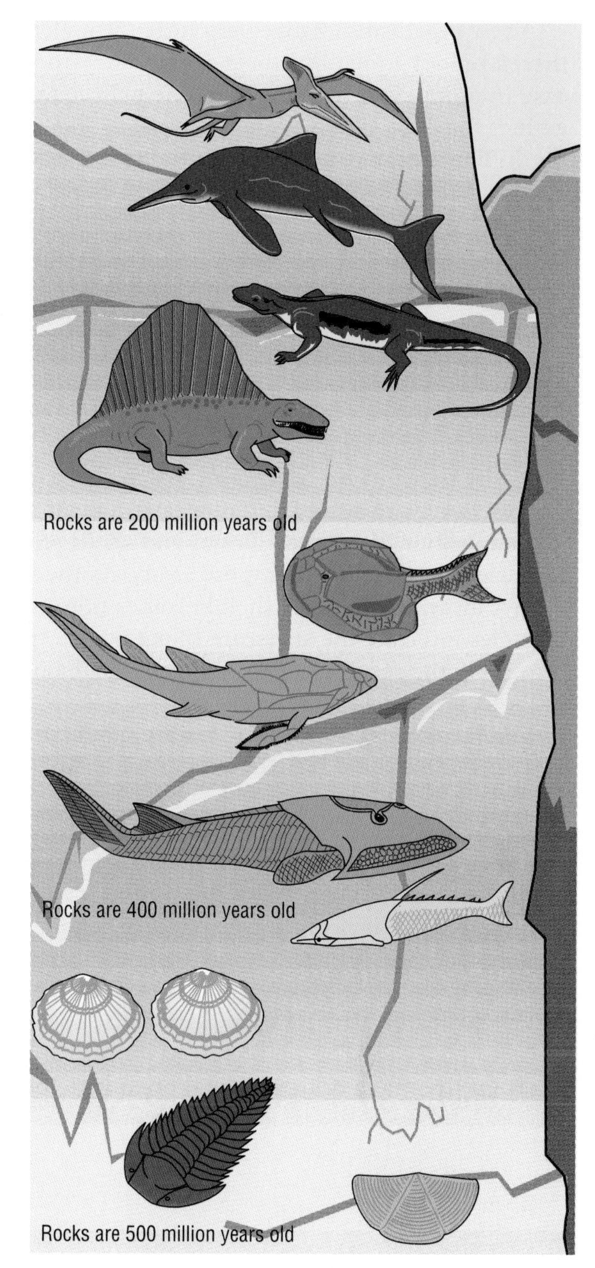

Figure F.4 Animals representative of those living at the time when the rock layers in which their fossils are preserved were deposited

What do the rocks of the Grand Canyon tell us about the evolution of life? Careful study of the fossils in successive layers of rock reveals that:

- representatives of the major invertebrate groups (see page 194) were in existence by 600 million years ago
- fish were the first vertebrates to appear about 500 million years ago
- reptiles appeared about 300 million years after the fish
- birds and mammals appeared more recently, too late to be represented as fossils in the rocks of the Grand Canyon.

The sequence of fossils also tells us that the older the fossils, the more likely they are to differ from present-day living things. From the fossil record we can conclude that organisms change through time. In other words, present-day living things are a product of evolution.

A few organisms change very little through time. Descendants therefore look much like their distant ancestors. The coelacanth is a case in point (Figure F.5). Everyone believed that this fish had been extinct for 200 million years. Then in 1938 a fisherman caught one in the Indian Ocean. Since then a number of specimens have been caught. The skeleton and structures of modern coelacanths differ little from the fossil remains of coelacanths that lived 400 million years ago. It seems that coelacanths have existed for all of this time in an environment where there is little pressure for evolutionary change.

Figure F.5 Coelacanths resemble the fossils of vertebrates that were among the first to live on land

Fossil formation

Dead organisms rapidly decompose (see page 141). Fossil formation therefore depends on conditions which replace organic material as it decays away with something more permanent, or which prevent decomposition altogether. Bones and shells are best preserved as fossils because they do not decompose easily. As the organic material gradually decays away, the bone or shell becomes porous. Mud and mineral particles infiltrate the structure filling up the spaces. If the mud turns into rock, the result is a fossil, which may be preserved for hundreds of millions of years (Figure F.6).

Similar processes can also fossilise plants. For example, the 'Petrified Forest' of Arizona, USA consists of tree trunks where the original organic material of the wood has been replaced by the mineral silica in a process of petrification. Originally the trees were buried under sediment, and water infiltrated silica particles into the trunks. The result was logs of rock (Figure F.7). Fine details have been preserved. For example, xylem tubes (see page 55) and annual rings are clearly visible.

Sometimes when an organism is rapidly buried in hardening mud or cooling volcanic ash and then decays, the space it occupied is filled with another material. The space acts as a **mould**. The replacement material takes on the shape of the original organism, forming a **cast**.

Figure F.6 Fossilised bones and tusk of a mammoth

Figure F.7 Arizona's petrified forest

Figure F.8 Plaster cast of the carcass of a baby woolly mammoth, whose frozen body was found in Siberia in 1977. Scientists believe this mammoth lived about 39 000 years ago. Woolly mammoths became extinct just a few thousands years ago

Do you think you could fancy mammoth steaks? Whole mammoths (a type of woolly haired elephant) have been found preserved in Siberia (Figure F.8). After the animals died, rapid freezing in sub-zero temperatures prevented decomposition. The bodies were in such a good state of preservation that it was possible for the scientists who discovered them to cook and eat mammoth steaks!

Extinction

We often read that species die out (become **extinct**) because of the harmful effects of human activities on the environment. Such extinctions usually occur over a relatively short period of time. Natural extinction is more long term. It makes room for new species to evolve and replace the previous ones.

Competition between species (see page 112) and environmental change are the causes of naturally occurring extinctions. For example, between 200 million years ago and 70 million years ago, an enormous variety of reptiles dominated the Earth (Figure F.9). However, about 70 million years ago many species became extinct. Today tortoises, turtles, snakes, lizards and crocodiles are virtually all that remain. Why so many species of reptile became extinct is not clear. Possible explanations include the following:

- The appearance of the first flowering plants replaced the vegetation on which herbivorous reptiles depended. The food webs (see page 120) of the reptilian world therefore collapsed.
- Some mammals that evolved from a group of reptiles about 200 million years ago were efficient thieves of reptilian eggs
- A meteor plunged to Earth causing catastrophic damage. The impact threw up dust and smoke that triggered global cooling, which most reptile species did not survive.

Whatever the cause, the mass extinction of most reptile species made way for mammals and birds to fill the vacant spaces. Pause and think for a moment. If the reptiles had not given way to mammals, we probably would not have evolved to read about these extraordinary events!

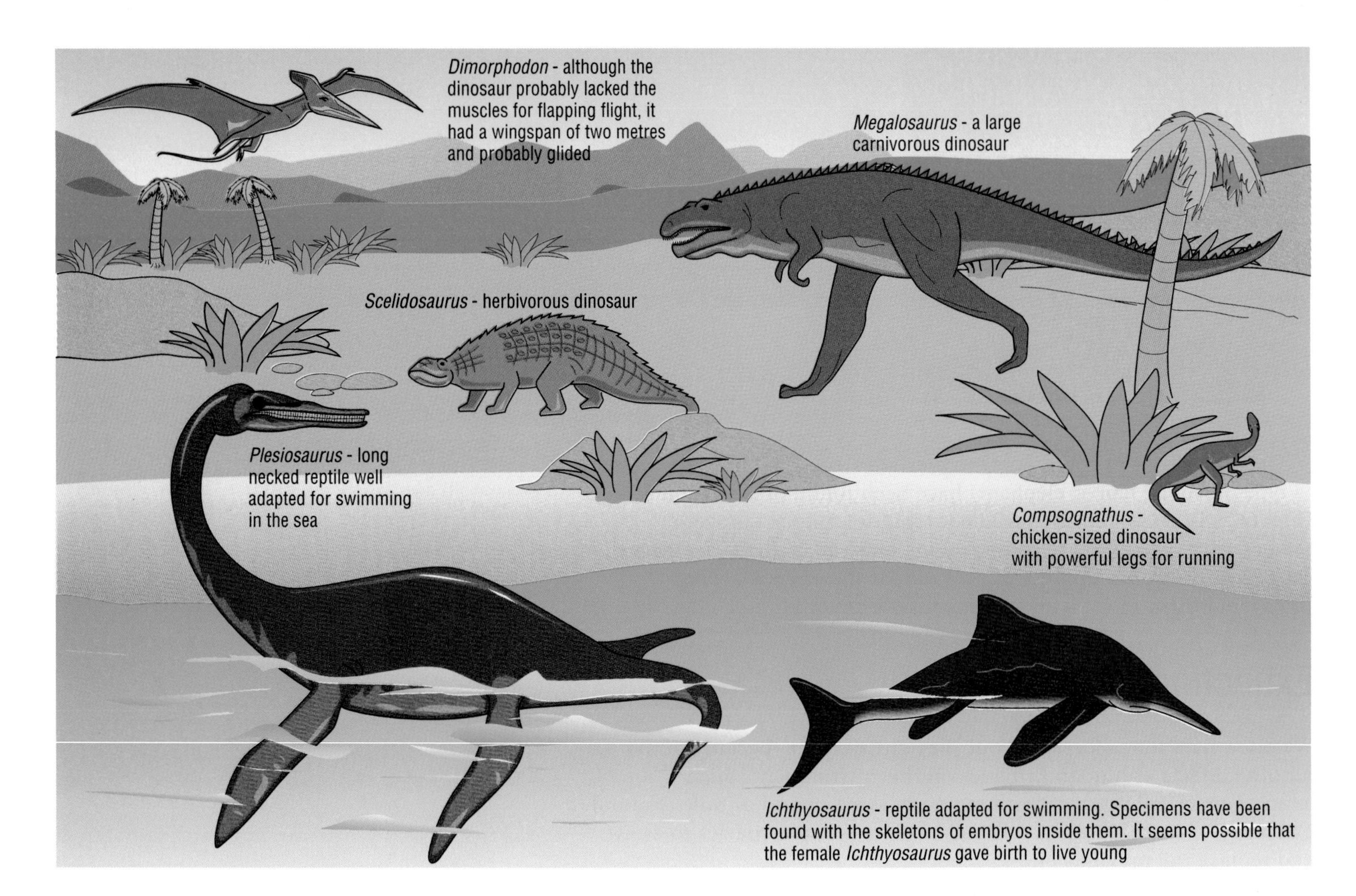

Figure F.9 Reptiles dominated land, sea and air 155 million years ago

Questions

1 Describe the different ways in which fossils can be formed.
2 Why do you think studying life in the past helps us to understand biology today?
3 What is an ancestor?
4 What is a descendant?
5 Give a brief account of some of the factors that might have contributed to the extinction of many types of reptile 70 million years ago. What single factor do you think might have been the most important? Give reasons for your choice.

Summary

- Fossils are the remains of organisms that lived long ago.
- Fossils provide evidence that evolution of species has occurred.
- Species eventually evolve into new species, or become extinct.
- Changes in the environment and competition are the causes of extinction.

Unit G
The great British naturalist

The British naturalist, Charles Robert Darwin (1809–82) (Figure G.1) collected evidence for the process of **evolution**.

In the nineteenth century, Darwin introduced a startling new theory to the world. It was an explanation of how all living organisms had come into existence. Darwin's theory was that all organisms were descended from either one or a very few simple forms of life. Over a period of millions of years, these original forms of life branched out. Gradually they developed new characteristics that helped them to survive. As a result of these changes all known species came into existence.

Darwin first explained his theories to a group of scientists in London in 1858. A year later, in 1859, Darwin published his book, *The Origin of Species*. In this book he showed that plants and animals could not possibly have been created in their present form. They must have gone through a long process of development and change – evolution.

Figure G.1 Charles Darwin (1809–82) – the great British naturalist

Furthermore, in *The Origin of Species*, Darwin also described the *method* of evolution. He showed not only that it was a fact but also how evolutionary processes actually worked.

As the years passed, more and more scientists accepted Darwin's ideas. Later, discoveries of chromosomes and genes confirmed his belief that all present-day organisms have descended from either one or a few original forms of life.

Figure G.2 HMS Beagle

Figure G.3 A cartoon of Darwin published in The London Sketch Book, *1874*

Darwin's whole life was devoted to scientific research. His interest started in his youth, when his hobbies were hunting for beetles, collecting minerals and watching birds. Darwin's father wanted him to become a doctor but Darwin changed the course of his studies after beginning medicine. He began studies to become a clergyman but found that this career did not suit him either.

Shortly after his graduation from Cambridge University, Darwin received an offer that was to change his whole life and also change the way the whole world thought about the origins of life.

He was invited to go on a trip around the world on the British Royal Navy ship HMS *Beagle* (Figure G.2). It was making a voyage of scientific discovery to survey little-known areas of the world. Darwin was offered the job of naturalist. Darwin was 22 when he sailed on the *Beagle* from Plymouth in 1831. He returned in 1836. (The voyage of the *Beagle* is shown in Figure G.4.) The voyage was a revelation to Darwin. It enabled him to study vast numbers of creatures that he had never known existed. Pressed in the rocks of remote countries, he discovered fossils of species which had since become extinct. These fossils showed Darwin that forms of life went through many changes. The patterns of life he saw around him and in the rocks had a profound effect on Darwin. They made him think that gradual changes in life took place over many thousands of years.

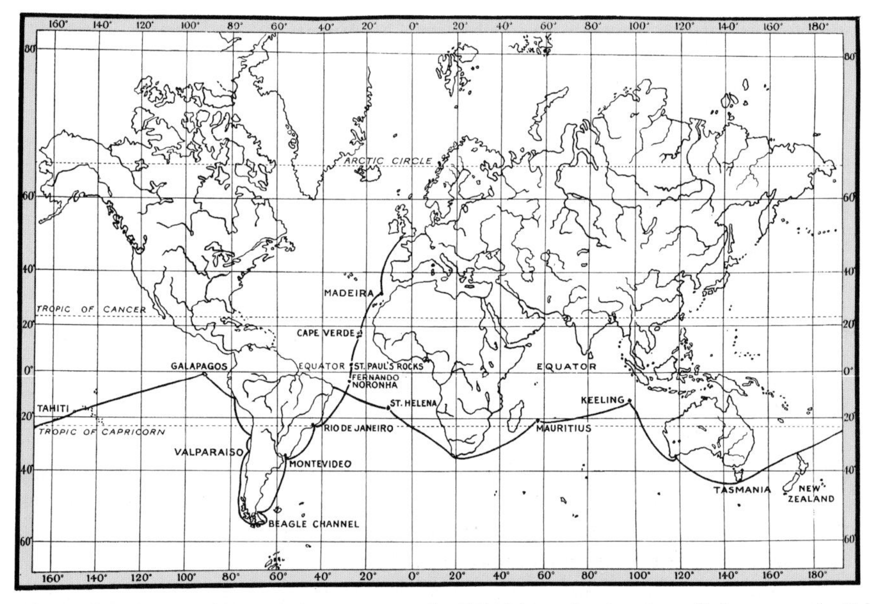

Figure G.4 A map of Darwin's voyage on the HMS Beagle *showing all the ports of call*

On the voyage, the *Beagle* visited South America. One of the many fossils Darwin found there was the extinct *Glyptodon* (Figure G.5(a)). This strange beast had a bony armour covering its skeleton. Most other observers would have left it at that, but Darwin compared it with other known animals. Figure G.5(b) shows the giant armadillo which Darwin knew a lot about. Why, he wondered, was *Glyptodon* so much like the giant armadillo, yet different in detail? He concluded that they must somehow be related, and one way in which this could have happened was that an ancestral stock of *Glyptodon* had changed to produce another species; that is, evolved.

Darwin began to form his ideas of evolution in the Galapagos Islands in the Pacific. When he returned home his notebooks were filled with his observations and his boxes were filled with specimens. For the next twenty years he continued his studies of evolution. Then, when he had reached

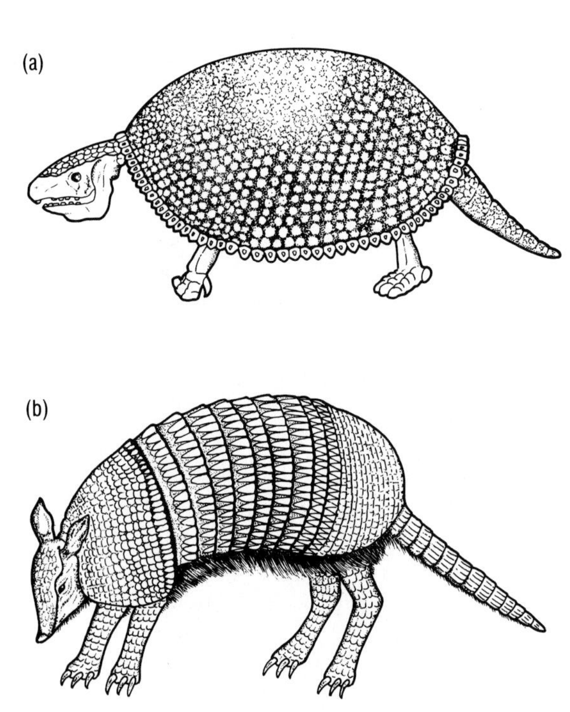

Figure G.5 Darwin compared the extinct Glyptodon *(a), found in fossil form, with the giant armadillo (b). He concluded that they must be related; in other words the armadillo had evolved from the* Glyptodon

conclusions as a result of his careful observations, he wrote down his ideas in *The Origin of Species*.

Darwin produced seventeen books on a large variety of scientific subjects. He wrote all but two of them in the study of his home in the village of Downe, 20 miles from London. His study was packed with books, papers, stuffed animals and specimens from his voyage.

Darwin conducted endless experiments in his study and in the greenhouse in his garden. He studied the way in which orchids attract bees and the way sundew plants trap fruit-flies. Watching colonies of earthworms during the night-time, he discovered that the worms react to light. His interests covered every kind of living organism. He wrote books on a variety of subjects ranging from the earthworm to coral reefs, including a four-volume work on barnacles.

But Darwin's most famous work was his theory of evolution. It revolutionised the science of biology. It brought new ideas into the study of religion and astronomy, of history and psychology. The truths about the history of life which Darwin perceived have affected every branch of science. Each one of these disciplines has benefited from his work.

An island laboratory

Charles Darwin was 26 years old when HMS *Beagle* took him to the Galapagos Islands. He had already been on board for over three years. He had explored jungles and grasslands of South America. He had climbed the high mountains of the Andes, and had filled his notebooks with thousands of observations about plants and animals.

These observations had convinced him of the truth of his theory of evolution. However, he had not yet worked out how the process actually took place. It was in the Galapagos Islands that he found some of the clues he was searching for.

The Galapagos are a cluster of lonely volcanic islands in the Pacific Ocean, lying on the Equator. They are about 650 miles from the mainland of South America. Although they were discovered in 1535, they had remained uninhabited for three hundred years. The only visitors were a few buccaneers and crews of whaling ships.

The name, Galapagos, comes from the Spanish *Galapago* which means giant tortoise. These animals (Figure G.6) live on the islands and were an important feature in Darwin's ideas of evolution.

Darwin remained on the islands for only five weeks. They were, however, among the most important weeks of his life, for he realized that these isolated islands held the key to the whole mystery of evolution. Because the Galapagos were isolated from the rest of the world, he was able to study the creatures that lived there as if he and they were in a laboratory. He called the islands 'a little world in itself'. Figure G.7 shows Darwin's route around the islands and some of the species he studied in his investigations.

Darwin was puzzled by the fact that living creatures inhabited the Galapagos at all. Most of the animals were unique and could not be found anywhere else in the world. The islands were of volcanic origin, so he was sure that they had not originated there. Also, some of the island creatures were very similar to those found on the mainland. Yet the islands were more than 600 miles from the mainland. How, then, had these creatures managed to reach the Galapagos? Darwin's first idea was that a land bridge might have once connected the islands to the mainland. If this theory were true, then the animals could have reached the islands by crossing the bridge. He gave up

Figure G.6 A Galapagos tortoise

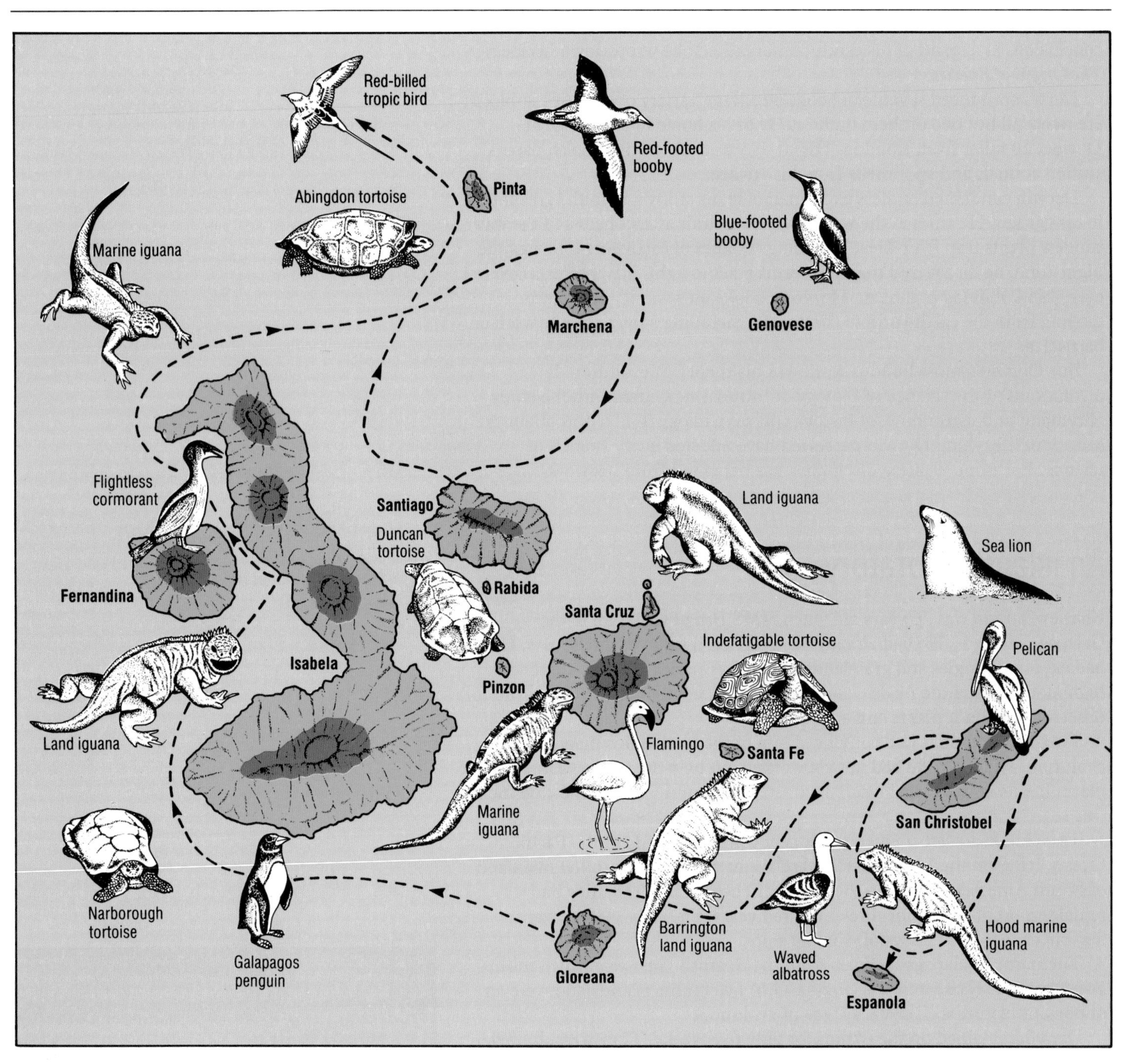

Figure G.7 The route of the HMS Beagle *took Darwin to four of the Galapagos Islands. He spent just over a month there putting together the foundation stones of his theory of natural selection*

this idea as soon as he noted that frogs and toads were absent from the island and hardly any mammals lived there. If there had been a land bridge then these types of animals would have reached the islands and thrived there.

Darwin's conclusion was that life had arrived from the mainland by way of the sea or air (Figure G.8). This would explain the abundance of reptiles, but the absence of amphibians. Reptiles have eggs protected by shells but amphibian eggs only have jelly surrounding them. Animals like lizards and tortoises would have survived the ocean journey while frogs and toads would have died.

Darwin performed many experiments to test his theory that life had reached the Galapagos in a hit-or-miss fashion across the ocean. His aim was to discover which kinds of eggs and seeds could survive the long journey by

Figure G.8 How some animals reached the Galapagos Islands

found that some would still germinate. He grew plants from seeds dropped by birds and from seeds embedded in the mud on birds' feet (Figure G.10). The results confirmed his belief that plants had come to the islands from the mainland.

During his time on the Galapagos, Darwin was particularly impressed by the enormous variety of reptiles and birds that he found on the islands. He noticed that the individual types of various species differed from one island to another. Since the climate and soil on different islands were similar, how and why had they arisen?

It was the vice-governor of the Galapagos who first called Darwin's attention to the way in which animals inhabiting different islands varied from each other. He said that by looking at a tortoise he could tell which island it came from. However, it was not the tortoises but the finches which put Darwin on the track that led him to the discovery of how different species of creatures come into existence.

Darwin had observed thirteen different varieties of finches in the Galapagos (Figure G.11). He suggested that these must have all descended from some birds which had flown to the islands or had been blown there in a storm.

There were no other land birds there. Therefore, without enemies or rivals, the finches were able to develop freely into the different species. On each island, the special features that were best suited to their environment were passed on from one generation to the next. In some finches, slender curved beaks for probing flowers had evolved; in others, strong parrot-like beaks useful for cracking seeds had survived.

It was this discovery which gave Darwin his first clue to the central part of his theory of evolution. In his book, *The Origin of Species*, he explained that there were two main factors which decide how creatures change and develop. The first is the physical surroundings in which they live. Organisms that have survived have gradually developed the characteristics they need to exist in their environment. The second factor is the other organisms living in the environment. Survivors develop characteristics needed to conquer their prey and avoid their enemies.

Although the climate and terrain were similar on all the Galapagos Islands, the animals and plants had begun to develop differently. Each slight change in one led to changes in others, so the process of evolutionary change

Figure G.9 Seeds being dispersed by the wind – is this how some plant varieties reached the Galapagos Islands?

Figure G.10 Seeds can also be dispersed in the mud on bird's legs

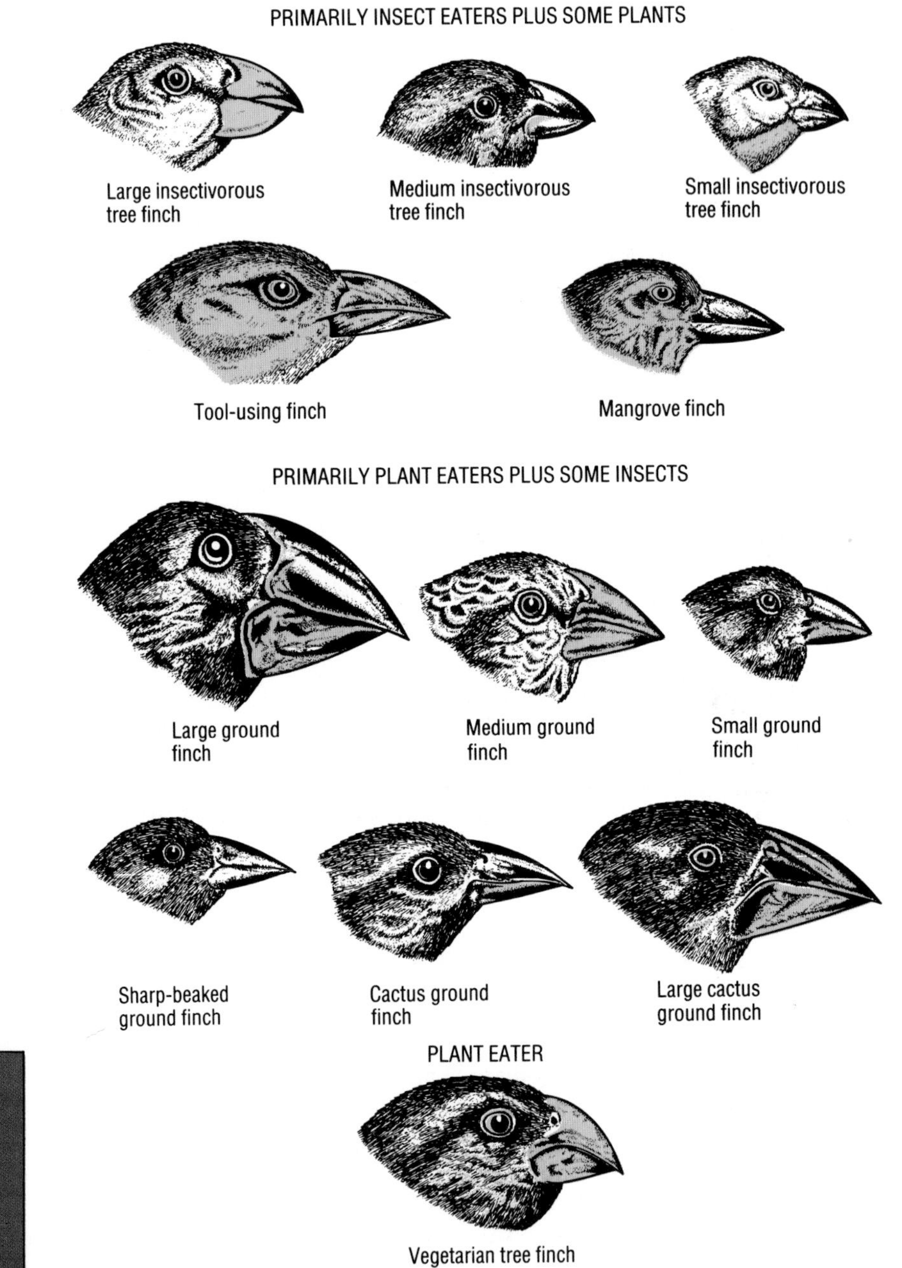

Figure G.11 The thirteen varieties of finch found on the Galapagos Islands. They vary in their beak shape, diet and plumage

Figure G.12 The Darwin memorial

continued until the differences had become so great that Darwin was able to make one of his most important discoveries.

In 1957 a group of scientists working with international conservation groups recommended that some of the islands be set apart as sanctuaries. A unique 'laboratory' of evolution now exists on the islands – The Charles Darwin Memorial Station (Figure G.12). Here, many scientists of all nationalities strive to answer questions relating to evolution.

water or air (Figure G.9). He kept seeds in water for several months and

Natural selection

Figure G.13 The black-headed gull has evolved to be dependent on our rubbish tips for food

We can now develop the idea of natural selection in the way that Charles Darwin himself worked it out. He used three facts and two deductions from them.

The first fact was stated in 1798 by Thomas Malthus in *Essays on Population*. Malthus explained that animals and plants have a tendency to multiply at a geometrical rate. That is, in numbers that run 2,4,8,16,32, and so on. In other words, the offspring always tend to be more numerous than parents. We have only to look at the human 'population explosion' to see this in action.

The second fact is that while all living organisms can increase at a geometrical rate, they seldom do. Few species, apart from humans and some animals and plants dependent on us, have been observed to increase so rapidly for very long. The species that have done so have often been presented with new opportunities by humans. For example, the black-headed gulls feeding on our refuse dumps (Figure G.13) and rats feeding in our sewers. Another example was seen in the rabbit which was introduced to Australia. There, they had no natural enemies and in a few years became a national nuisance (Figure G.14).

Figure G.14 A plague of wild rabbits was caused by humans importing a few to Australia

From these two facts, Darwin deduced a 'struggle for existence'. More accurately, competition for a chance to reproduce. Almost everywhere in nature, organisms produce more young than can possibly survive to the age at which they can reproduce. They must compete for food and for all their other needs. For example, we are not completely covered with flies despite their enormous breeding rate.

The third fact is that all living things vary. We have seen this in our studies of genetics. Darwin hence deduced a mechanism of **natural selection**. The principle of natural selection states that the competition is for existence between individuals which vary among themselves. Thus some individuals must be more likely to succeed than others. Those with favourable variations will be more likely to survive and reproduce than those with unfavourable variations.

A great deal of variation is inherited. Favourable inheritable variations have a better chance of being passed on than unfavourable variations.

Natural selection is the principal agent of evolution. It is not a physical force; it has no purpose; it is a process that occurs completely by chance, like evolution itself. It is a process that has made, through millions of years, the human brain, the bird's eye and the bat's ear – all from simple cells.

Summary

- Evolution occurs via natural selection.
- Individual organisms within a species might show a wide range of variation.
- Predation, disease and competition cause some individuals to die.
- Other individuals, which carry variations more suited to the environment, are more likely to survive and breed successfully.
- A great deal of variation is inherited. Favourable inheritable variations are more likely to be passed on to the next generation.

Extension questions

1 The Galapagos Islands and their wildlife are famous for their influence on Charles Darwin's theory of natural selection. The Hawaiian Islands were even more remote and were never visited by Darwin. However, similar observations can be made on many animals, including insects that live there. The Hawaiian Islands are about 3000 km from the mainland. The table below shows the number of species in certain insect groups in the British Isles and the Hawaiian Islands. Study the data and answer the questions below.

Group	**Number of species**	
	British Isles	**Hawaiian Islands**
Fruit-flies (*Drosophila*)	32	500
Other flies (Order *Diptera*)	5950	0
Total insect species	21833	6500

a What two types of animals found on the Galapagos can be compared with these observations on the Hawaiian Islands?
b Suggest an evolutionary explanation of the differences between the British Isles and the Hawaiian Islands in the number of species of:
i total insects,
ii fruit-flies.
c The Galapagos Islands and the Hawaiian Islands are about 400 km and 3000 km from the nearest mainland, respectively. Explain how this has been important in the evolution of animals and plants on both groups of islands.
d On the Hawaiian Islands there are no native amphibians or reptiles. There are two species of native mammals – a seal and a bat. Compare this with the situation on the Galapagos Islands and explain why there are not more native vertebrates present.

2 Charles Darwin suggested that natural selection is a way in which species either adapt to changing conditions or become extinct.
a State the meaning of:
i adapt,
ii extinct.
b The peppered moth exists as a light-coloured form and a dark-coloured form. Both are eaten by birds. Before the *Clean Air Act* of 1956, both light and dark moths were released in two different areas and as many as possible were recaptured. One group was released near an industrial area where soot and dust covered trees. The other was released in a non-industrial area. The percentages of moths recaptured in the two areas are given in the table below.

Area	**Percentage recaptured**	
	Light-coloured type	**Dark-coloured type**
Non-industrial	12.5	6.3
Industrial	13.1	27.5

i Use this data to explain 'natural selection'.
ii Since the *Clean Air Act* of 1956, industrial areas have become much less polluted. Explain how this might have influenced the number of light-coloured moths surviving in these areas.
c Explain why it might be an advantage for the species to produce dark mutations occasionally, even in a clean environment.

Unit H
Variation and mutation

No two living things are exactly the same. If you look casually at a group of organisms of one species, you might think that they were all alike. However, anyone who has to care for animals soon learns to distinguish between them. Furthermore, plants can vary just as much as animals.

A child tends to resemble brothers, sisters and parents more closely than more distant relatives; and resembles relatives more closely than unrelated friends. These resemblances are due mainly to **heredity**. The resemblance of identical twins is due totally to heredity. The differences that enable us to tell one twin from another are due to the influences of the environment.

Separation of chromosomes in meiosis (page 205) and the random way in which they recombine also produces variation. Even the simplest of organisms have hundred of genes; complex organisms have millions. A tremendous amount of variation is possible from this number. Furthermore, crossing over of parts of chromosomes increases the possibility of variation by making new combinations of genes. The mixing of chromosomes at fertilisation ensures that all individuals are unique. Other factors besides genes affect the development of an organism. A pea plant may inherit tall genes, but will still not grow in poor soil. Regardless of how many times different genes are combined into new arrangements, the results are all variations on an existing pattern.

A change in the genes themselves – a **mutation** – will give a new type of pattern. One of the first mutations to be studied was one which was observed in the fruit-fly, *Drosophila melanogaster*. It was found that in a strain of pure-bred red-eyed flies, there was one with white eyes (Figure H.1). This was caused by a sudden change in one of the many genes controlling eye colour. Since then, hundreds of fruit-fly mutations have been found and studied.

Figure H.1 A mutant fly (left) and a normal fly (right). The mutant type has white eyes and shorter wings than the normal fly

What is a mutation?

Mutation involves a change in the chemical structure of a gene. Mutations can result from a mistake in gene duplication. For example, in replication the DNA might pick up the wrong base and so produce a new arrangement. Since genes control the manufacture of proteins, the new gene might make a different protein, or be unable to make a protein at all. This could result in the change of a body characteristic or even the introduction of a new characteristic.

Most mutations are recessive and are masked by dominant normal genes. Thus, except for some sex-linked genes, mutations do not show up until two of the mutant genes occur together in the same individual. The individual is said to be **homozygous** for the mutant gene. Some mutations produce only minor changes in body chemistry. If it is an unfavourable change for the organism it will be lost; the organism with the mutation is unlikely to live long enough to have offspring which will inherit the mutant gene. Sooner or later, however, the mutation will occur again. A few examples of mutations are shown in Figures H.2–H.5.

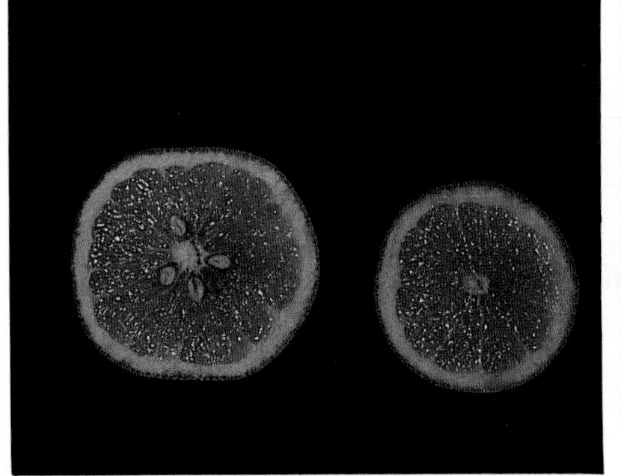

Figure H.2 A normal fruit with seeds and a mutation without seeds

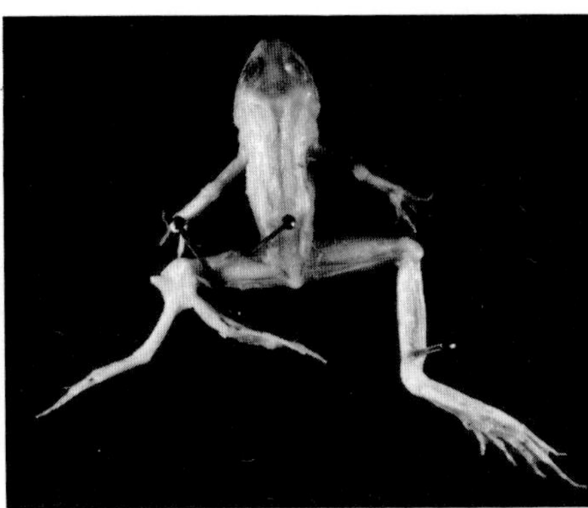

Figure H.3 A five-legged frog

Figure H.4 A mutant breed of cat with no hair, born to an ordinary black and white cat in Canada in 1966

Figure H.5 Two-toed Africans from the Zambesi valley

Although we speak of mutant genes and normal genes, the genes we now call normal were once mutants. Because they were favourable, they have been passed on from generation to generation and have become part of the normal collection of genes. Mutation is all part of the process of evolution.

Radiation and genetics

All living things are exposed to a certain amount of background radiation. The two sources of this radiation are:

1 Cosmic rays from space.

2 Radioactive materials in the Earth's crust.

The background radiation is, in part, responsible for the mutations that occur in all organisms.

Consider radiation as being a series of rapidly-fired mini-bullets. Then think of these bullets hitting the molecules that make up organisms. These molecules will be damaged; if DNA is one of them, then the code which is normally inherited can be altered or destroyed.

In fact, radiation is a series of sub-atomic particles (mini-bullets) which leaves some atoms with enormous energy. We say that radiation is **ionising** because when the particles hit an atom they strip off an electron (ionisation) and release a large amount of energy. Thus any biological molecule hit by radiation is destroyed. This, in itself, is enough to cause considerable damage to a living cell.

It has been known for a long time that an increase in radiation causes an increase in the mutation rate. In other words, it causes an increase in the rate at which genes are altered.

Indeed, this knowledge has been used in genetic experiments to produce mutations in experimental organisms. This treatment has produced some useful plant varieties. It is also known that there is no level of radiation that is so low that no change in mutation rate occurs – radiation always has an effect.

These findings have real significance for us in our nuclear and technological age. The human race has increased the radiation levels in several ways. First, there has been an increase in the use of X-rays for medical purposes. Then there was an increase in the testing of nuclear weapons (Figure H.6). This brought a whole new meaning to the word 'fallout'. The dumping of radioactive waste from nuclear power stations is also a problem which has not yet been solved.

It must be emphasised here that this increase in radiation is only a fraction of the naturally occurring background radiation. However, it must also be emphasised that **any** increase will increase the mutation rate.

Figure H.6 These soldiers were unaware of the dangers to which they were being exposed. What damage would the radiation from this blast have caused their bodies?

How damage is done

Radiation may damage a cell by either direct or indirect action. The result of **direct** action is damage to a molecule; eg damage to a molecule of DNA by the passage through it of an electron or other atomic particle. **Indirect** action is the change in a large molecule of DNA brought about by highly active pieces of molecules (**free radicals** – **ions**). Free radicals are formed by the action of radiation on water molecules. Although these free, highly active pieces of molecules are very short-lived, they can do a lot of damage to any large molecules that are near them. It is in this way that radiation brings about the mutation of genes.

The different types of radiation

From a biological point of view, the most important damaging radiations are **alpha** particles, **beta** particles, **X-rays** and **gamma** rays.

Alpha particles spread their energy very quickly and cause the production of an enormous number of ions. There is little danger from alpha particles unless they get inside the body. The skin easily stops the particles entering from external sources. However, if an organism takes in food exposed to alpha radiation, hence contaminating the body cells, the alpha particles become a serious hazard. The most active tissues in the body are blood-forming bone marrow, the liver, testes and ovaries. Once alpha particles get into these tissues they will do two kinds of harm.

1 The marrow and liver cells will suffer damage – this will shorten the life of the person.
2 The damage done to the sex organs will be inherited via DNA and passed on to future generations.

Products from nuclear reactors are one source of alpha particles and pose a serious problem. For example, nuclear waste contains plutonium-239 which can easily enter food chains. Once in the body, it accumulates in the bone and is suspected of causing anaemia and leukaemia. Leukaemia is a particularly distressing form of cancer which affects blood cells in the bone marrow. Leukaemia occurs when mechanisms which control blood-cell manufacture break down. An imbalance in the correct proportions of red cells, white cells and platelets then occurs (Figure H.7).

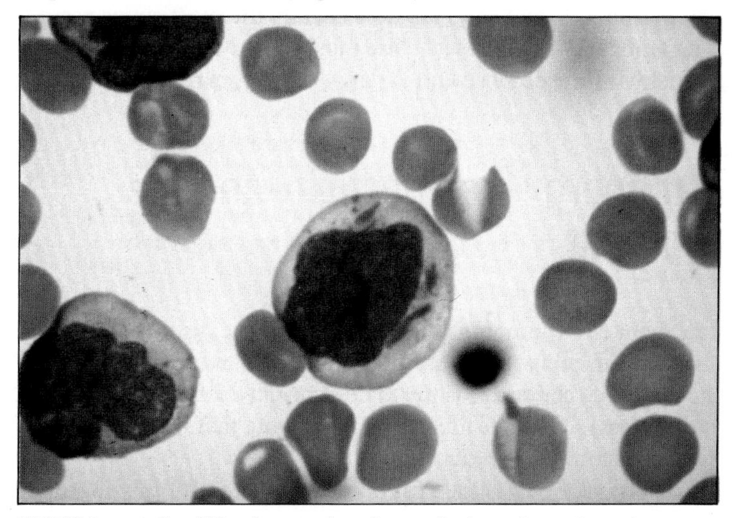

Figure H.7 This sample of blood was taken from a leukaemia patient: three abnormal blood cells can clearly be seen

Ultraviolet light and skin cancer

Ultraviolet (UV) light is part of the spectrum of radiation emitted by the Sun. Over-exposure to sunlight (and therefore to UV light) is a major risk factor for a type of skin cancer called **melanoma.** UV light striking each skin cell can damage DNA in the nucleus, causing a gene mutation. If mutations affect the genes controlling the rate at which cells divide, then cell division can run out of control. In other words, the cells become malignant and a mass of cancerous cells (**tumour**) develops. Scientists are worried that thinning of the ozone layer in the atmosphere (page 108) increases our exposure to UV and therefore increases the risk of melanoma developing. Recently, the occurrence of melanoma worldwide has increased significantly.

Chemicals causing mutations

Chemical substances (called **carcinogens**) in cigarette smoke and food damage genes, causing them to mutate. The effect is to increase the activity of genes which stimulate cell division. Cells proliferate and contribute to the development of cancer.

Figure H.8 *The Sellafield reprocessing plant in Cumbria*

Figure H.10 *Collecting grass samples for analysis near a nuclear power station*

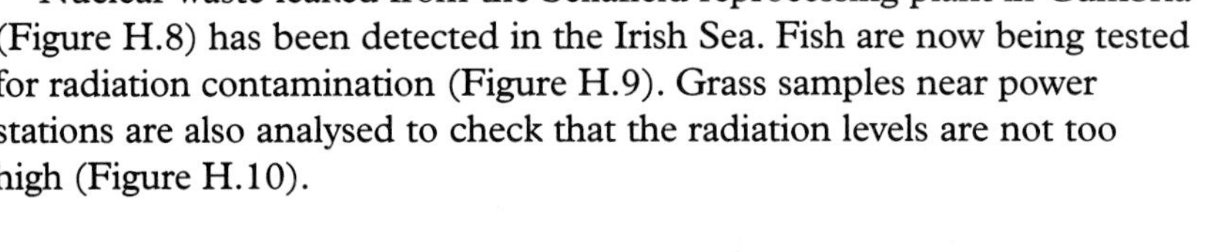

Nuclear waste leaked from the Sellafield reprocessing plant in Cumbria (Figure H.8) has been detected in the Irish Sea. Fish are now being tested for radiation contamination (Figure H.9). Grass samples near power stations are also analysed to check that the radiation levels are not too high (Figure H.10).

Figure H.9 Collecting samples of fish off the west Cumbrian coast for monitoring by British Nuclear Fuels at Sellafield

Beta particles are high-speed electrons and can penetrate the skin. Internally, they are also a hazard. Radioactive fallout contains beta particles as well as alpha particles. Strontium-90 and caesium-137 have attracted considerable attention because they readily accumulate in living organisms. For example, fish caught in the North Sea have been found with caesium-137 in their flesh. Strontium-90 is similar to calcium and becomes localised in bones. Emissions can damage the bone marrow and cause anaemia and leukaemia. Caesium-137 is absorbed by all cells.

X-rays and gamma rays have great penetrating power. They are a major hazard when their source is outside the body. Great care must be taken to minimise exposure when X-rays are used. Again, the most active tissues – in developing embryos, for instance – are most sensitive. X-raying pregnant women is risky for the embryo and is rarely, if ever, carried out.

Nuclear energy, radioactivity and the environment

One of our main environmental concerns today is the disposal of radioactive nuclear waste from power stations. The problem of extremely long **half-lives** of radioactive waste chemicals is the main consideration when deciding on the method of disposal. The half-life period is the time that it takes for the radioactive emissions to be reduced by half.

Table K.1 on page 152 gives a list of the half-lives of some important radioactive elements. Some of these may enter our environment as a result of human activities.

Radiation and you

Can nuclear wastes be buried at sea?

The production of highly radioactive waste is a problem with all technologies which use radioactive materials, including nuclear power stations. This waste must be disposed of safely, in places from where it cannot be recovered.

Three main places for the disposal of solid, high-level waste have been suggested. One is the surface of the deep ocean floor, others are under the surface of land and under the surface of the ocean floor. Whichever method or combination of methods is used, there will be routes by which radiation will return to the human environment.

If leakage from containers stored on or under the ocean bed takes place, pollution by radiation will occur and could affect us. Figure H.11 shows the possible routes the radiation could then take as it returns to our environment.

The total radiation that affects the human environment is given in Table H.1 and is shown diagrammatically in Figure H.12.

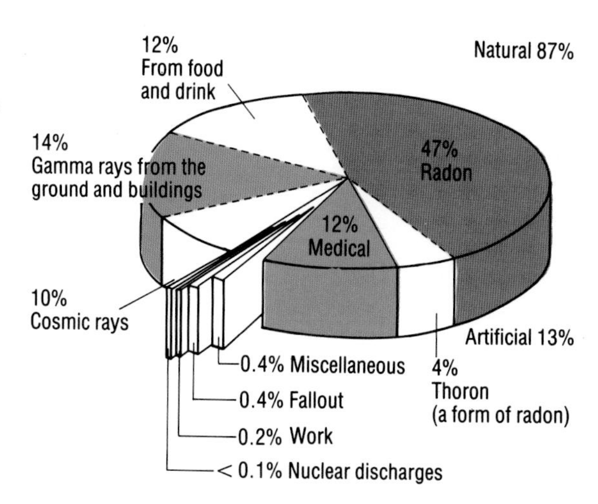

Figure H.12 The health hazard posed by the various emitters of ionising radiations depends on the location of the source. Internally, alpha and beta sources are the greatest danger, while X-rays and gamma rays are more dangerous from an external source

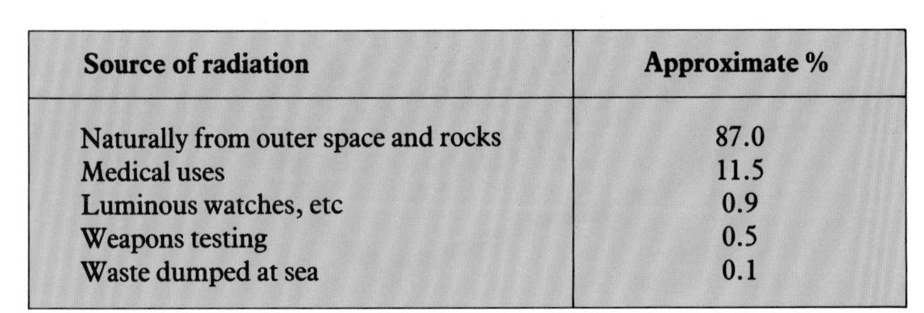

Source of radiation	Approximate %
Naturally from outer space and rocks	87.0
Medical uses	11.5
Luminous watches, etc	0.9
Weapons testing	0.5
Waste dumped at sea	0.1

Table H.1

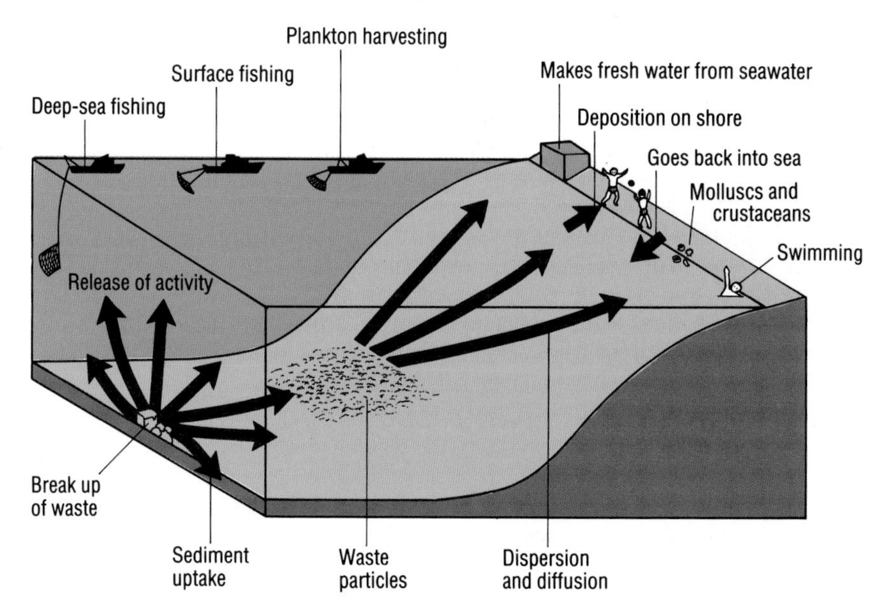

Figure H.11 Radiation routes from nuclear waste stored on the surface of the ocean floor to humans

The great nuclear debate

Case study – **Chernobyl**

In 1986 an accident at the Soviet Union's largest nuclear power station released a cloud of radioactive material high into the atmosphere. The winds then blew this cloud across Poland and Scandinavia. These countries were showered with radioactive chemicals. The direct risks to health were inhalation and skin irradiation, but there were also other indirect risks. For example, there were restrictions on the consumption of fresh vegetables and milk in several countries.

In March 1989, the following headline appeared in a science magazine.

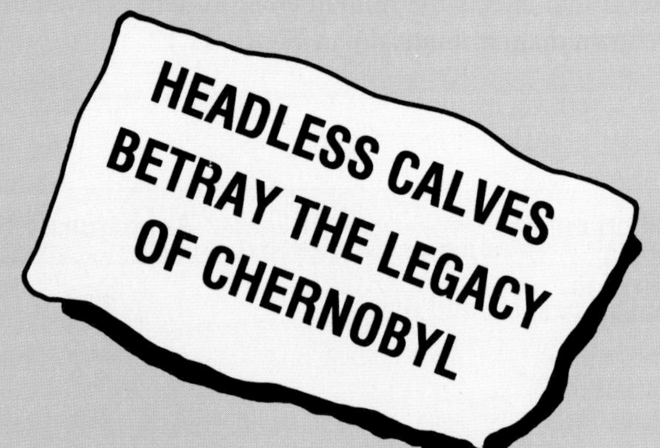

This is a summary of the article:
The effects of the world's worst nuclear accident at Chernobyl are becoming clear at farms just outside the 30-kilometre exclusion zone around Chernobyl. There has been an alarming increase since 1986 in the number of deformed animals being born on a farm 50 km from Chernobyl. This farm, which has 350 cows and 87 pigs, has records of only 3 deformed pigs born in the five years before the accident and no deformed calves. In 1987, 37 pigs and 27 calves were born with serious deformities. During the first nine months of 1988, a further 41 deformed pigs and 35 deformed calves were born. Some calves lacked heads, limbs, eyes or ribs. Most of the pigs had deformed skulls.

Radiation at the farm was 148 times higher than the background level. Food is still delivered to the area from elsewhere. However, local people still consume home-produced milk, fruit and vegetables. Cattle still eat locally produced fodder. Concern is now growing about the possible effects on people. For instance, women from the area are advised not to have children. The average annual number of new cancer cases, especially of the lip, has doubled since the accident.

Farmers use pressurized cabins on tractors to protect themselves from radioactive dust from the fields. Medical workers have found the thyroid glands of more than half of the children were affected by radiation.

Questions

a Explain how radioactive materials get into milk and vegetables.
b Besides the potential hazards of accidents in nuclear power stations, state another problem which arises from generating electricity by nuclear means rather than by using tidal or solar energy.
c One of the radioactive materials was iodine-131. Explain the link between this and its accumulation in the thyroid glands of people affected by the radiation.
d Give a genetic explanation for the birth of deformed animals that took place two years after the accident.

Summary

- New forms of genes result from changes (mutations) in existing genes.
- Mutations occur naturally, but the chance of mutations occurring is increased by:
  – exposure to ionising radiations e.g. UV light, X-rays, radiation from radioactive substances
  – some chemical substances
- Human use of nuclear technology can result in serious accidents.

Extension questions

1. Explain the harmful effects of radiation to humans.
2. Use the information given and write balanced arguments for and against the use of nuclear power stations.

Unit I
Mendel and inheritance

Figure I.1 Gregor Mendel – the 'father of genetics'

It might seem strange to you now but the whole science of genetics really began with a monk experimenting with pea seeds in the 1860s. Using this unlikely material and a great deal of patience, this Augustinian monk, Gregor Mendel, set out the first principles of inheritance in 1865. For his unique contribution to science, he is often called the 'father of genetics' (Figure I.1).

Although Mendel performed very few experiments that had not already been done before, he succeeded where others had failed. His success was due to the unusual combination of skills he brought to the task. He was trained in mathematics as well as in biology. With this background, he planned experiments that at the time were novel in three respects:

1 Instead of studying a small number of offspring from one mating, Mendel used many identical matings. As a result he had enormous numbers of offspring to study.
2 As a result of having large numbers to study, he was able to apply statistical methods to analyse the results.
3 He limited each cross to a single difference, a single pair of contrasted characters, at a time. People who had carried out genetic crosses with plants and animals before Mendel did not concentrate on one character at a time. Pure-breeding stock was not always used (pure-breeding individuals receive similar genes from both parents). Characters were often masked by each other.

Mendel selected garden peas for his experiments because he knew that they possess many varieties (Figure I.2). The plants are easy to cultivate and cross, and the generation time is reasonably short. Finally, and of great importance, the plants are self-pollinating. The significance of this requires an explanation.

Figure I.2 Pea seeds – an unlikely plant for the foundation of genetics

Let us consider the structure of the pea flower (Figure I.3) and see how fertilisation usually takes place. Pollen from the anthers falls onto the stigma of the same flower. This happens before the flower opens fully. Pollen tubes develop from the pollen grains. These tubes carry the male gamete to the female gamete in the ovule.

If you want to pollinate one pea flower with pollen from another plant, you must remove the anthers from the flower before its own pollen is mature. Later, when the stigma is ready to receive pollen, you can dust it with pollen

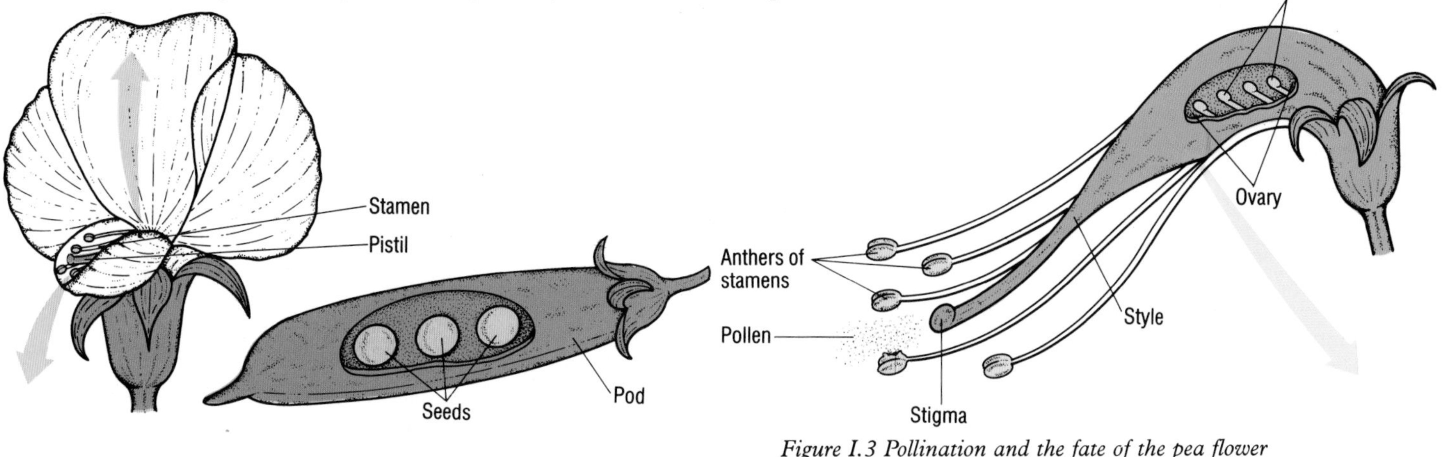

Figure I.3 Pollination and the fate of the pea flower

taken from some other pea flower of your choice. In this way the parents of the next generation can be controlled.

Mendel made sure that the plants he started with were all **pure-bred** for each character he was studying. He did this by letting the plants fertilise themselves for a number of generations. The offspring of each generation were studied to make sure they were all like one another and like the parent plant. Then Mendel made hundreds of crosses by dusting the pollen of one kind of plant on to the stigmas of plants of another kind. For example, he pollinated plants from a type whose seeds were always round, with pollen from a type whose seeds were always wrinkled.

In every case he found that all the offspring resembled one of the parents and showed no sign of the character of the other parent. Thus all the crosses between plants with round seeds and plants with wrinkled seeds produced offspring whose seeds were always round. One character seemed to 'dominate' the other. Mendel therefore called this character the **dominant** character.

Mendel's results – a summary

Mendel used seven pairs of contrasted characters (Figure I.4).

Seed shape	Round seed dominant to wrinkled seed
Seed colour	Yellow seed dominant to green seed
Seed coat colour	Coloured seed coat dominant to white seed coat
Pod shape	Inflated pod dominant to wrinkled pod
Pod colour	Green pod dominant to yellow pod
Flower position	Axial flowers dominant to terminal flowers
Stem length	Long stem dominant to short stem

Figure I.4 Dominance in seven pairs of characters in garden peas

Characters selected	**F1**	**F1 Self-pollination**	**F2**	**Ratio**
Round × wrinkled seeds	all round	round × round	5474 round 1858 wrinkled	2.95:1
Yellow × green seeds	all yellow	yellow × yellow	6022 yellow 2001 green	3.01:1
Coloured × white seed coat	all coloured	coloured × coloured	705 coloured 224 white	3.15:1
Inflated × wrinkled pods	all inflated	inflated × inflated	882 inflated 299 wrinkled	2.96:1
Green × yellow pods	all green pods	green × green	428 green 152 yellow	2.82:1
Axial × terminal flowers	all axial	axial × axial	651 axial 207 terminal	3.14:1
Long × short stems	all long	long × long	787 long 277 short	2.84:1

Figure I.5 Mendel's results with two generations of garden peas

When a cross is made between two plants which have been pure-bred for contrasted characters (eg, round seed versus wrinkled seed), the parents are the P1 generation. The offspring are the first filial generation, F1. These symbols will help us to follow the steps in Mendel's thinking.

He let the F1 plants pollinate themselves. This produced the F2 (second filial) generation. One of the contrasted characters appeared in 75% of the F2 generation, while 25% of the other character reappeared. Since the other character had receded into the background for a generation, he called it **recessive**. He also noted that there were no 'in-between' forms. The new seeds were either round or wrinkled, or tall or short, etc – just like the P1.

Mendel's results are shown in Figure I.5. The 3:1 ratio is obvious in all cases.

Figure I.6 Mendel used garden peas for his famous experiments

Inheritance of a single character (monohybrid inheritance)

Mendel made his greatest contribution to genetics by explaining his observations. He began by using symbols to represent the characters he was dealing with. His mathematical training encouraged him to use symbols rather than written descriptions.

Mendel assumed that the character for tall plants was caused by a dominant factor. He used a capital T to symbolise this element. The character for short plants, the only alternative to tall plants, was caused by a recessive factor, t. Basically, we use the same idea today.

In 1910, many years after Mendel's work, the genetic factor was given the name **gene**.

Next, Mendel assumed that every plant had a pair of genes for each character. These pairs of genes are called **alleles**. An allele is one of two alternative forms of a gene which may occupy the same part of a chromosome. For example, if T represents tall and t represents short, TT or Tt or tt can be the allelomorphic pair. T and t are at the same place on corresponding pairs of chromosomes. Mendel was convinced of the existence of alleles because some parent plants with the dominant gene produced some offspring with the recessive gene. Therefore, every F1 plant must have each type of gene. The F1 plant could be represented by Tt. Mendel also assigned a pair of genes per character even to plants which are true breeding. A plant from parents that bred true for tall plants was therefore TT. Similarly, a plant from parents that bred true for short plants was tt. These paired symbols representing the genes of an organism are its **genotype**. When an organism has identical factors in its genotype (eg TT or tt), it is known as **homozygous**; if it has different factors (eg Tt), it is known as **heterozygous**. Using this logic Mendel was now able to test his hypothesis about genes. If he knew the genotype of each parent, he could predict the kinds and proportions of gametes each parent could produce. From this, he could predict the kinds and proportions of the offspring.

If every plant had a pair of genes for each character, was there any rule about how these genes were passed on to the next generation? Mendel thought about the short plants that appeared in the F2 generation. These could not carry the dominant gene, T. They must have received the recessive t from the F1. Remember the F2 was produced by self-pollinating the F1.

The next question was: How frequently do gametes that carry t occur among all the gametes produced by the F1 Tt parents? Mendel reasoned back from the proportions of tt short plants in the F2. These amounted to one-quarter of the entire F2 generation. So the frequency of t in the eggs and male gametes produced by the F1 generation should be one-half (since one-half multiplied by one-half = one-quarter). This means that half the gametes of a Tt plant would carry T; the other half, t.

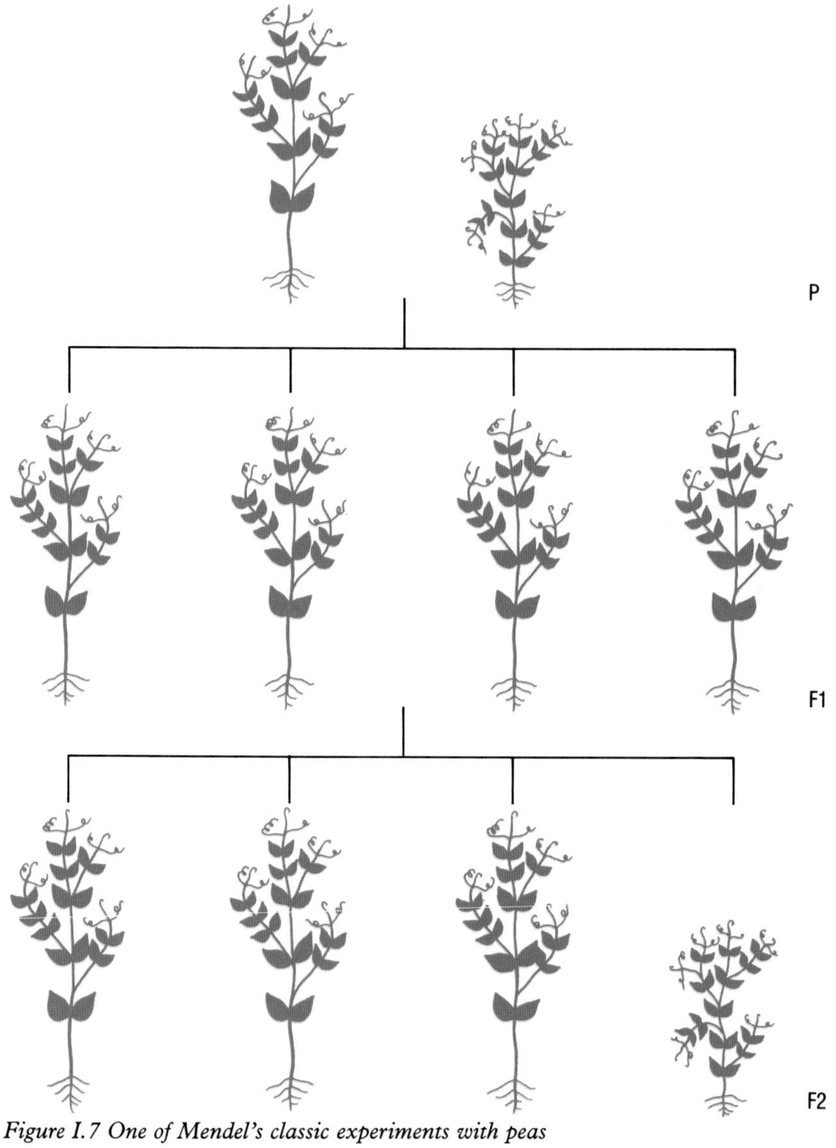

P

F1

F2

Figure I.7 One of Mendel's classic experiments with peas

Figure I.7 shows one of Mendel's classic experiments with peas. If a pure-breeding tall variety was crossed with a pure-breeding short variety, all the offspring were tall. When two of these plants were crossed, three tall offspring were produced for every short one. Mendel saw that these results could be explained if the characteristics were inherited as 'particles'. Each plant had two of these particles, and one was dominant to the other. In this case the tall character (T) was dominant to short (t). His interpretation of the results is shown in Figure I.8.

Mendel thus arrived at a general rule:

'The two members of each pair of genes must separate when gametes are formed, and only one of each pair can go to a gamete.'

If a parent is TT its gametes will all inherit one or other of its T genes, but not both. If the parent is tt, its gametes will all inherit one of its t genes. If the parent is Tt, roughly half of the gametes will inherit its T gene, and the other half will inherit its t gene.

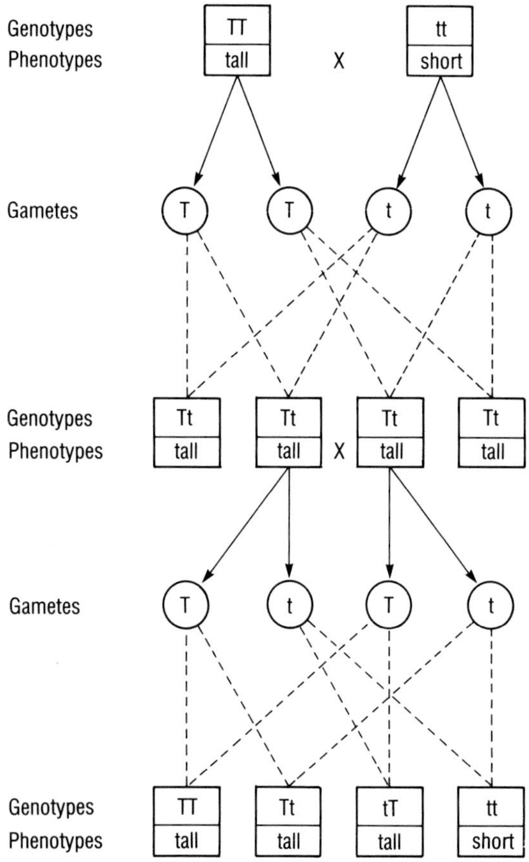

Genotypes TT tt
Phenotypes tall X short

Gametes T T t t

Genotypes Tt Tt Tt Tt
Phenotypes tall tall X tall tall

Gametes T t T t

Genotypes TT Tt tT tt
Phenotypes tall tall tall short

Figure I.8 A graphic explanation of Mendel's results as shown in Figure I.7

Extension questions

1 Gregor Mendel formulated his first law of genetics which states 'Of a pair of contrasted characters, only one can be represented in a gamete by its germinal unit'.
  a Give the modern name for 'germinal unit'.
  b State where these germinal units are found in the gametes.
  c A red-haired woman marries a brown-haired man, and all the children are brown haired. Explain this genetically.

2 Figure I.9 is part of a family tree showing the distribution of brown eyes and blue eyes.
  a Which numbers represent a man who is homozygous for blue eyes?
  b Which numbers represent a woman who must be heterozygous?
  c Which of the eye colours is controlled by a dominant gene?
  d Which part of the family tree enables you to answer part **c**?

3 In humans, the gene for tongue-rolling, R, is dominant to the gene for the inability to roll the tongue, r.
  a Using these symbols, give the genotypes of the children that could be born from a marriage between a heterozygous father and a non-tongue-rolling mother.
  b State whether the children would be tongue-rollers or not.

4 A boy was interested to see how eye colour was inherited in his family, so he wrote down the colour of the eyes of all his family. The information and his deductions are given in Figure I.10. Unfortunately, he could not finish it. Fill in the gaps after copying the table.
  a Write the correct words for the numbers 1–17.
  b From the results, state the dominant eye colour.
  c Also from the results, state what recessive means.
  d How many factors for eye colour do the gametes shown contain?

Summary

- Genetics is the study of the ways in which characteristics can be inherited from parents to offspring.
- Gregor Mendel was responsible for establishing some fundamental principles of genetics.
- Certain characteristics are controlled by one pair of genes. The genes of a pair are called alleles.
- An allele that controls the development of a characteristic when present on just one of the chromosomes is a dominant allele.
- An allele that controls the development of characteristics only if it is present on both chromosomes is a recessive allele.
- If both chromosomes in a pair contain the same allele of a gene, the individual is homozygous for that gene.
- If both chromosomes in a pair contain different alleles of a gene, the individual is heterozygous for that gene.

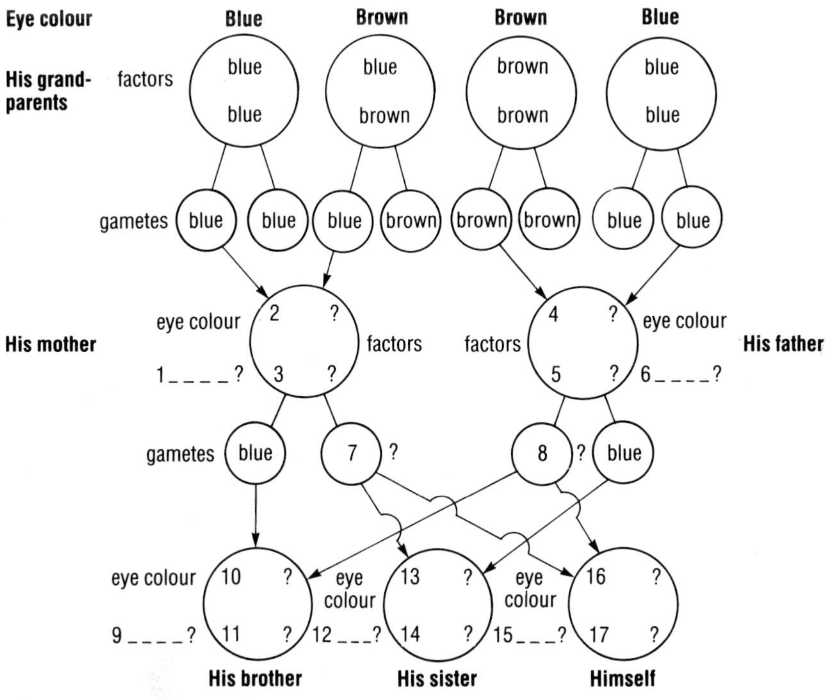

Figure I.10 Copy and complete this table

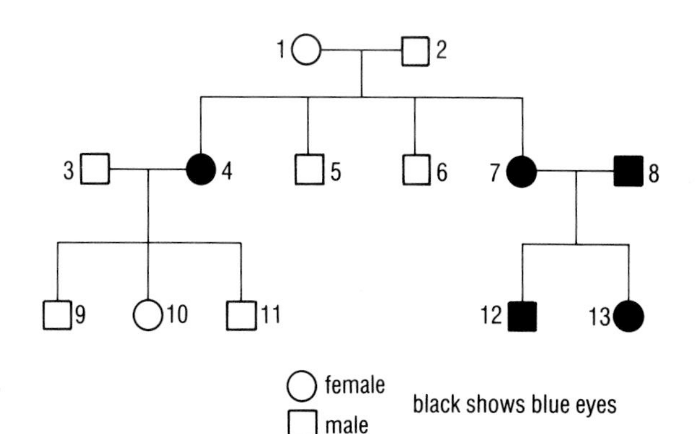

Figure I.9 Distribution of eye colour in a family

Unit J
Chromosomes and genes

An organism's characteristics are passed down from generation to generation by genes. But what are genes, where can they be found and what do they actually do?

Life is a series of complex chemical reactions. It is not surprising, therefore, that early in the history of genetics, scientists knew that genes must be chemicals. Proteins are the most essential chemicals found in living cells. Hence, the early geneticists guessed that genes probably existed as protein contained in the chromosomes in the nucleus of cells.

By 1950, it became clear that it was not the protein in chromosomes that passed on the code of life from generation to generation. It was another chemical component of cells: nucleic acids. These are some of the largest and by far the most fascinating of all life's molecules. Two forms are known: **deoxyribonucleic acid (DNA)** which is found in all chromosomes, and **ribonucleic acid (RNA)** which is found in the cytoplasm and nuclei of cells. It was DNA that carried the genetic code. The structure of the DNA molecule was discovered in 1953 by the American, James Watson, and the English scientist, Francis Crick, working at the Cavendish Laboratory in Cambridge (Figure J.1).

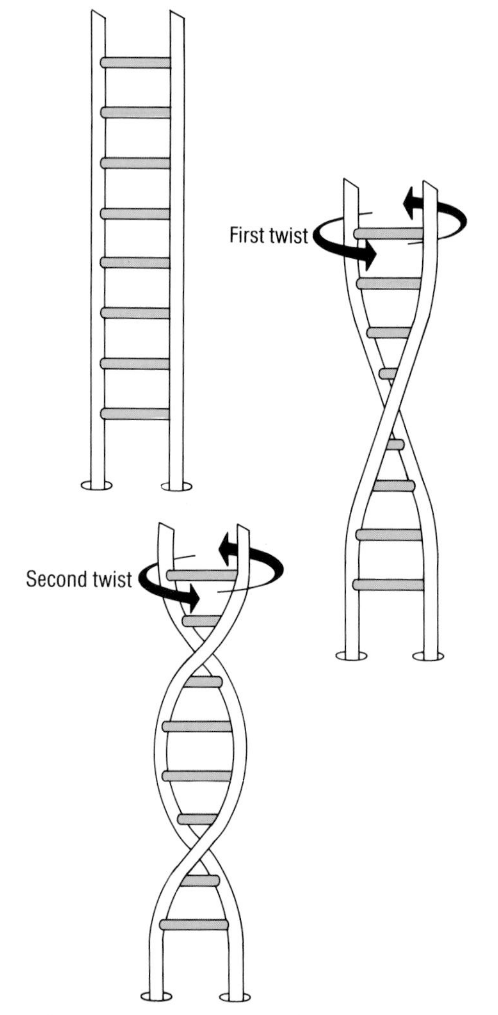

Figure J.2 The double helix of DNA

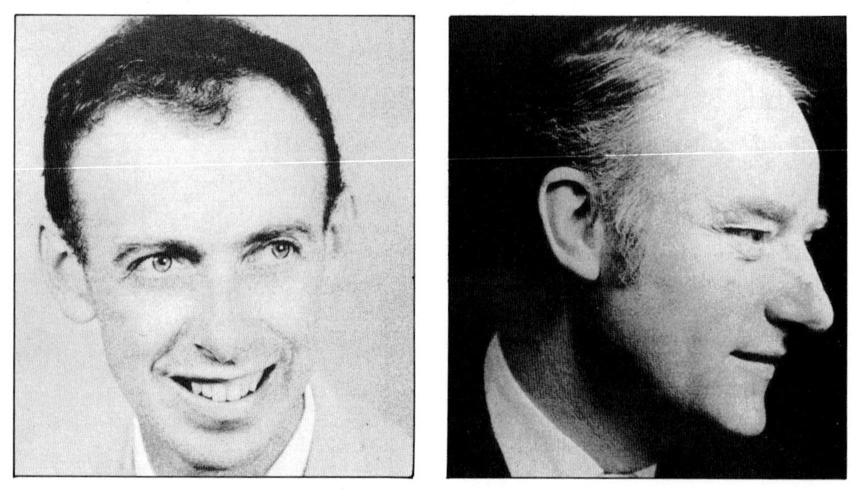

Figure J.1 James Watson (left) and Francis Crick (right) discovered the structure of the DNA molecule in 1953

Nucleic acids, like proteins, are made of many units strung together. DNA has a 'ladder-like' structure of two long sugar-phosphate chains joined together by connecting bases ('rungs'). The ladder is twisted to form a three-dimensional double helix as shown in Figure J.2. Figure J.3 shows a model of a strand of DNA.

Watson and Crick discovered that the rungs of the ladder are made of four different types of bases: **guanine, cytosine, thymine** and **adenine**. The bases fit together as shown in Figure J.4. Guanine only pairs with cytosine, and thymine only pairs with adenine. The differences between one DNA molecule and another – or one gene and another – depends on the pattern of

Figure J.3 A model of DNA – the spheres represent molecules

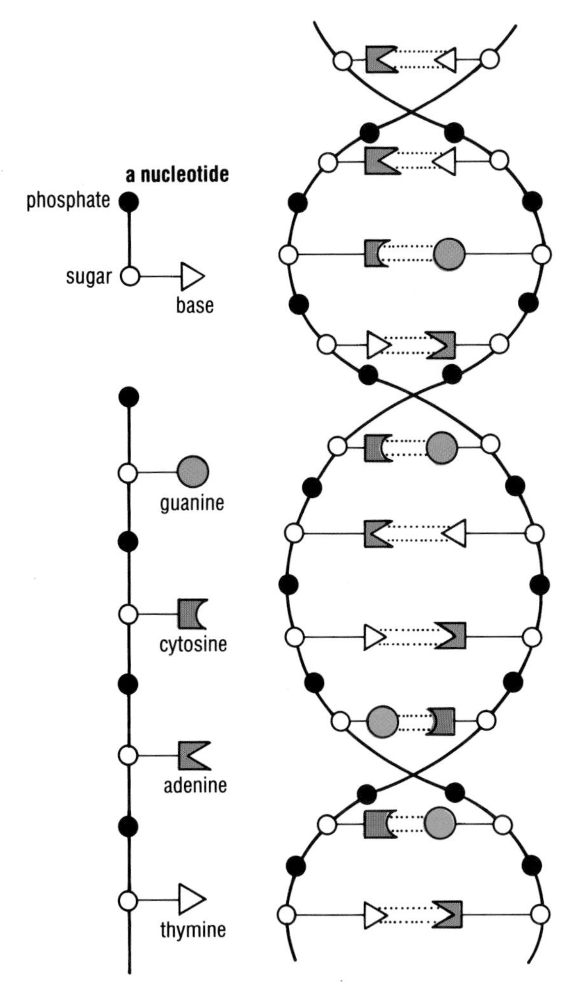

Figure J.4 The structure of DNA

these base pairs. This is how genes produce certain effects in organisms. For example, the order of base pairs which produces blue eyes is different to the order of base pairs which produces brown eyes.

There is an almost unlimited number of possible arrangements of base pairs, if you consider that a strand of DNA can be over 10 000 base units long. Therefore, there is an almost unlimited number of possible genes in plants and animals. The DNA contains the genetic code which tells an organism how to develop.

How do genes work?

How do genes control the development of an organism? How can such microscopic particles have such a staggering effect on life?

Genes consist of DNA. DNA expresses a code that determines which chemical reactions take place in a cell and at what speed. It does this by determining which proteins are made (**synthesised**) in the cell. The growth and development of a cell is determined by the type and speed of the chemical reactions taking place within it. Hence, by controlling protein synthesis, DNA controls the life of the cell, and hence the development of the organism.

DNA controls protein synthesis

Proteins are made of building blocks called **amino acids**. The amino acids are linked together in chains. The different ways in which different amino acids are linked together determines the type of protein synthesised. DNA is able to regulate how the amino acids are arranged: the types and arrangement of the bases in the DNA molecule act as a code that determines which amino acids are linked together. Thus, by determining the form and arrangement of these basic building blocks, DNA controls protein synthesis.

Sex chromosomes in humans

Figure J.5 shows the chromosomes of a man and a woman. In each case, a photograph of all the chromosomes from the nucleus of a body cell has been cut up and the chromosomes arranged into similar pairs and in order of size.

Of the 23 pairs of chromosomes in each photograph, the chromosomes in each of 22 pairs are similar in size and shape in both the man and the woman. Notice, however, that the chromosomes of the 23rd pair in the man are different from the chromosomes in the 23rd pair in the woman. These are the **sex chromosomes**. The larger chromosomes are called the X chromosomes; the smaller chromosome is called the Y chromosome.

Since the body cells of a woman each carry two X chromosomes, meiosis (see page 205) can only produce eggs containing an X chromosome. Each body cell of a man, however, carries an X chromosome and a Y chromosome, so meiosis produces two types of sperm cells. Of the sperm cells produced, 50% carry an X chromosome, and 50% carry a Y chromosome. A baby's sex depends on whether the egg is fertilised by a sperm carrying an X chromosome or one carrying a Y chromosome (Figure J.6). The birth of almost equal numbers of girls and boys in the population overall is governed by the production of equal numbers of X and Y sperms at meiosis.

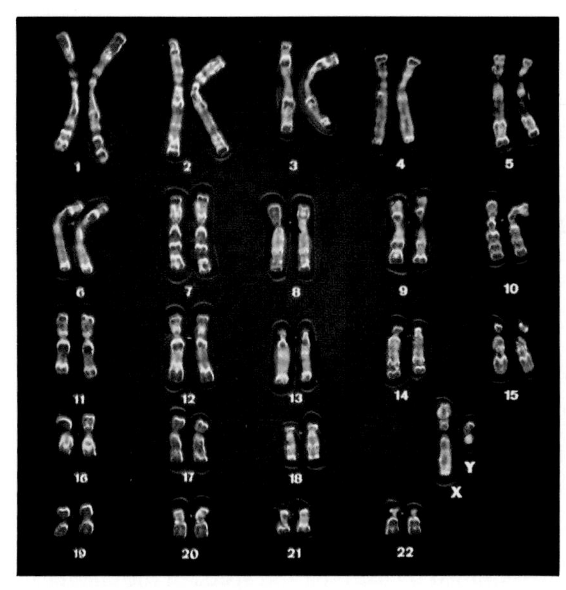

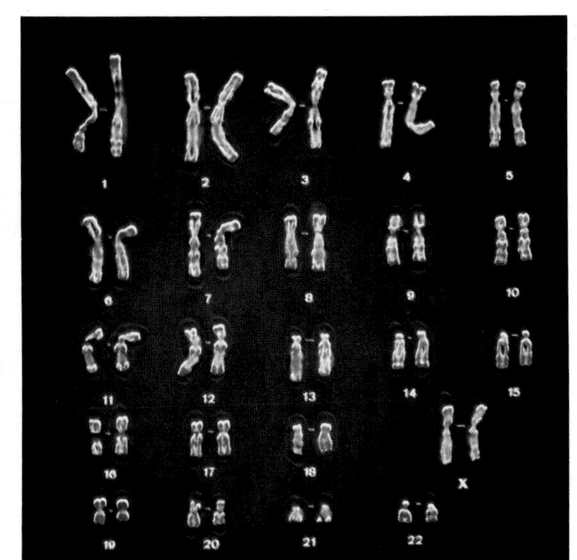

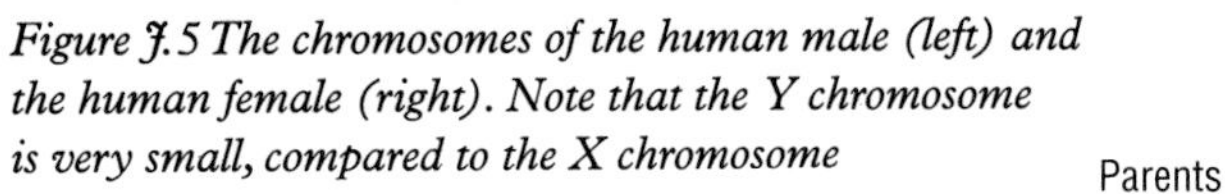

Figure J.5 The chromosomes of the human male (left) and the human female (right). Note that the Y chromosome is very small, compared to the X chromosome

Figure J.6 Inheritance of sex in humans

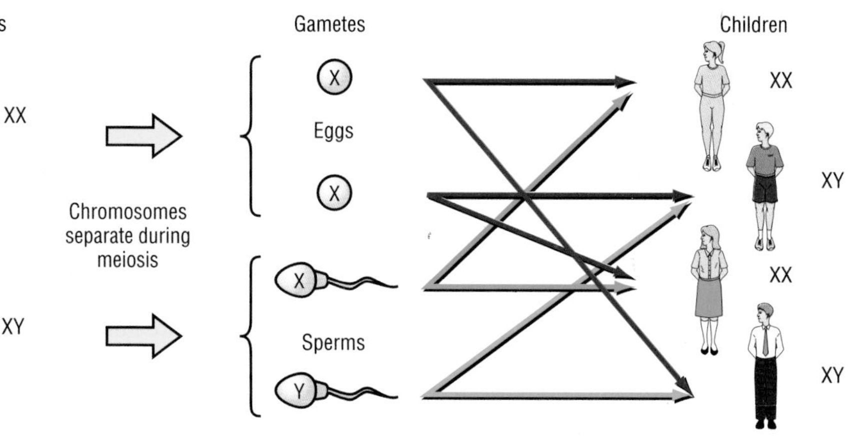

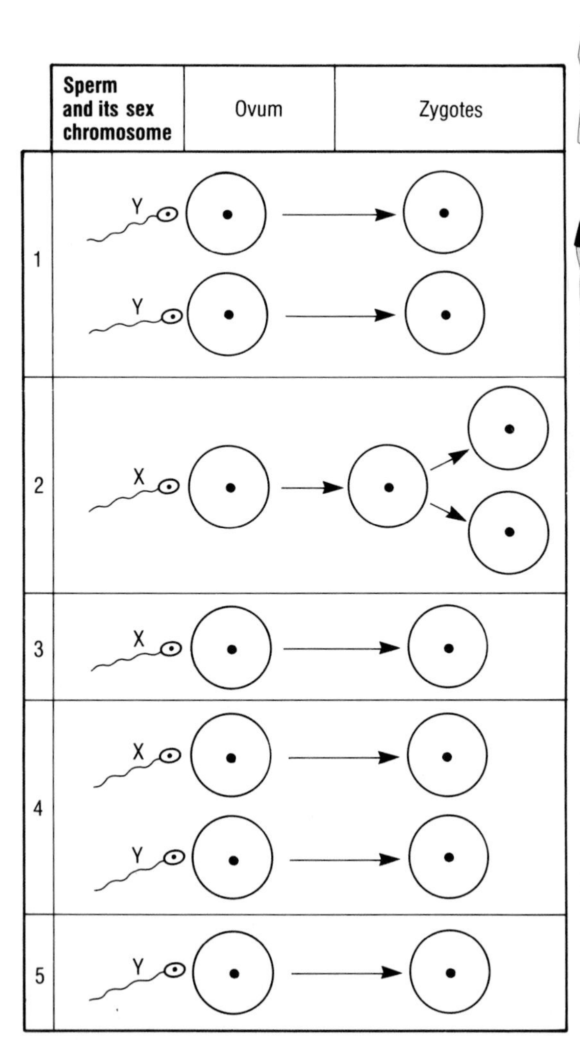

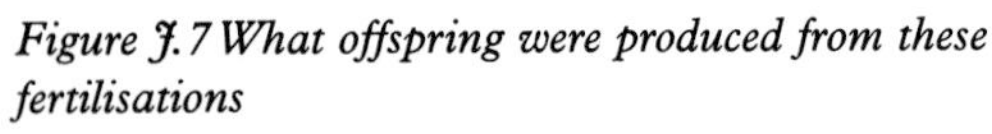

Figure J.7 What offspring were produced from these fertilisations

Summary

- Chromosomes are carriers of genetic information in the form of genes.
- Genes are collections of complex molecules, the structure of which was discovered by Watson and Crick.
- Genes work by controlling the way in which key molecules, proteins, are made in cells.
- The presence of special sex chromosomes determines the sex of an organism.

Question

1 Select a letter from the list below to show the children produced by the fertilisations given in Figure J.7.
Children produced:
A – twins, one boy and one girl.
B – identical twin girls.
C – twin boys, one with brown eyes, the other with blue eyes.
D – one boy.
E – one girl.

Unit K
When things go wrong

You know by now that we are far from understanding all of the mutations that interfere with the way the body works. It is possible to classify at least two genetic effects of mutations. One group consists of mutations that stop the body making certain essential proteins; eg enzymes, hormones or proteins making up the structure of cells.

The second group is related to the first in that it often involves the lack of certain enzymes. However, the main problem is the accumulation in the body of chemicals which normally would be broken down. An example of this type is the accumulation of a chemical called phenylalanine in the blood. It leads to the condition called **phenylketonuria** (PKU).

Sickle-cell anaemia

Sickle-cell anaemia is an often fatal condition which is quite common in West Africans. The distribution of the disease is shown in Figure K.1. If there is a low level of oxygen in the blood, the red blood cells of a person suffering from this disorder collapse into a sickle shape (Figure K.2) and may form blockages in blood vessels.

The disease is inherited as a single mutant recessive gene. If a child inherits the gene from both parents it has only a 20% chance of surviving into adulthood. The gene that controls the formation of haemoglobin is not formed properly. Therefore the haemoglobin in the sickle-shaped cells is not very good at carrying oxygen around the body.

If this is so important you might think that, over thousands of years, the mutant gene would have disappeared from human chromosomes because

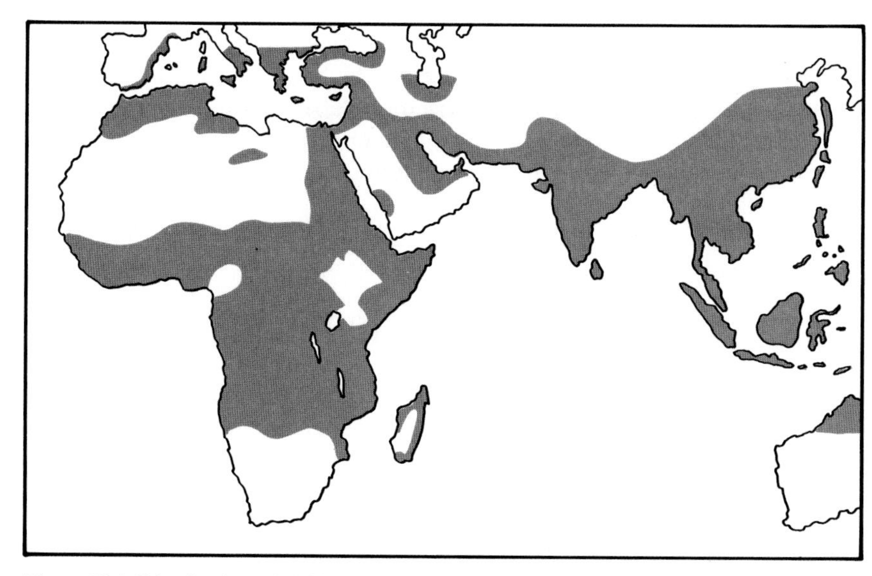

Figure K.1 Distribution of sickle-cell anaemia – up to 30% of the population may carry the sickle gene

Figure K.2 Deformed red blood cells in sickle-cell anaemia

♀ gametes ♂	HbS	Hb
HbS	HbSHbS	HbSHb
Hb	HbHbS	HbHb

Figure K.3(a) Parents HbSHb × HbSHb

♀ gametes ♂	C	c
C	CC	Cc
c	cC	cc

Figure K.3(b) Parents Cc × Cc

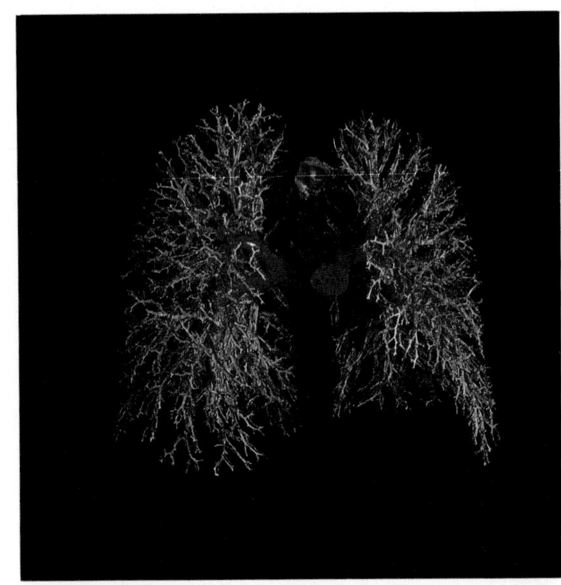

Figure K.4 The respiratory tree

carriers would die before having children. However, under certain circumstances, it is an advantage to have some sickle-shaped red blood cells: the malarial parasite is less likely to affect people who are heterozygous for the mutant gene.

For example, let us assume that the gene for haemoglobin is Hb and that the gene for sickle-cell haemoglobin is HbS. Then a normal person will have the haemoglobin genotype HbHb. A person with the sickle-cell disorder will have the genotype HbSHbS and will usually die. However, a person with the genotype HbSHb can still survive and will also be resistant to malaria.

If two people, each with the genotype HbSHb, have a child, then the predicted genotype of the child is given in Figure K.3(a).

It can be seen that there is a one-in-four chance of the child dying with the sickle-cell disorder (HbSHbS). There is a one-in-two chance of the child being a carrier of the sickle-cell gene but probably surviving (HbSHb). There is a one-in-four chance of the child being perfectly normal for the haemoglobin gene (HbHb).

Cystic fibrosis – a recessive problem

Possibly the most common human genetic disorder is **cystic fibrosis**. It is an inherited condition which affects the pancreas and the bronchioles of the lungs. It results in failure of the pancreas to function properly, intestinal obstruction, and sweat-gland and salivary-gland malfunction. Cystic fibrosis is one of the most common fatal diseases of childhood and is inherited as a recessive gene in the following way.

Let C represent the gene for a normal pancreas and bronchioles, and c be the recessive gene for cystic fibrosis.

A person suffers from the disease only if he or she has two genes for cystic fibrosis; ie the genotype cc. A person with the genotype Cc is a carrier of the disorder, but does not suffer from the disease. There is, therefore, a one-in-four chance of a child suffering from the complaint if two carriers have a child (Figure K.3(b)).

The condition occurs once in about 2000 births. It accounts for between 1 and 2% of admissions to children's hospitals. Figure K.4 shows the enormous number of very fine tubes of the lungs called **bronchioles**. In people suffering from cystic fibrosis, these tubes become blocked with mucus and have to be cleared regularly. Many children die as a result of pneumonia, although some survive to adulthood.

Huntington's chorea – a dominant problem

In 1872 an American doctor, Fraser Roberts, wrote the following observation in his notebook. 'The boy George Huntington, driving through a wooded lane in Long Island while accompanying his father on professional rounds, suddenly saw two women, mother and daughter, both tall and very thin, both bowing, twisting and grimacing.' Fifty years later George Huntington made a major contribution to science by describing and studying the inherited disease, now known as **Huntington's chorea**.

Huntington's chorea is an inherited disease characterised by involuntary muscular movement and mental deterioration. The age of onset is about 35 years. The majority of those affected can therefore produce a family before being aware of their own condition. It is transmitted by a dominant gene and both sexes can be equally affected. An estimate of its frequency is roughly 5 in 100 000 and affected people are heterozygous.

Theoretically, two parents could be affected and a homozygous child could be produced. The effect of this would probably be lethal. The gene is so rare that it is most unlikely to occur in the homozygous condition.

Let HC be the gene for Huntington's chorea and hc be the normal gene.

Parents HChc × hchc

♀ gametes ♂	hc	hc
HC	HChc	HChc
hc	hchc	hchc

Summary

- Genetic disorders can be caused by altered genes or chromosomes.
- Cystic fibrosis is caused by a recessive allele.
- Huntington's chorea is caused by a dominant allele.
- Sickle-cell anaemia is caused by a recessive allele. Carrying this allele can bring resistance to malaria.

Other genetic disorders

Key: D = dominant SL = sex-linked
R = recessive

Achondroplasia – Dwarfism due to short limbs (D)
Agamma globulinemia – Antibody formation disrupted (R,SL)
Aniridia – Blindness due to absent or rudimentary iris (D)
Cleft palate with harelip (D)
Colour blindness – Inability to distinguish red and green (R,SL)
Congenital clubfoot – Malformation of feet (D)
Epilepsy – Convulsions, abnormal brain functioning (R)
Gout – Painful swelling of joints due to abnormal uric acid levels (D)
Muscular dystrophy (shoulder girdle) – Degeneration of the muscles of the shoulder girdle in youth and adulthood (D)
Muscular dystrophy (hip girdle) – Progressive degeneration of the hip girdle in early childhood (R,SL)
Myopia – Near-sightedness (R,SL)
Nystagmus – Quivering of the eyeballs (D)
Optic atrophy – Degeneration of the optic nerves leading to blindness (D)
Pernicious anaemia – Abnormal blood corpuscles (R)
Piebaldness – Spotting of body due to unpigmented areas (D)
Polydactyly – Extra fingers and/or toes (D)

Extension questions

1 A blood disorder leading to a form of anaemia has been found to be inherited. Investigation of the medical records of a particular family gave the following information, although in some cases the records gave no information regarding the presence or absence of the symptoms.

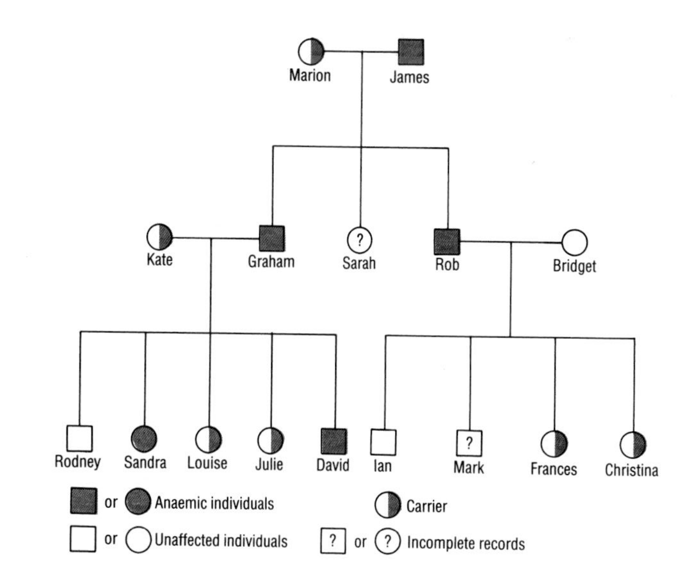

a What is meant by the term 'carrier'?
In your answers to each of the following questions explain your reasoning clearly.
b What is the likely genotype of Mark?
c What is the probability that Sarah is a carrier?
d If Ian and Anna (who has the same genotype as Julie) were to have children, what proportion of their sons might be expected to be anaemic?

2 Sickle-cell anaemia is caused by a mutant gene HbS. The normal gene is Hb. The homozygous condition of the mutant gene is lethal but it is possible for the heterozygotes to survive.

a Show the genotypes of children of:

i a normal parent and one carrying the sickle-cell gene,

ii two parents, both carrying the sickle-cell gene.

b The percentages of adults who are heterozygous for sickle-cell anaemia in different areas are shown in circles on the map of parts of Africa below. The distribution of the malarial parasite is also shown. Explain the distribution of the sickle-cell gene and that of the malarial parasite. Suggest an explanation for the distribution of the malarial parasite shown on the map.

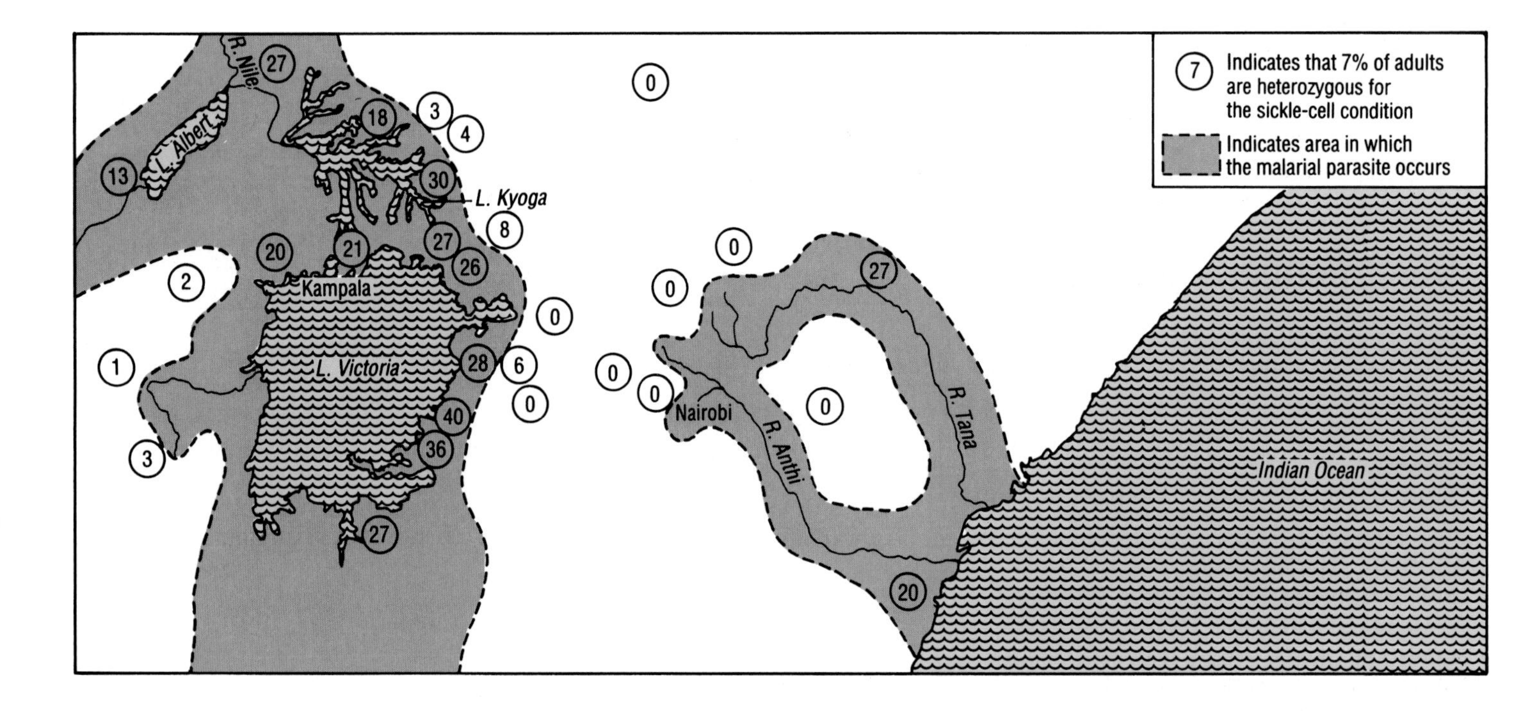

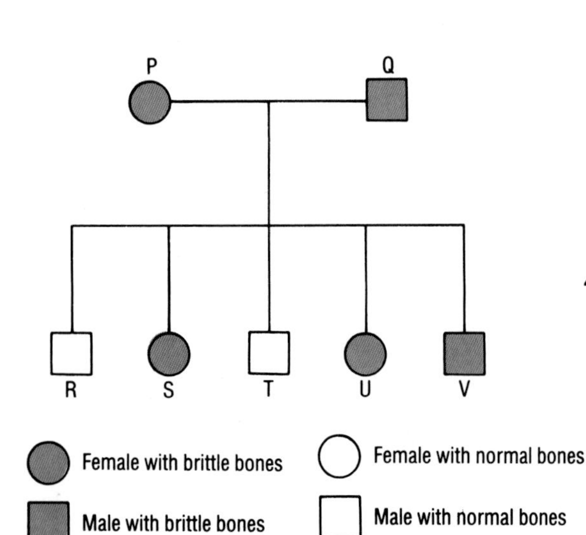

3 Huntington's chorea is a rare but very serious inherited disease caused by a dominant gene, HC. The effects of the disease do not appear until the age of 35 years.

A woman of 25 years of age is planning to start a family. However, the woman's father has Huntington's chorea and is heterozygous for the condition. There is no history of the disease in her mother's family.

a What are the chances of the woman having Huntington's chorea?

b **i** What advice might be given to the woman by a genetic counsellor about the desirability of starting a family?

ii By means of a diagram, explain the reasoning on which the advice is based.

4 Humans sometimes grow up with bones that are brittle and break easily. This condition is passed on from parents to their children, as shown in the family tree.

a What is the sex of the child labelled U? Does the child have brittle bones?

b Which is recessive, normal or brittle bones?

c Are the two parents homozygous or heterozygous?

d The gene for brittle bones is B and the gene for normal bones is b. What are the possible genotypes for children S and T?

Unit L
Reproduction

Growing up

As children develop into adults their bodies grow and change. The proportions of the body alter and during the teenage years the reproductive organs mature. As this happens (during **puberty**) the bodies of girls and boys become less alike. They start to show **secondary sexual characteristics.** Girls develop breasts and their body shape changes; hair grows in the armpits and around the reproductive area. Boys' voices change or 'break'; hair grows on the body and face; the penis and testes are enlarged. All these changes are controlled by hormones produced by glands in the body (see page 92).

Female reproductive organs

Figure L.1 shows the organs of reproduction in the female body. The **ovaries** contain eggs even in small girls but it is only as the body matures that eggs become ready for reproduction. In women the reproductive organs have a separate opening behind the urethra. The **uterus** (womb) opens at the **cervix** into the **vagina** which leads to the outside.

The menstrual cycle

In most women one egg (**ovum**) is released from an ovary (Figure L.1) each month. This is **ovulation**. It happens about 14 days before a period or **menstruation**. The ripening of each egg and its release are controlled by two hormones from the pituitary gland (page 92). Ovulation continues regularly until a woman is about 50, when the **menopause** (change of life) occurs. After ovulation the uterus wall begins to thicken. If the egg has not been fertilised it dies. Eventually, the lining or **endometrium** of the uterus breaks down and is shed during the menstrual period. The period lasts about three to seven days, and then the endometrium grows again.

If the egg is **fertilised** by a **sperm** as it passes along the **oviduct**, the fertilised egg remains in the uterus during development. Menstrual periods cease until after the birth. Figure L.2 summarises the menstrual cycle.

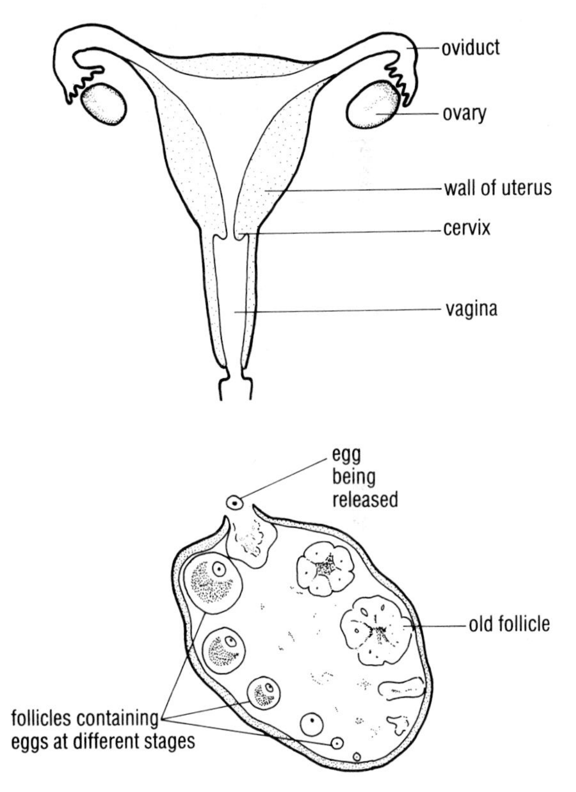

Figure L.1 The female reproductive system, with one ovary enlarged

	About 5 days	About 9 days		About 14 days
Hormones	Gonadotrophin from the pituitary	Oestrogens from ovary	Ovulation (egg is released)	Progesterone from follicle
Uterus lining	Menstrual period	Endometrium grows		Endometrium thickens
Ovum		Ovum develops in follicle within ovary		

Figure L.2 The menstrual cycle

The development of the endometrium and the menstrual cycle are brought about by hormones from the pituitary and ovary. Called **gonadotrophins**, the pituitary hormones control the development of eggs within the follicles of each ovary. **Oestrogens** from the ovaries are produced while each egg develops. This leads to the beginning of the growth of the endometrium. After ovulation, the **follicle** that released the egg produces another hormone, **progesterone**. This stimulates thickening of the endometrium and the development of the blood vessels in it.

Summary

- Hormones control growth, sexual development and the menstrual cycle.

Unit M
Controlling fertility

The fertility rate gives the average number of children a woman has throughout her childbearing life. In the UK, on average each woman bears 1.8 children. The '0.8' takes into account women who are unable to have children (**infertile**) and those who choose not to have them. In either event, hormones may help women control their fertility.

Infertility

Couples are usually thought to be infertile if they have regular unprotected sexual intercourse for 12 months without pregnancy occurring. Of all couples who are infertile (10%), about 35% are due to female infertility. Absence of (or infrequent) ovulation is one of the causes. Stimulating ovulation is the aim of treatment. Drugs that increase the levels of gonadotrophins from the pituitary (see page 256) induce ovulation in 80% of treated women.

Choice: no children

The **contraceptive pill** contains one or both of the hormones oestrogen and progesterone (Figure M.1). The concentration of the hormones in the blood stops the ovaries from producing eggs.

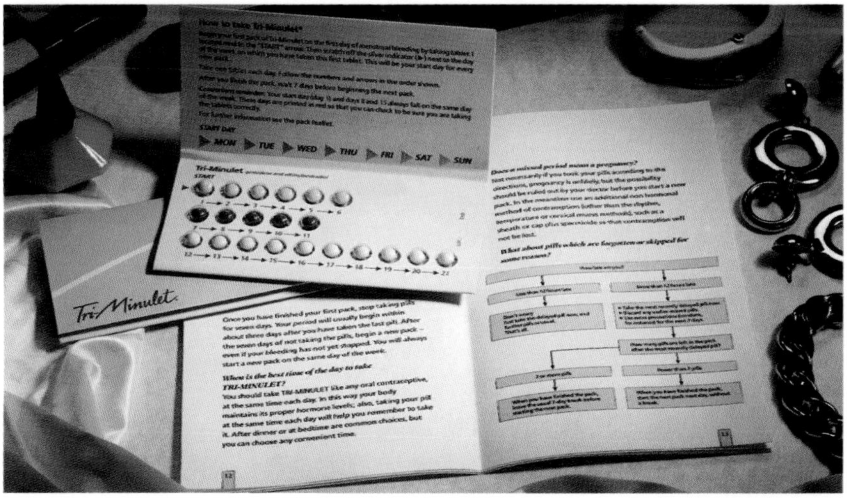

Figure M.1 The woman takes a pill every day for 21 days of her menstrual cycle. When she stops taking the pill menstruation (see page 255) begins. The woman begins taking the pill again on day one of her next menstrual cycle

If the woman forgets to take a pill on one day then protection is not complete and another form of contraception must be used until the woman's next menstrual cycle begins.

Another type of pill is the morning-after pill. It delivers a large dose of hormones which prevents implantation of the embryo. The morning-after pill must be taken within three days of intercourse to be effective.

Activity M.1

Use books and pamphlets to write a short account of one of the following: twins; ways of helping infertile couples to have a baby; advantages, disadvantages and reliability of each method of contraception; diseases associated with sexual intercourse.

More about hormones, reproduction and fertility

The human female usually produces one mature egg each month from the onset of puberty (age 11–14 years) to the beginning of the menopause (see page 255). The different events of this monthly (menstrual) cycle are summarised in Figure M.2. They occur regularly in sequence throughout the woman's fertile time of life, although they may become more erratic during the menopause. Egg production stops altogether, usually by about the age of 50.

Different hormones play their part in the development and release of mature eggs, as we have seen. Figure M.2 shows how they work together. The changes that the hormones bring about prepare the uterus to receive an egg if it is fertilised. If the egg is not fertilised, the production of oestrogen and progesterone tails off and the thick lining of the uterus begins to break down. The release of blood and tissue through the vagina is called menstruation and, as we mentioned on page 255, is what is meant by 'having a period'. It lasts for several days. A new menstrual cycle then begins.

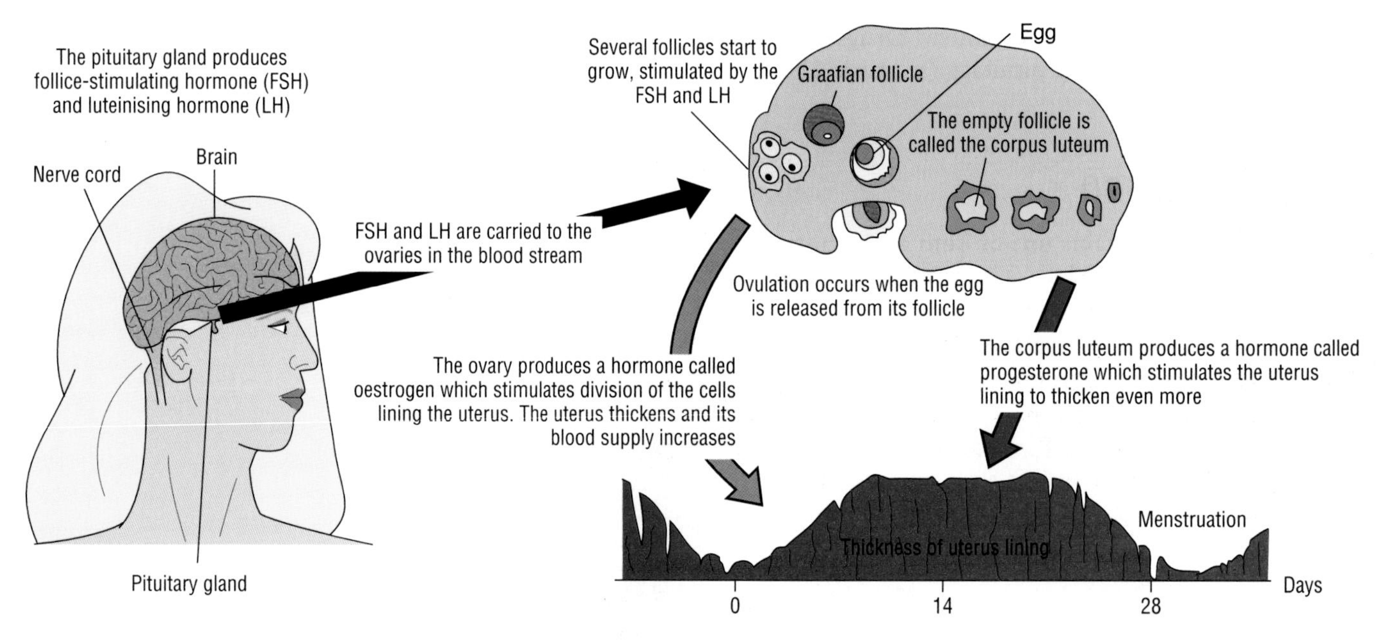

Figure M.2 The hormones of the menstrual cycle; a sharp increase in the level of luteinising hormone (LH) causes ovulation. The intervals of time for each stage vary from woman to woman

Notice the role of follicle stimulating hormone (FSH) in Figure M.2. The hormone is one of the gonadotrophins (the other is luteinising hormone – LH) referred to on page 256. Insufficient FSH is one of the causes of infertility in women. Treatment aims to stimulate egg development by raising the level of FSH in the body.

Tablets of the anti-oestrogen drug **clomiphene** make tissues insensitive to oestrogen, including the pituitary gland. Notice in Figure M.2 that oestrogen inhibits the production of FSH. Treatment with clomiphene prevents this effect of oestrogen and the pituitary continues to produce FSH. Levels of FSH increase with the result that eggs develop normally.

The treatment for infertility just described aims to *promote* egg development by increasing the level of FSH in the woman's body. Hormonal contraceptives aim to *prevent* egg development by inhibiting FSH production. They contain oestrogen and progesterone or progesterone alone. Oestrogen inhibits FSH production and reduces LH secretion, preventing ovulation. The progesterone inhibits LH secretion further. Hormonal contraceptives taken as pills are called **oral** (by mouth) **contraceptives**. There are different types of pill:

- The **combined oral contraceptive** (COC) combines small amounts of oestrogen and progesterone. The combination is called the 'low dose' pill.
- The **progesterone only pill** (POP) called the 'mini-pill' is suitable for women who are breast feeding or who may be at risk if they take the COC. The POP allows a short monthly bleed.

Injectable hormone contraceptives contain synthetic progesterone. Once injected the hormone is slowly released within the body over two or three months. Injectables are useful for women who find it difficult to take the pill every day or experience problems with other forms of contraception.

Summary

- Hormones are used to treat infertility and prevent pregnancy.
- Understanding how the effects of hormones fit together in the events of the menstrual cycle helps us use hormones to our advantage.

Extension questions

1 Figure M.3 shows how the thickness of the uterus lining and the levels of two hormones, A and B, made in the ovaries, vary during the menstrual cycle. The first month is a normal menstrual cycle but fertilisation occurs during the second month.

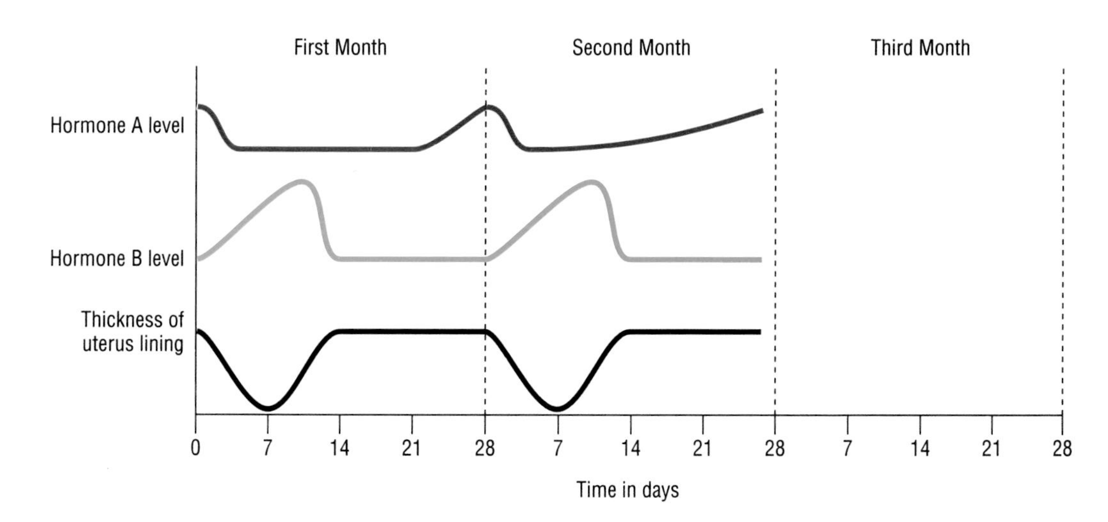

Figure M.3

a In the first month, when is the most likely time of ovulation?

b When, during the second month, is fertilisation most likely to have occurred?

c Describe what happens to the uterus in the third month.

d What will happen to the levels of hormones A and B in the third month?

e What is the role of hormones A and B in the first month?

f A type of contraceptive pill contains both hormones A and B.

i How does this type of pill work as a contraceptive?

ii A woman using the contraceptive pill usually takes it each day for 21 days and then takes a pill which does not contain hormones for the next seven days. What is the advantage of taking the pill containing hormones for 21 days only?

End-of-module questions

QUESTION ONE

1 In the human body it is the X and Y chromosomes that determine sex. The diagram shows how this happens.

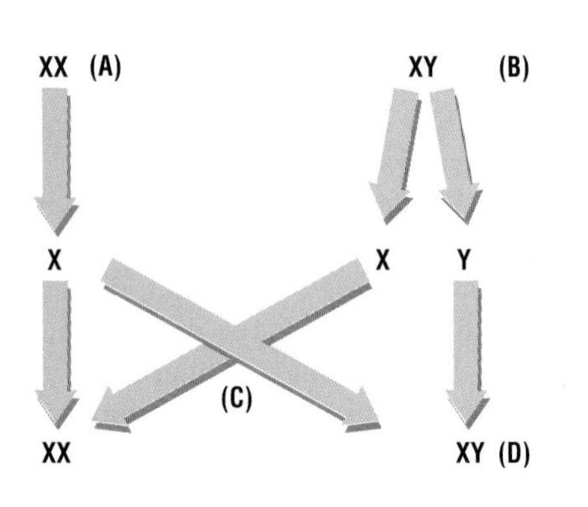

Choose letters from the diagram for each of the items 1–4 in the table.

1	the joining of the sex cells
2	the male sex cells
3	the mother
4	the baby boy

(E/F)

QUESTION TWO

The sentences below are about inheritance. Choose words from the list for each of the items 1–4 in the sentences.

genes **chromosomes** **characteristics** **nucleus**

A cell ____1____ contains ____2____.
These carry ____3____that control the ____4____ of the body.

(C/D)

QUESTION THREE

The diagrams show how to take a cutting.

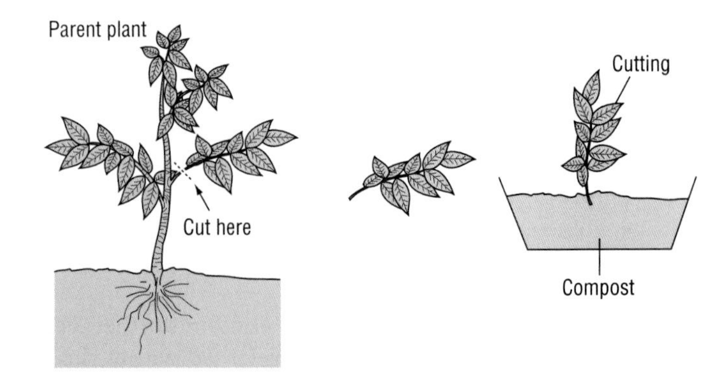

Choose from the list **two** statements that are true about cuttings.

cuttings prefer dry conditions to grow properly
cuttings are a quick way of producing new plants
cuttings are an expensive way of producing new plants
the cutting will be genetically identical to the parent plant
cuttings are a good way of introducing variety into a species

(E/F)

QUESTION FOUR

Meiosis is a form of cell division, which takes place in the body. Choose from the list **two** statements that are true about meiosis.

the resulting cells have the same number of chromosomes as the original cell
two cells result from the division of one cell
the resulting cells are all different from one another
four cells result from the original cell
this type of cell division results in growth of the body

(A/B)

QUESTION FIVE

Evolution occurs through natural selection. Choose from the list **two** statements that are true.

individuals best suited to their environment survive
individuals generally show a narrow range of variation
predation, disease and competition reduce the effect of natural selection
most mutations are generally useful to organisms
the greater the dose of radiation, the greater the rate of mutation

(A/B)

QUESTION SIX

Many species have become extinct. The drawing is of a Dodo, which is a bird that became extinct over 100 years ago.

Give two reasons why species become extinct.

(C/D)

QUESTION SEVEN

Why do animals often look quite like their parents?

(E/F)

QUESTION EIGHT

Fossils are formed from the remains of plants and animals that lived many years ago.

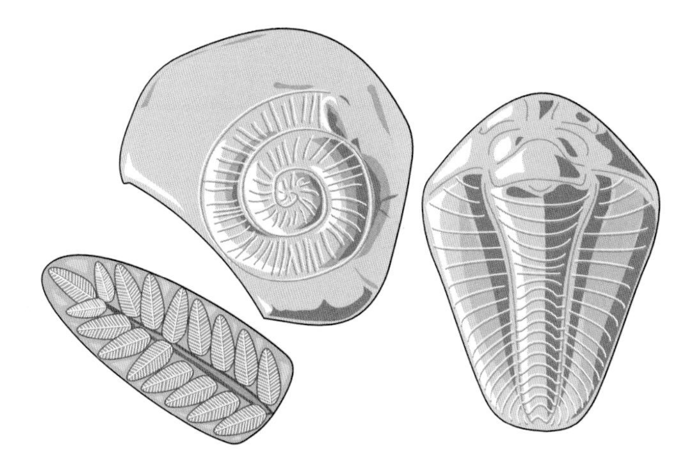

Describe two ways in which fossils can be formed.

(C/D)

QUESTION NINE

How do hormones control the release of an egg from the ovary, in a woman?

(A/B)

QUESTION TEN

Some diseases can be inherited.

1 State one symptom of Huntington's chorea.
2 State one symptom of cystic fibrosis.
3 Two parents decide to have a child. The father carries the gene for cystic fibrosis, but the mother does not have the gene. What is the chance of their baby being born with the condition? Use a genetic diagram to help you find the answer.
4 What is meant by the term 'allele'?
5 What is meant by the term 'homozygous'?

(A/B)

QUESTION ELEVEN

Sickle-cell anaemia is a disorder of red blood cells. Two parents are both carriers of the condition.

1 What is the chance that the couple's child will have sickle-cell anaemia? You could use a genetic diagram to help find your answer.
2 When can it be an advantage to an individual to carry the sickle-cell anaemia gene?
3 How does DNA control characteristics?

(A/B)

Answers to End-of-module questions

Humans as organisms

QUESTION ONE
1 gullet 2 stomach
3 small intestine 4 large intestine (4)

QUESTION TWO
1 nutrition 2 reproduction
3 respiration 4 excretion (4)

QUESTION THREE
1 salivary glands 2 large intestine
3 small intestine 4 stomach (4)

QUESTION FOUR
1 nucleus 2 cytoplasm
3 tissue 4 cell membrane (4)

QUESTION FIVE
1 white cells 2 plasma
3 platelets 4 red cells (4)

QUESTION SIX
they have muscular walls
the walls contain elastic fibres (2)

QUESTION SEVEN
they ingest microbes
they produce antibodies (2)

QUESTION EIGHT
muscles between the ribs contract
the diaphragm muscles contract (2)

QUESTION NINE
energy is released
lactic acid is produced (2)

QUESTION TEN
sugar can diffuse through membranes
the greater the difference in concentration the faster the rate of diffusion (2)

QUESTION ELEVEN
1 B 2 C 3 A 4 D (4)

QUESTION TWELVE
1 A 2 C 3 D 4 B (4)

QUESTION THIRTEEN
1 A 2 C 3 D 4 B (4)

QUESTION FOURTEEN
1 Axes labelled (1) with reasonable scales (1).
Points correctly plotted (2).
Line of best fit (1).
2 Answer between 165 and 195 seconds or between 225 and 240 seconds gives (1). Answer between 195 and 225 gives (2).
3 Heart rate is higher in order to move the blood faster (1), to carry more oxygen (1) and food, or glucose, (1) to the cells for respiration (1).
4 Haemoglobin in red cells (1) joins with oxygen to form oxyhaemoglobin (1).

QUESTION FIFTEEN
1 Axes labelled (1) with reasonable scales (1).
Points correctly plotted (1).
Line of best fit (1).
2 Answer 18 or 19 breaths per minute, or 23 or 24 breaths per minute gives (1). Answer between 20 and 22 breaths per minute gives (2).
3 oxygen + glucose $\rightarrow$ carbon dioxide + water + energy (5)
4 Breathing rate increases during exercise in order to increase the oxygen supply (1) to muscles and tissues (1), to increase the energy released (1).
5 Breathing rate remains high because anaerobic respiration during exercise (1) results in an oxygen debt (1). Incomplete breakdown of glucose produces lactic acid (1) which must be oxidised (1) to carbon dioxide and water (1).

Maintenance of life

QUESTION ONE
1 cell membrane 2 cell wall
3 nucleus 4 chloroplast (4)

QUESTION TWO
1 receptors in the tongue 2 receptors in the skin
3 receptors in the eye 4 receptors in the ear (4)

QUESTION THREE
1 phosphate 2 nitrate
3 sugars 4 potassium (4)

QUESTION FOUR
1 Sun 2 oxygen
3 chlorophyll 4 sugar (4)

QUESTION FIVE
an increase in the wind speed
an increase in the temperature of the air (2)

QUESTION SIX
tobacco smoke can result in heart damage
solvents can cause damage to the lungs, liver and brain (2)

QUESTION SEVEN
more sugar in the cell
the pressure in the cell being low (2)

QUESTION EIGHT
1 C **2** D
3 A **4** B (4)

QUESTION NINE
1 B **2** D
3 A **4** C (4)

QUESTION TEN
1 C **2** A
3 D **4** B (4)

QUESTION ELEVEN
1 D **2** B
3 A **4** C (4)

QUESTION TWELVE
1 At 9.00 a.m. it is light, so the plant is photosynthesising (1). Therefore its stomata are open (1) in order to take in carbon dioxide (1).
2 Drop in rate might be due to any of the following:
drop in wind speed
Sun shaded by cloud
temperature drop
rise in humidity (3)
3 Adaptations that reduce water loss include:
thick waxy layer on leaves
leaves reduced to spines or needles to reduce surface area
fewer stomata
4 root (1), or root hairs (2)

QUESTION THIRTEEN
1 Axes labelled (1) with reasonable scales (1).
Points correctly plotted (2).
Line of best fit (1).
2a Answer between 1.1 and 1.2, or between 1.3 and 1.4, gives (1). Answer between 1.2 and 1.3 gives (2).
b Answer 16.00 gives (1). Answer after 12.00 but before 16.00 gives (2).
3 Photosynthesis requires light (1), so sugar is produced during the day (1). No sugar is produced at night (1). Sugar is used up in respiration (1) during both day and night. The more light there is (the greater the light intensity), the more sugar is present (1).

Environment

QUESTION ONE
1 fungus **2** prey
3 producer **4** predator (4)

QUESTION TWO
1 fat **2** respire
3 microbes **4** photosynthesis (4)

QUESTION THREE
beech leaves
beech nuts (2)

QUESTION FOUR
carbon dioxide
light (2)

QUESTION FIVE
sulphur dioxide
nitrogen oxides (2)

QUESTION SIX
caterpillar
aphids (2)

QUESTION SEVEN
some energy is lost as waste materials
a lot of energy is lost as heat (2)

QUESTION EIGHT
1 Examples include:
in sewage treatment, to break down human waste
in the formation of compost, to break down garden waste (4)
2 Microbes decompose the body (1). Protein is converted to ammonium compounds by nitrifying bacteria (1). The ammonium compounds are then converted to nitrate (1).
3 The build up of nitrogen fertilisers in rivers and lakes can lead to eutrophication (1). Plants grow more rapidly (1). As the plants die, microbes break down the plant material (1). The respiration of these microbes uses up large amounts of oxygen from the water (1), leaving less available for the animals living there (1). The animals could suffocate and die.

QUESTION NINE
1 40 deer (1)
2 1940 (1)
3 Possible reasons include:
change in habitat
change in climate
disease
increased hunting by humans
competition with other animals for food or shelter (3)

QUESTION TEN

1 September (1)
2 28 mm (1)
3 February or March (1)
4 a Microbes (1) break down the leaf litter (1).
b Warm (1) and moist (1) conditions.

Inheritance and selection

QUESTION ONE

1 C **2** B
3 A **4** D (4)

QUESTION TWO

1 nucleus **2** chromosomes
3 genes **4** characteristics (4)

QUESTION THREE

cuttings are a quick way of producing new plants
the cutting will be genetically identical to the parent plant (2)

QUESTION FOUR

the resulting cells are all different from one another
four cells result from the original cell (2)

QUESTION FIVE

individuals best suited to their environment survive
the greater the dose of radiation, the greater the rate of mutation (2)

QUESTION SIX

Reasons include:
environmental change
new predators
disease
competition (2)

QUESTION SEVEN

Information about body characteristics is passed on from parents to offspring through sexual reproduction (1). The information is carried as genes (1).

QUESTION EIGHT

Some fossils were formed from parts of plant or animal remains that did not decay (1) because conditions were not suitable e.g. hard parts of the remains, such as bones (1). Other fossils were formed as parts of the remains became replaced by other materials (1), such as minerals. These materials then solidified as layers of sediment/sedimentary rocks built up on top of the remains (1).

QUESTION NINE

A hormone called FSH (1) is secreted by the pituitary gland (1). This causes an egg to mature in an ovary (1). It also stimulates the ovaries (1) to produce a second hormone called oestrogen (1), which inhibits FSH production (1). Oestrogen also stimulates the pituitary gland to produce another hormone called LH (1), which stimulates the release of the egg from the ovary (1).

QUESTION TEN

1 Symptoms include:
nervous disorder
brain shrinks
lack of co-ordination (1)

2 Symptoms include:
disorder of cell membranes
production of thick mucus (1)

3

Let C represent the normal gene and c be the recessive gene for cystic fibrosis.
Parents CC (♀) × Cc (♂)

♀ gametes ♂	C	C
C	CC	CC
c	Cc	Cc

There is 0% chance that the couple's baby will have cystic fibrosis. (4)

4 An allele is a pair of genes (1) controlling some characteristic (1).
5 An individual is homozygous for an allele when both genes in the pair are the same (2).

QUESTION ELEVEN

1

Let Hb represent the gene for normal haemoglobin and HbS be the gene for sickle-cell haemoglobin.
Parents HbSHb × HbSHb

♀ gametes ♂	HbS	Hb
HbS	HbSHbS	HbSHb
Hb	HbHbS	HbHb

There is a 1 in 4, or 25%, chance that the child will have sickle-cell anaemia. (4)

2 The sickle-cell gene can give an individual some protection against malaria, so where malaria is common it can be an advantage to carry the gene. (1)
3 DNA molecules carry coded information as a series of bases (1). The order of the bases determines the order of amino acids in proteins made in the cell (1). The order of the amino acids in a protein determines its shape and therefore its function. Many different kinds of proteins are encoded in DNA to build an organism with particular characteristics (1).

Index